Other Titles in This Series

169 S. G. Gindikin and E. B. Vinberg, Editors, Lie Groups and Lie Algebras: E. B. Dynkin's Seminar
168 V. V. Kozlov, Editor, Dynamical Systems in Classical Mechanics
167 V. V. Lychagin, Editor, The Interplay between Differential Geometry and Differential Equations
166 O. A. Ladyzhenskaya, Editor, Proceedings of the St. Petersburg Mathematical Society, Volume III
165 Yu. Ilyashenko and S. Yakovenko, Editors, Concerning the Hilbert 16th Problem
164 N. N. Uraltseva, Editor, Nonlinear Evolution Equations
163 L. A. Bokut', M. Hazewinkel, and Yu. G. Reshetnyak, Editors, Third Siberian School "Algebra and Analysis"
162 S. G. Gindikin, Editor, Applied Problems of Radon Transform
161 Katsumi Nomizu, Editor, Selected Papers on Analysis, Probability, and Statistics
160 K. Nomizu, Editor, Selected Papers on Number Theory, Algebraic Geometry, and Differential Geometry
159 O. A. Ladyzhenskaya, Editor, Proceedings of the St. Petersburg Mathematical Society, Volume II
158 A. K. Kelmans, Editor, Selected Topics in Discrete Mathematics: Proceedings of the Moscow Discrete Mathematics Seminar, 1972–1990
157 M. Sh. Birman, Editor, Wave Propagation. Scattering Theory
156 V. N. Gerasimov, N. G. Nesterenko, and A. I. Valitskas, Three Papers on Algebras and Their Representations
155 O. A. Ladyzhenskaya and A. M. Vershik, Editors, Proceedings of the St. Petersburg Mathematical Society, Volume I
154 V. A. Artamonov et al., Selected Papers in K-Theory
153 S. G. Gindikin, Editor, Singularity Theory and Some Problems of Functional Analysis
152 H. Draškovičová et al., Ordered Sets and Lattices II
151 I. A. Aleksandrov, L. A. Bokut', and Yu. G. Reshetnyak, Editors, Second Siberian Winter School "Algebra and Analysis"
150 S. G. Gindikin, Editor, Spectral Theory of Operators
149 V. S. Afraĭmovich et al., Thirteen Papers in Algebra, Functional Analysis, Topology, and Probability, Translated from the Russian
148 A. D. Aleksandrov, O. V. Belegradek, L. A. Bokut', and Yu. L. Ershov, Editors, First Siberian Winter School "Algebra and Analysis"
147 I. G. Bashmakova et al., Nine Papers from the International Congress of Mathematicians, 1986
146 L. A. Aĭzenberg et al., Fifteen Papers in Complex Analysis
145 S. G. Dalalyan et al., Eight Papers Translated from the Russian
144 S. D. Berman et al., Thirteen Papers Translated from the Russian
143 V. A. Belonogov et al., Eight Papers Translated from the Russian
142 M. B. Abalovich et al., Ten Papers Translated from the Russian
141 H. Draškovičová et al., Ordered Sets and Lattices
140 V. I. Bernik et al., Eleven Papers Translated from the Russian
139 A. Ya. Aĭzenshtat et al., Nineteen Papers on Algebraic Semigroups
138 I. V. Kovalishina and V. P. Potapov, Seven Papers Translated from the Russian
137 V. I. Arnol'd et al., Fourteen Papers Translated from the Russian
136 L. A. Aksent'ev et al., Fourteen Papers Translated from the Russian
135 S. N. Artemov et al., Six Papers in Logic
134 A. Ya. Aĭzenshtat et al., Fourteen Papers Translated from the Russian
133 R. R. Suncheleev et al., Thirteen Papers in Analysis
132 I. G. Dmitriev et al., Thirteen Papers in Algebra
131 V. A. Zmorovich et al., Ten Papers in Analysis

(Continued in the back of this publication)

Lie Groups and Lie Algebras:
E. B. Dynkin's Seminar

Dedicated to E. B. Dynkin
on the occasion of his seventieth birthday

American Mathematical Society

TRANSLATIONS

Series 2 • Volume 169

Advances in the Mathematical Sciences – 26

(*Formerly Advances in Soviet Mathematics*)

Lie Groups and Lie Algebras:
E. B. Dynkin's Seminar

S. G. Gindikin
E. B. Vinberg
Editors

American Mathematical Society
Providence, Rhode Island

ADVANCES IN THE MATHEMATICAL SCIENCES
EDITORIAL COMMITTEE

V. I. ARNOLD
S. G. GINDIKIN
V. P. MASLOV

Translation edited by A. B. SOSSINSKY

1991 *Mathematics Subject Classification.* Primary 22Exx, 57Sxx; Secondary 14J15, 15A24, 17A99, 42A38, 44A12, 53C65, 57R45, 58C27.

ABSTRACT. The notion of Dynkin diagrams are currently used in many areas of mathematics. The present book, whose publication is timed to the 70th birthday of E. B. Dynkin, contains papers of the former participants of Dynkin's seminar at Moscow University. It includes works on Lie groups and Lie algebras, cyclic homology, singularity theory, and representation theory, in which Dynkin diagrams play an important role.

This book is of interest to researchers and graduate students working in the Lie groups, Lie algebras, and related areas.

Library of Congress Catalog Card Number: 91-640741
ISBN 0-8218-0454-5
ISSN 0065-9290

Copying and reprinting. Material in this book may be reproduced by any means for educational and scientific purposes without fee or permission with the exception of reproduction by services that collect fees for delivery of documents and provided that the customary acknowledgment of the source is given. This consent does not extend to other kinds of copying for general distribution, for advertising or promotional purposes, or for resale. Requests for permission for commercial use of material should be addressed to the Assistant Director of Production, American Mathematical Society, P. O. Box 6248, Providence, Rhode Island 02940-6248. Requests can also be made by e-mail to reprint-permission@math.ams.org.

Excluded from these provisions is material in articles for which the author holds copyright. In such cases, requests for permission to use or reprint should be addressed directly to the author(s). (Copyright ownership is indicated in the notice in the lower right-hand corner of the first page of each article.)

© Copyright 1995 by the American Mathematical Society. All rights reserved.
The American Mathematical Society retains all rights
except those granted to the United States Government.
Printed in the United States of America.

⊚ The paper used in this book is acid-free and falls within the guidelines
established to ensure permanence and durability.
♻ Printed on recycled paper.
10 9 8 7 6 5 4 3 2 1   00 99 98 97 96 95

Contents

Preface	ix
On the Work of E. B. Dynkin in the Theory of Lie Groups F. I. Karpelevich, A. L. Onishchik, and E. B. Vinberg	1
Matrix Vieta Theorem Dmitry Fuchs and Albert Schwarz	15
Integral Geometry on Real Quadrics Simon Gindikin	23
Dynkin Diagrams in Singularity Theory S. M. Gusein-Zade	33
Variations on the Triangular Theme A. A. Kirillov	43
Vector Fields and Deformations of Isotropic Super-Grassmannians of Maximal Type A. L. Onishchik and A. A. Serov	75
A_∞ Algebras and the Cohomology of Moduli Spaces Michael Penkava and Albert Schwarz	91
On Hamburger's Theorem I. Piatetski-Shapiro and Ravi Raghunathan	109
An Analogue of M. Artin's Conjecture on Invariants for Nonassociative Algebras Vladimir L. Popov	121
On Reductive Algebraic Semigroups E. B. Vinberg	145
Crystal Bases and the Problem of Reduction in Classical and Quantum Modules D. P. Zhelobenko	183

Therefore, it can be said that the first version of the Lie groups seminar was a part of the 1947–1955 seminar.

Although Dynkin was interested in both Lie groups and in probability from his student times, at the beginning of the 1950s his work was concentrated mainly around Lie groups. His publications in this area stoped in 1955, and at this moment he presumably switched completely to the probability theory. It is only at the beginning of 1960 that he returned for some time to Lie groups in connection with his work on Brownian motion and the Martin boundary of symmetric spaces. It is interesting to notice that this work lies on the intersection of his two interests.

The outstanding results of E. B. Dynkin on Lie groups are well known. However, I am certain that the survey prepared for this volume by Karpelevich, Onishchik, and Vinberg will allow the reader to understand Dynkin's contribution to this theory even better.

Dynkin organized his Lie group seminar in 1957, when he practically stopped working in this area. This makes the success of this seminar even more remarkable. Here is the sequence of events, as I recollect them. In the fall of 1956, Dynkin announced a one-year course in differential manifolds and Lie groups. At the end of the fall semester he suddenly declared that in the spring semester the course will become a seminar.

Mainly, this was a seminar for second- and third-year undergraduate students. Among the participants I remember A. Kirillov, E. Vinberg, M. Freidlin, M. Shur, and V. Tutubalin (the last three soon switched to probability.) Most of us knew each other from our school years, when we used to attend the same mathematical circles and mathematical competitions.

In the spring of 1957, we mainly learned how to read mathematical books, give talks, and solve problems. At the end of the semester, most of the participants prepared an essay, which sometimes could have been considered as a small scientific work.

The situation had changed dramatically in the fall of 1957. Dynkin invited his students of the previous generation to the seminar, F. Berezin and F. Karpelevich among them. A. Onishchik then a graduate student, was an even earlier participant of the seminar. Soon afterward I. Pyatesky-Shapiro, who just returned from Kaluga, where he had worked after graduating from Moscow Pedagogical University, started to attend the seminar. At about the same time the seminar was joined by A. Schwarz who just entered the graduate school, and by G. Tyurina and D. Fuchs, who were in their junior undergraduate year.

This blend of mathematicians of different generations produced wonderful results. It is difficult to overestimate the importance of this for us, the junior participants of the seminar. We found ourselves involved in the process of mathematical research. Problems and results were discussed in a very preliminary format. Here are some examples that are closer to me. Ilya Pyatesky-Shapiro spoke about his plans and results related to automorphic forms and related problems in the theory of bounded symmetric domains. This is how we first heard about Siegel domains. At some point, the speaker asked Onishchik, who had the reputation of an expert on homogeneous complex manifolds, whether it had already been proved that each homogeneous bounded domain is symmetric. The answer was that it was almost proved. After some discussion we came to the conclusion that the remaining part of the proof — either semisimplicity or unimodularity of the group — should not pose a serious obstacle. Soon afterwards (I am tempted to say in just a week) Ilya, somewhat hesitantly, was demonstrating to us an example of a nonsymmetric homogeneous domain in $\mathbb{C}^4$.

I also recall team attempts to include the exceptional domains into the general classification scheme for Siegel domains. An exceptional cone for the first domain was soon found, but no one could figure out what to do with the second domain, until Ilya finally found its place among the Siegel domains of the second type related to the light cone. It was no wonder to all of us that soon afterwards Vinberg and I found ourselves involved in the development of the theory of nonsymmetric homogeneous domains.

Another frequently discussed topic was the problem of the boundary of a symmetric space. Pyatesky-Shapiro was interested in the corresponding constructions in his attempts to find compactifications of fundamental domains of discrete groups. Dynkin approached the problem from the point of view of Martin boundaries. Finally, one of the most important chapters in the theory of symmetric spaces, the Karpelevich compactification theory, was developed.

Those were happy years for all of us. The seminar was an important part of our "mathematical house." We spent a lot of time together. I can recall joint weekend trips to the country (together with Dynkin's seminar on probability). We have continued our friendly relations for some 40 years, and we all are very grateful today to Evgeniĭ Borisovich Dynkin, who did so much at the start our mathematical life.

S. G. Gindikin   Rutgers University

On the Work of E. B. Dynkin in the Theory of Lie Groups

F. I. KARPELEVICH, A. L. ONISHCHIK, AND E. B. VINBERG

E. B. Dynkin began working in the theory of Lie groups and Lie algebras in his student years. His work in this field was relatively brief but extremely intense. This can be seen from the list of his main achievements during that period: the invention of simple roots and "Dynkin diagrams", the discovery of explicit formulas for the Campbell-Hausdorff coefficients, the description of maximal subalgebras of simple complex Lie algebras, the classification of semisimple subalgebras of exceptional complex Lie algebras, the description of primitive homology and cohomology classes of the classical compact Lie groups.

The notion of simple root of a semisimple Lie algebra was invented by Dynkin in 1944 when he was studying the work of H. Weyl and Van der Waerden in order to prepare, at I. M. Gelfand's request, a survey on the structure and classification of semisimple Lie algebras for the latter's seminar. The theory of simple roots is presently widely known, so we limit ourselves to some brief remarks. Suppose $\mathfrak{h}$ is a Cartan subalgebra of a semisimple Lie algebra $\mathfrak{g}$ and let Δ be the root system of the Lie algebra $\mathfrak{g}$ with respect to $\mathfrak{h}$. Note that in the work of Dynkin the dual space $\mathfrak{h}^*$ is identified with $\mathfrak{h}$ by means of the Killing form, so that $\Delta \subset \mathfrak{h}$. In the real linear span E of the system Δ the lexicographic ordering related to a fixed basis is introduced; this allows to distinguish the subsystem Δ_+ of positive roots. A positive root is called *simple* if it cannot be presented as the sum of two positive roots. It turns out that the system Π of positive roots is a basis in $\mathfrak{h}$, and any positive root can be expressed in terms of simple roots with nonnegative integer coefficients. Further, the system Π may be used to construct a system of generators for the Lie algebra $\mathfrak{g}$, so that the relations between these generators are entirely determined by the angles between the simple roots and the ratios of their lengths. Thus the system Π, regarded as a system of vectors in Euclidean space E, determines uniquely (up to isomorphism) the Lie algebra $\mathfrak{g}$. The system Π may be represented by means of a graph whose vertices correspond to simple roots and are painted in two colors depending on their lengths, while the edges are supplied with multiplicities 0, 1, 2, or 3 in accordance with the angle between the corresponding roots, which can only assume the values $\pi/2$, $2\pi/3$, $3\pi/4$, or $5\pi/6$. (In our time instead of

---

1991 *Mathematics Subject Classification.* Primary 22-03, 01A65, 01A70.

Key words and phrases. Lie group, Lie algebra, simple root, Dynkin diagram, regular subalgebra, Dynkin index, primitive element.

©1995, American Mathematical Society

painting the graph in two colors, one supplies some of the edges with arrows.) This graph is connected if and only if the Lie algebra $\mathfrak{g}$ is simple. The classification of semisimple complex Lie algebras is thereby reduced to a combinatorial problem, whose solution can be obtained from simple facts in Euclidean geometry (see [1, 5]). Thus a simple and elegant exposition of Killing–Cartan results on the classification of complex semisimple Lie algebras was obtained.

At the present time, the method of simple roots is the basis of the structural theory of Lie algebras, of algebraic groups, of Kac-Moody algebras, and other similar algebraic objects. The graph representing the simple root system of a semisimple Lie algebra $\mathfrak{g}$ is called the *Dynkin diagram* of this Lie algebra.

While he was a graduate student of Moscow University, Dynkin worked on other questions of Lie group theory. We have in mind his study of the formal power series

$$\Phi(x, y) = \log(e^x e^y)$$

in two noncommuting variables x, y, where the exponential and logarithmic series are defined by the usual formulas

$$e^x = \sum_{n=0}^{\infty} \frac{x^n}{n!}, \qquad \log x = \sum_{n=1}^{\infty} (-1)^{n-1} \frac{1}{n} (x-1)^n.$$

It is easy to see that

$$(1) \qquad \Phi(x, y) = \sum (-1)^{n-1} \frac{1}{n} \frac{1}{p_1! q_1! \cdots p_n! q_n!} x^{p_1} y^{q_1} \cdots x^{p_n} y^{q_n},$$

where the sum is taken over all systems of nonnegative integers $(p_1, q_1, \ldots, p_n, q_n)$ satisfying the conditions $p_i + q_i > 0$, $i = 1, \ldots, n$. As Campbell and Hausdorff showed at the turn of the century, this series can be expressed in terms of x, y using only linear operations and the commutation operation $[u, v] = uv - vu$. Written in this form, the series (1) expresses (in canonical coordinates) the multiplication operation in an arbitrary Lie group via the commutation operation in the corresponding Lie algebra. An algorithm for computing this expression was given, but the explicit form of the expression remained unknown until Dynkin found an unexpectedly simple answer to the problem (see [2]).

This answer is based on the following simple algebraic theorem. Let $A(x_1, \ldots, x_m)$ be the free associative algebra with generators $x_1, \ldots, x_m$ over an arbitrary field K of characteristic zero and $L(x_1, \ldots, x_m)$ the set of elements of $A(x_1, \ldots, x_m)$ that can be expressed in terms of $x_1, \ldots, x_m$ using only linear operations and the commutation operation (actually $L(x_1, \ldots, x_m)$ is the entire free Lie algebra generated by $x_1, \ldots, x_m$). Then the element $Q \in A(x_1, \ldots, x_m)$ of the form

$$Q = \sum_{i_1 \ldots i_m} a_{i_1 \ldots i_m} x_{i_1} \cdots x_{i_m},$$

where $a_{i_1 \ldots i_m} \in K$, belongs to $L(x_1, \ldots, x_m)$ if and only if

$$Q = \sum_{i_1, \ldots, i_m} \frac{1}{m} a_{i_1 \ldots i_m} [[[x_{i_1}, x_{i_2}], \ldots], x_{i_m}].$$

This implies that the required expression of the series (1) in terms of the commutator can be obtained by replacing, in each term, the multiplication operation by commutation and dividing by the degree. This theorem allowed, in particular, to generalize the correspondence theorems between local Lie groups and Lie algebras, and it was proved by Sophus Lie over the fields of real and complex numbers in the finite-dimensional case, and then by Birkhoff in the Banach case, to the case of Banach local Lie groups and Lie algebras over an arbitrary complete normed field of zero characteristic. These results constituted Dynkin's PhD (candidate dissertation), which he defended successfully in 1948 (see [7]).

The method of simple roots discussed above was the basis of the following stage of Dynkin's research dealing with the structure of semisimple Lie algebras, their automorphisms, linear representations and subalgebras. Here Dynkin succeeded in obtaining new fundamental results, and also considerably simplified the statements and proofs of theorems known earlier.

In the paper [9], the method of simple roots is applied to the study of finite-dimensional linear representations of a complex Lie algebra $\mathfrak{g}$. According to E. Cartan, an irreducible linear representation is determined by its highest weight. In Dynkin's theory, the highest weight Λ of a linear representation φ is specified by the so-called "numerical marks"

$$\Lambda_\alpha = \frac{2(\Lambda, \alpha)}{(\alpha, \alpha)}, \qquad \alpha \in \Pi,$$

where Π is the system of simple roots and $(\cdot, \cdot)$ is the Killing form of the Lie algebra $\mathfrak{g}$. The expressions Λ_α, $\alpha \in \Pi$, can assume arbitrary nonnegative integer values. Dynkin established several properties of the system of all weights of an irreducible representation φ with highest weight Λ. Any weight M of the representation φ can be written in the form $M = \Lambda - \alpha_1 - \cdots - \alpha_k$, where $\alpha_i \in \Pi$, $i = 1, \ldots, k$, and $\Lambda - \alpha_1 - \cdots - \alpha_i$ for any $i = 1, \ldots, k$ is a weight of the representation φ. The number k is uniquely determined, and Dynkin calls it the *floor* of the weight M (the floor of the highest weight Λ is 0). It is proved in [9] that the system of weights of the representation φ is "spindle-like". This means that if the number of floors is $r + 1$ and s_k is the number of weights living on the k th floor (each weight is counted according to its multiplicity), then

$$(2) \qquad s_k = s_{r-k}, \qquad s_0 \leq s_1 \leq \cdots \leq s_{r_0},$$

where $r_0 = [r/2]$. The proof of relations (2) is based on the construction of an important three-dimensional subalgebra in $\mathfrak{g}$ (later called the *principal subalgebra* by Dynkin, see [16]). Independently this subalgebra was constructed in 1950 by Siebenthal.

It is interesting to note that the numbers s_k for the adjoint representation, as established by Steinberg [48] and Kostant [40], are closely related to the Betti numbers of the corresponding Lie group. Namely, the Poincaré polynomial of the latter has the form

$$\prod_{i=1}^{n}(1 + t^{2k_i+1}),$$

where $n = \operatorname{rk} \mathfrak{g}$ and each natural number $k < r_0$ appears in the system $k_1, \ldots, k_n$ with multiplicity $s_{r_0-k} - s_{r_0-k-1}$ (note that in the case considered $s_{r_0} = s_{r_0-1} = n$).

In the same paper Dynkin obtained simple formulas for the number $r + 1$ of floors of a weight system by expressing it in terms of the numerical marks of the highest weight of the irreducible representation φ, and established that if a representation admits an invariant bilinear form, then it is symmetric for even r and skew-symmetric for odd r. This simple criterion immediately leads to the orthogonality and symplecticity conditions of a linear representation obtained by A. I. Mal'tsev in 1944 (see [44]) separately for each type of simpe Lie algebra.

In the paper [12] Dynkin develops a unified approach to the description of the automorphism group of semisimple complex Lie algebras. The theorem found in this paper claims, in particular, that there exists a one-to-one correspondence between the connected components of the automorphism group of a semisimple Lie algebra $\mathfrak{g}$ and the automorphisms of the Dynkin diagram of this Lie algebra. Here the identity automorphism corresponds to the group of interior automorphisms, while the nonidentical automorphisms correspond to different components consisting of exterior automorphisms. For example, if the Dynkin diagram has no symmetries other than the identity, then the Lie algebra $\mathfrak{g}$ has no exterior automorphisms. Let us note that the connected components of the automorphism group of simple complex Lie algebras were described earlier by F. V. Gantmakher [33] by an exhaustive search of all simple algebras.

The papers [10, 11, 13–17] deal with the classification of subalgebras of complex semisimple complex Lie algebras. A subalgebra $\tilde{\mathfrak{g}}$ of a semisimple complex Lie algebra $\mathfrak{g}$ is called *regular* if there exists a basis in $\tilde{\mathfrak{g}}$ consisting of elements of the Cartan subalgebra $\mathfrak{h}$ and of root vectors e_α of the algebra $\mathfrak{g}$ (α are the roots of the Lie algebra $\mathfrak{g}$ with respect to $\mathfrak{h}$). In this case the root system $\tilde{\Sigma}$ of the Lie algebra $\tilde{\mathfrak{g}}$ is naturally identified with a certain part of the system of root vectors Σ of the algebra $\mathfrak{g}$. The paper [10] (for a more detailed exposition, see [16]) contains a classification of semisimple regular subalgebras of semisimple complex Lie algebras.

It is convenient to specify a regular subalgebra $\tilde{\mathfrak{g}}$ of a Lie algebra $\mathfrak{g}$ by means of the system $\tilde{\Pi}$ of its simple roots. It can be proved that the subalgebra $\tilde{\mathfrak{g}}$ can be extended to a semisimple subalgebra $\mathfrak{g}'$ of maximal rank in $\mathfrak{g}$ (such subalgebras are always regular), whose system Π' of simple roots contains $\tilde{\Pi}$. The system $\tilde{\Pi}$ can be obtained by deleting one or more roots from Π'. Thus the problem is reduced to finding all semisimple subalgebras of maximal rank. In order to describe the system $\tilde{\Pi}$ of simple roots of such subalgebras, Dynkin proposed a simple and elegant algorithm that consists of the following.

First, suppose that the Lie algebra $\mathfrak{g}$ is simple. To its system of simple roots $\Pi = \{\alpha_1 \ldots \alpha_n\}$, where n is the rank of the algebra $\mathfrak{g}$, let us add the lowest root δ, i.e., the smallest root with respect to the order described above. We obtain the system $\hat{\Pi} = \{\alpha_1 \ldots \alpha_n, \delta\}$, which is called the *extended system of simple roots* of the algebra $\mathfrak{g}$. Deleting any root from $\hat{\Pi}$, we obtain system Π_1, again consisting of n roots, which is the system of simple roots of some semisimple (not simple in general) subalgebra $\mathfrak{g}_1$ of maximal rank n. The transformation of the system Π into Π_1 is called an *elementary transformation* of the system Π. In the general case, when the Lie algebra $\mathfrak{g}$ is not necessarily simple, by an elementary transformation of the system Π we mean an elementary transformation of one of its irreducible components. In these terms, we have the following theorem. The system of simple roots $\tilde{\Pi}$ of any semisimple subalgebra $\tilde{\mathfrak{g}}$ of maximal rank in $\mathfrak{g}$ may be obtained from the system Π of simple roots of the Lie algebra $\mathfrak{g}$ by means of a sequence of elementary transformations. Using this result, in [16] Dynkin listed all regular semisimple subalgebras (up to conjugation) of all simple complex Lie algebras. Note that somewhat earlier, Borel and de Siebenthal [30] published a paper in which they classified connected subgroups of maximal rank in semisimple compact Lie groups using different methods; this is equivalent to the classification of regular semisimple subalgebras in semisimple complex Lie algebras.

It is obvious that a semisimple subalgebra of maximal rank can be a maximal subalgebra only in the case when its system of simple roots can be obtained from Π via a single elementary transformation. In the paper [16] it was erroneously stated that, conversely, any proper subalgebra obtained in this way is maximal. This mistake was corrected by Tits [49].

It turned out that the extended systems of simple roots and the *extended Dynkin diagrams* that represent them were to play a significant role in other questions in the theory of Lie groups and Lie algebras. The extended Dynkin diagram of a semisimple complex Lie algebra $\mathfrak{g}$ describes the fundamental polyhedron of the extended Weyl group, which may be used to classify globally all connected Lie groups having $\mathfrak{g}$ for their tangent algebra (see [27]), or to classify all automorphisms of finite order of the algebra $\mathfrak{g}$ (see [34, 35]). The latter classification is closely related to the fact that the extended simple root system of a Lie algebra $\mathfrak{g}$ is the simple root system of a certain infinite-dimensional Lie algebra, namely, the affine Kac–Moody algebra associated to $\mathfrak{g}$.

Over a hundred years ago S. Lie put forward the classification problem for *primitive local transformation groups*, i.e., transitive local Lie groups of transformations that do not leave any nontrivial fiber space invariant. In the language of Lie algebras this is equivalent to the classification of maximal subalgebras in finite-dimensional Lie algebras. V. V. Morozov reduced this problem to the case when the underlying Lie algebra is simple. Moreover, in 1943, in his dissertation he proved that any nonsemisimple maximal subalgebra of a semisimple complex Lie algebra is regular (see also [44]) and gave an explicit list all nonsemisimple maximal subalgebras in simple complex Lie algebras. (In 1951 this classification was simplified by F. I. Karpelevich, who showed [36] that any nonsemisimple subalgebra in a semisimple complex Lie algebra belongs to the class of subalgebras now called *parabolic* and described this class in terms of simple roots.) In 1950 Dynkin obtained a description of all maximal subalgebras in the classical Lie algebras (see [11, 15]). The most difficult case occurs when the subalgebra is simple and irreducible (i.e., does not leave any proper subspace invariant); in this case the result obtained may be expressed as follows: almost any irreducible simple complex subalgebra of the complete Lie algebra $\mathfrak{gl}_n(\mathbb{C})$ is maximal in one of the classical Lie algebras $\mathfrak{sl}_n(\mathbb{C})$, $\mathfrak{so}_n(\mathbb{C})$, and, for even n, in $\mathfrak{sp}_n(\mathbb{C})$. The exceptions to this rule (which are 4 series and 14 separate subalgebras) were explicitly listed. In the paper [15] Dynkin also found all the reducible maximal subalgebras of the classical complex Lie algebras. This work was defended in 1950 as a doctoral dissertation.

In 1951 Dynkin succeeded in solving a very difficult problem of classifying semisimple subalgebras of the exceptional complex Lie algebras [16]. In particular, all the semisimple maximal subalgebras were listed; together with the results described above, this gave a complete solution to the S. Lie problem in the case of the field of complex numbers. An important role in this classification was played by the notions of R-subalgebra and S-subalgebra. A subalgebra $\tilde{\mathfrak{g}}$ of a semisimple Lie algebra $\mathfrak{g}$ is said to be an *R-subalgebra* if it is contained in a proper regular subalgebra of the Lie algebra $\mathfrak{g}$, and an *S-subalgebra* otherwise. If $\mathfrak{g}$ is a classical complex Lie algebra, then the class of its proper R-subalgebras is contained in the class of reducible subalgebras and almost always concides with this class. Any S-subalgebra is semisimple, and any semisimple subalgebra is an S-subalgebra of some semisimple regular subalgebra. Hence the classification of of all semisimple subalgebras of semisimple Lie algebras reduces to the classification of S-subalgebras.

This ideology was applied by Dynkin, in particular, to the classification of three-dimensional simple algebras; this classification, according to Morozov's theorem [43] (also see [45]) is equivalent to the classification of nilpotent elements. In order to solve the conjugation problem for three-dimensional simple subalgebras, Dynkin introduced the notion of characteristic. The characteristic of a three-dimensional simple subalgebra is a semisimple element, and two three-dimensional simple subalgebras are conjugate if and only if their characteristics are conjugate. Since the conjugacy problem for semisimple elements has a simple solution, this solves the conjugacy problem for three-dimensional simple subalgebras.

The methods developed by Dynkin for the classification of three-dimensional simple subalgebras of semisimple Lie algebras, and the tables of three-dimensional simple subalgebras of exceptional simple Lie algebras that he obtained in [16] became the basis of numerous further studies. A. G. Elashvili [31] found the centralizers of the nilpotent elements of exceptional simple Lie algebras, and in the process corrected several (not very significant) errors in Dynkin's tables.

From the point of view of the theory of invariants, the nilpotent elements in the space V of a linear representation of a semisimple complex Lie algebra G are elements that have an orbit whose closure contains 0. (In the case of an adjoint representation, this is equivalent to the element of the Lie algebra $\mathfrak{g}$ being nilpotent in the ordinary sense.) According to the Hilbert–Mumford theorem, for any nilpotent element $e \in V$ there exists an element $h \in \mathfrak{g}$ such that

$$\lim_{t \to +\infty} (\exp th)e = 0.$$

In general, there are many such elements h, and we can choose the one for which $(\exp th)e$ tends to 0 most rapidly for a fixed length of h. In the case of an adjoint representation, this will be the characteristic of the three-dimensional simple subalgebra corresponding to e. In the general case this provides an approach to the classification of nilpotent elements, and for a certain class of linerar groups introduced by Vinberg, the method of characteristics combined with the ideology of S-subalgebras yields the complete classification of nilpotent elements. For example, in the paper by Vinberg and Elashvili [51], the classification of trivectors in 9-dimensional space was obtained in this way.

For the cycle of papers on the classification of subalgebras in semisimple Lie algebras Dynkin was granted the prize of the Moscow Mathematical Society in 1951.

Let us note two results obtained by Dynkin in connection with the classification of subalgebras, but of interest on their own.

In the paper [15], the following theorem deals with the decomposition of the Kronecker (tensor) product of irreducible linear representations of a semisimple complex Lie algebra $\mathfrak{g}$ into irreducible components. Suppose R_N denotes an irreducible representation with highest weight N, and $P(\Lambda, M)$ is the set of highest weights of irreducible components of the representation $R_\Lambda \otimes R_M$. In the weight lattice, let us introduce a partial order by setting $\Lambda \geq M$ if $\Lambda - M$ is a linear combination of simple roots with nonnegative coeffficients. It is known that the set $P(\Lambda, M)$ has a unique maximal element with respect to this order, namely $\Lambda + M$. Dynkin described [15] the maximal elements of the set $P(\Lambda, M) \setminus \{\Lambda + M\}$. It turned out that they are precisely the weights of the form $\Lambda + M - \alpha_1 - \cdots - \alpha_k$, where $\alpha_1, \ldots, \alpha_k$ are simple roots such that in the sequence $\Lambda, \alpha_1, \ldots, \alpha_k, M$ the pairs of orthogonal elements are those and only those that are neighbors, except perhaps Λ and M. In this case the irreducible components of the representation $R_\Lambda \otimes R_M$ (as well as those of $R_{\Lambda+M}$) are of multiplicity 1.

This theorem implies, in particular, that the representation $R_\Lambda \otimes R_M$ is irreducible if and only if the algebra $\mathfrak{g}$ can be decomposed into the direct sum of two ideals so that the representation R_Λ is trivial on one of them and R_M is trivial on the other.

In the paper [16] an important invariant of homomorphisms of simple Lie algebras is introduced. Let $\varphi : \tilde{\mathfrak{g}} \to \mathfrak{g}$ be a homomorphism of simple complex Lie algebras and let $\widetilde{(\cdot, \cdot)}$, $(\cdot, \cdot)$ be the Killing forms of the Lie algebras $\tilde{\mathfrak{g}}$ and $\mathfrak{g}$ respectively. Then $(\varphi(x), \varphi(y)) = j(\varphi)\widetilde{(x, y)}$ for all $x, y \in \tilde{\mathfrak{g}}$, where $j(\varphi) \in \mathbb{C}$. It turns out that $j(\varphi)$ is an integer, which is positive if $\varphi \neq 0$. It is called the *index* (or the *Dynkin index*) of the homomorphism φ.

In 1952 Dynkin's first papers dealing with the topology of compact Lie groups appeared. As shown by Hopf and Samelson in their well-known papers published in 1941, the dual cohomology algebra and homology algebra (or Pontryagin algebra) of a compact Lie group over the field of rational numbers are Grassman algebras whose free generators may be taken to be *primitive*, i.e., orthogonal to the decomposable elements of the dual algebra. But already in 1935, Pontryagin had computed the homology algebras of the classical Lie groups, and explicitly constructed certain integer-valued cycles determining systems of free generators in those algebras (see [46]). In the papers [20, 21, 22] Dynkin showed that these generators are primitive. It is also proved there that for the primitive generators of the cohomology algebra or the Pontryagin algebra of any compact group one can always choose integer-valued cohomology (respectively homology) classes. In the papers [23, 24] free systems of integer-valued primitive generators of the rational cohomology of the classical groups were also indicated. Here the relationship between the symmetric and skew-symmetric invariants of the adjoint group of an arbitrary compact Lie group was indicated; it became clear later that this relationship is due to the transgression operation in the principal bundle of G.

This remarkable relationship is the following (see [24]). Suppose $\mathfrak{g}$ is the tangent Lie algebra of the Lie group G. To any symmetric p-linear form ξ on $\mathfrak{g}$ let us assign the skew-symmetric $(2p-1)$-linear form η according to the rule

$$\eta(x_1, \ldots, x_{2p-1}) = \text{Alt}\,\xi(x_1, [x_2, x_3], \ldots, [x_{2p-2}, x_{2p-1}]),$$

where Alt stands for alternation. It turns out that the assignment $\sigma : \xi \mapsto \eta$ maps the algebra S_G of invariant symmetric forms on $\mathfrak{g}$ into the algebra A_G of invariant skew-symmetric forms. Here $\text{Ker}\,\sigma$ coincides with the ideal of decomposable elements in the free commutative algebra S_G, while $\text{Im}\,\sigma$ coincides with the space of primitive elements of the algebra A_G (if one identifies the latter algebra in the usual way with the algebra $H^*(G, \mathbb{R})$) of real-valued cohomology of the Lie group G). Thus σ defines a one-to-one correspondence between the free generators of the polynomial algebra S_G and the primitive free generators of the Grassmann algebra $H^*(G, \mathbb{R})$, the generators of degree p corresponding to the primitive class of dimension $2p-1$. For example, if G is simple and non-Abelian, then, taking for ξ the Killing form

$$\xi(x, y) = \text{tr}(\text{Ad}\,x\,\text{Ad}\,y),$$

which is the only (up to proportionality) invariant symmetric bilinear form on $\mathfrak{g}$, we see that $H^*(G, \mathbb{R})$ contains a unique primitive generator η of degree 3, and we have

$$\eta(x, y, z) = \xi(x, [y, z])$$

(in the case $\mathfrak{g} = \mathfrak{so}_3$ the form η is the "mixed product" of the vectors x, y, z in the three-dimensional Euclidean space).

In the papers [23, 24] the following question was also studied. It is well known that any homomorphism of connected compact Lie groups $f : H \to G$ induces an algebra homomorphism in cohomology $f^* : H^*(G, \mathbb{R}) \to H^*(H, \mathbb{R})$ that takes primitive elements to primitive ones. If $x_1, \ldots, x_r$ and $y_1, \ldots, y_s$ are integer-valued primitive generators of the algebras $H^*(G, \mathbb{R})$ and $H^*(H, \mathbb{R})$, repectively, then

$$f^*(x_i) = \sum_{j=1}^{s} d_{ij} y_j, \quad i = 1, \ldots, r,$$

where the coefficients d_{ij} are integers. The problem is to find the integers d_{ij} (sometimes called *Dynkin coefficients*) from the homomorphism f, which may be specified, say, by the set of highest weights of the corresponding linear representation. This question is of fundamental importance for the study of the topology of the corresponding homogeneous space $G/f(H)$. The problem easily reduces to the case when the group G is simple, and for the classical groups, to the case when f is an irreducible linear representation. On the other hand, the map σ allows us to reduce the computation of the coefficients d_{ij} to the study of the homomorphism $(df)^* : S_G \to S_H$ induced by the differential $df : \mathfrak{h} \to \mathfrak{g}$ of the homomorphism f. Using this method, Dynkin proved that for an irreducible linear representation f with highest weight Λ we have $d_{ij} = P_{ij}(\Lambda)$, where P_{ij} are polynomials depending on H, but not on f, and also found an explicit (but very complicated) form for the polynomials P_{ij} in the case when $H = U_r$ is the unitary group. Somewhat simpler formulas for these polynomials were indicated later by R. Z. Rozenknop [47]. Note that if G and H are simple and the classes x_1, y_1 are of dimension 3, then d_{11} is nothing but the index of the homomorphism f, which is therefore a topological invariant.

The papers [27, 28, 29] bring together the algebraic and probabilistic research interests of Dynkin. They deal with the homogeneous space $E = SL_l(\mathbb{C})/SU_l$, which is a symmetric space of negative curvature with repect to the invariant Riemann metric. Let $\mathfrak{D}$ be the Laplace–Beltrami operator on E corresponding to the above-mentioned metric. In Dynkin's work the Green function of the operator $\mathfrak{D} - c$ (where c is a constant) is calculated, the Martin boundary of the space E corresponding to this operator is described, the structure of the set of nonnegative eigenfunctions of the operator $\mathfrak{D}$ is studied, as well as that of the set of bounded eigenfunctions. In the same papers the behavior as $t \to \infty$ of the trajectory of Markov processes of diffusion type with continuous trajectories in the space E invariant with respect to $SL_l(\mathbb{C})$.

The space E can be realized as the space of positive definite unimodular Hermitian matrices of order l with transformations of the form

$$(3) \qquad x \mapsto gxg^*, \quad x \in E, \; g \in SL_l(\mathbb{C}),$$

where g^* is the matrix transposed and complex-conjugate to g. All the characteristic numbers of any matrix $x \in E$ are positive. The tuple $(\rho_1, \ldots, \rho_l)$ of their logarithms, listed in nonincreasing order $\rho_1 \geq \cdots \geq \rho_l$, is denoted by $\rho(x)$. Clearly, we have $\rho_1 + \cdots + \rho_l = 0$. For any tuple of real numbers $(\rho_1, \ldots, \rho_l)$ let us put $\rho^2 = \rho_1^2 + \cdots + \rho_l^2$. The invariant Riemann metric in E is appropriately normed and the equation

$$(4) \qquad \mathfrak{D}f = cf,$$

where c is a constant, is considered. Let us put $\delta = (\delta_1, \ldots, \delta_l)$, where $\delta_j = (l+1-2j)/2$, $j = 1, \ldots, l$, and require that the constant c in equation (4) satisfy the condition

$$(5) \qquad c \geq -\delta^2.$$

(If this condition is not satisfied, then equation (4) has no nonzero nonnegative solutions [29].)

In [27, 29] an explicit formula for the Green function of equation (4) is derived. It turned out that this function can be expressed in terms of Bessel functions of an imaginary argument. Using the expression found for the Green function, Dynkin describes in detail the Martin boundary [42] of the space E corresponding to equation (4), studies the structure of the set of nonnegative solutions of this equation, and determines the minimal one of these solutions. Let $d_k(x)$, $x \in E$, $k = 1, \ldots, l$, the corner minor of order k of the matrix $x = (a_{ij})$, i.e.,

$$d_k(x) = \begin{vmatrix} a_{11} & \cdots & a_{1k} \\ \vdots & & \vdots \\ a_{k1} & \cdots & a_{kk} \end{vmatrix},$$

and T_c is the set of all tuples $(p_1, \ldots, p_l)$ of real numbers satisfying the conditions

$$p_1 \geq \ldots \geq p_l, \quad p_1 + \ldots + p_l = 0, \quad p^2 = c + \delta^2.$$

Let us put

$$(6) \qquad f_p(x) = \prod_{k=1}^{l-1} d_k(x)^{p_{k+1}-p_k-1}.$$

It is proved that the set of minimal nonnegative solutions of equation (4) coincides with the family of functions

$$(7) \qquad x \mapsto a f_p(u^* x u), \quad x \in E,$$

where a is a positive constant, $p \in T_c$, the matrix u is unitary and is defined up to multiplication from the right by an arbitrary diagonal matrix from U_l. Under these conditions the representation (7) is unique.

Using (7), Dynkin obtained an integral representation of an arbitrary nonnegative solution of equation (4). The paper [27] also contains an integral representation of any nonnegative general eigenfunction of all differential operators on E that commute with all isometries (3). In [29] nonzero bounded solutions of equation (4), which exist only for $c = 0$, are also studied and an integral representation for them is obtained.

To the operator $\mathfrak{D}$ corresponds a continuous random Markov process x_t of diffusive type (Brownian motion) in the space E. In [29] the behavior of the trajectories of this process are studied as $t \to \infty$. To state the result, let us present the matrix x_t in the form

$$(8) \qquad x_t = u_t \lambda_t u_t^*,$$

where $u_t \in U_l$ and λ_t is the diagonal martix with diagonal elements $e^{p_1}, \ldots, e^{p_l}$, $(p_1, \ldots, p_l) = p(x_t)$. Let us put

$$\tilde{p}_j = \frac{p_j}{|p(x_t)|},$$

and let $\tilde{\lambda}_t$ be the diagonal matrix with diagonal elements $e^{\tilde{p}_1}, \ldots, e^{\tilde{p}_l}$. Suppose

$$\tilde{x}_t = u_t \tilde{\lambda}_t u_t^*,$$

where the matrix u_t is the same as in (8).

It is established in [29] that the following limit almost surely exists

$$\lim_{t \to \infty} \tilde{x}_t = y.$$

The matrix y is random and belongs to the space E. The probability distribution of the matrix y was derived in [29]. In particular, it turned out that $p(y)$ is not random and equals δ.

Let us note that the method developed by Dynkin in [27, 29] was applied by E. V. Nolde to the spaces $E = O_l(\mathbb{C})/O_l$ and $E = Sp_{2l}(\mathbb{C})/Sp_l$. Many results of these papers were carried over by Karpelevich [37, 38, 39] and Furstenberg [32] to arbitrary symmetric spaces of noncompact type.

Dynkin's papers dealing with the Laplace–Beltrami operator and Markov processes on symmetric spaces were his last works related to the theory of Lie groups.

References

(A) Papers by Dynkin in the theory of Lie groups and Lie algebras

1. Dynkin, *Classification of simple Lie groups*, Mat. Sb. **18**. (1946), 347–352. (Russian)
2. _____, *Computation of the coefficients in the Campbell–Hausdorff formula*, Doklady Akad. Nauk SSSR **57** (1947), 323–326. (Russian)
3. _____, *An effective construction of a Lie group from a Lie algebra*, Uspekhi Mat. Nauk **2** (1947), 192. (Russian)
4. _____, *Structure of semisimple Lie algebras*, Uspekhi Mat. Nauk **2** (1947), 59–127; English transl., Amer. Math. Soc. Transl. (1), vol. 9, Amer. Math. Soc., Providence, RI, 1962, pp. 328–469.
5. _____, *Normed Lie algebras and analytic groups*, Uspekhi Mat. Nauk **3** (1948), 157–158. (Russian)
6. _____, *On the representation of the series $\log(e^x e^y)$ in noncommuting x and y via the commutators*, Mat. Sb. **25** (1949), 155–162. (Russian)
7. _____, *Normed Lie algebras and analytic groups*, Uspekhi Mat. Nauk **5** (1950), 135–186; English transl., Amer. Math. Soc. Transl. (1), vol. 9, Amer. Math. Soc., Providence, RI, 1962, pp. 470–542.
8. _____, *On the inclusion relation for irreducible subgroups of matrix groups*, Uspekhi Mat. Nauk **5** (1950), 214. (Russian)
9. _____, *Certain properties of the system of weights of linear representations of semisimple Lie groups*, Dokl. Akad. Nauk SSSR **71** (1950), 221–224. (Russian)
10. _____, *Regular semisimple subalgebras of semisimple Lie algebras*, Dokl. Akad. Nauk SSSR **73** (1950), 877–880. (Russian)
11. _____, *Maximal subgroups of semisimple Lie groups and the classification of primitive transformation groups*, Dokl. Akad. Nauk SSSR **75** (1950), 333–336. (Russian)
12. _____, *Automorphisms of semisimple Lie algebras*, Dokl. Akad. Nauk SSSR **76** (1951), 629–632. (Russian)
13. _____, *Inclusion relations between irreducible groups of linear transformations*, Dokl. Akad. Nauk SSSR **78** (1951), 5–7. (Russian)
14. _____, *On semisimple subalgebras of semisimple Lie algebras*, Dokl. Akad. Nauk SSSR **81** (1951), 987–990. (Russian)
15. _____, *Maximal subgroups of classical groups*, Trudy Moskov. Mat. Obshch **1** (1952), 39–166; English transl., Amer. Math. Soc. Transl. (2), vol. 6, Amer. Math. Soc., Providence, RI, 1957, pp. 245–378.
16. _____, *Semisimple subalgebras of semisimple Lie algebras*, Mat. Sb **30** (1952), 349–462; English transl., Amer. Math. Soc. Transl. (2), vol. 6, Amer. Math. Soc., Providence, RI, 1957, pp. 111–244.
17. _____, *Topological invariants of linear representations of the unitary group*, Dokl. Akad. Nauk SSSR **85** (1952), 697–699. (Russian)
18. _____, *Relashionship between the homology of a compact Lie group and its subgroups*, Dokl. Akad. Nauk SSSR **87** (1952), 333–336. (Russian)
19. _____, *Topological invariants of subgroups of the group of unitary matrices*, Uspekhi Mat. Nauk **7** (1952), 142–143. (Russian)
20. _____, *On the topology of compact Lie groups*, Uspekhi Mat. Nauk **8** (1953), 174–175. (Russian)
21. _____, *Construction of primitive cycles in compact Lie groups*, Dokl. Akad. Nauk SSSR **91** (1953), 201–204. (Russian)
22. _____, *The homology of compact Lie groups*, Uspekhi Mat. Nauk **8** (1953), 73–120; English transl., Amer. Math. Soc. Transl. (2), vol. 12, Amer. Math. Soc., Providence, RI, 1959, pp. 251–300.
23. _____, *Homological characteristics of homomorphisms of compact Lie groups*, Dokl. Akad. Nauk SSSR **91** (1953), 1007–1009. (Russian)
24. _____, *Topological characteristics of homomorphisms of compact Lie groups*, Mat. Sb **35** (1954), 129–173; English transl., Amer. Math. Soc. Transl. (2), vol. 12, Amer. Math. Soc., Providence, RI, 1959, pp. 301–342.
25. _____, *Compact Lie groups in the large*, Uspekhi Mat. Nauk **10** (1955), 3–74; English transl., Amer. Math. Soc. Transl. (2), vol. 21, Amer. Math. Soc., Providence, RI, 1962, pp. 119–192, (jointly with A. L. Onishchik).
26. _____, *Lie groups*, Mathematics in the USSR for 40 years,, vol. 1, Fizmatgiz, Moscow, 1959, pp. 213–227. (Russian)

27. _____, *Nonnegative eigenfunctions of the Laplace–Beltrami operator and Brownian motion in certain symmetric spaces*, Dokl. Akad. Nauk. SSSR **141** (1961), 288–291; English transl. Soviet Math. Dokl. **2** (1961).
28. _____, *Markov processes and problems in analysis*, Proc. Intern. Congr. Math. (Stockholm 1962), Uppsala, 1963, pp. 38–52; English transl., Amer. Math. Soc. Transl. (2), vol. 31, Amer. Math. Soc., Providence, RI, 1963, pp. 1–24.
29. _____, *Brownian motion in certain symmetric spaces and nonnegative eigenfunctions of the Laplace–Beltrami operator*, Izvestiya Akad. Nauk SSSR **30** (1966), 455–478; English transl., Amer. Math. Soc. Transl. (2), vol. 72, Amer. Math. Soc., Providence, RI, 1968, pp. 203–228.

(B) Other publications

30. A. Borel and J. de Siebenthal, *Sur les sous-groupes fermés de rang maximum des groupes de Lie compacts connexes*, Comment. Math. Helv **23** (1949), 200–221.
31. A. G. Elashvili, *Centralizers of nilpotent elements in semisimple Lie algebras*, Trudy Tbil. Mat. Inst **46** (1975), 109–132. (Russian)
32. H. Furstenberg, *A Poisson formula for semisimple Lie groups*, Ann. Math. **77** (1963), 335–386.
33. F. P. Gantmacher, *Canonical representation of automorphisms of a complex semi-simple Lie group*, Mat. Sb **5** (1939), 101–146. (Russian)
34. S. Helgason, *Differential geometry, Lie groups, and symmetric spaces.*, Academic Press, New York and London, 1984.
35. V. G. Kac, *Infinite dimensional Lie algebras*, Third ed., Cambridge Univ. Press, Cambridge and New York, 1990.
36. F. I. Karpelevich, *On nonsemisimple maximamal subalgebras of semisimple Lie algebras*, Dokl. Akad. Nauk SSSR **76** (1951), 775–778. (Russian)
37. _____, *Geodesic lines and harmonic functions on symmetric spaces*, Dokl. Akad. Nauk SSSR **124** (1959), 1199–1202. (Russian)
38. _____, *Nonnegative eigenfunctions of the Laplace–Beltrami operator on symmetric spaces of nonpositive curvature*, Dokl. Akad. Nauk SSSR **151** (1963), 1274–1276; English transl. Soviet Math. Dokl. **4** (1963).
39. _____, *The geometry of geodesic lines and eigenfunctions of the Laplace–Beltrami operator on symmetric spaces*, Trudy Moskov. Mat. Obshch **14** (1965), 48–185; English transl. in Trans. Moscow Math. Soc. **1965** (1967).
40. B. Kostant, *The principal three-dimensional subgroup and the Betti numbers of a complex simple Lie group*, Amer. J. Math. **81** (1959), 973–1032.
41. A. I. Maltsev, *On semisimple subgroups of Lie groups*, Izv. Akad. Nauk SSSR **8** (1944), 143–174; English transl., Amer. Math. Soc. Transl. (1), vol. 9, Amer. Math. Soc., Providence, RI, 1962, pp. 172–213.
42. R. S. Martin, *Minimal positive harmonic functions*, Trans. Amer. Math. Soc. **49** (1941), 137–162.
43. V. V. Morozov, *On the nilpotent element of a semisimple algebra*, Dokl. Akad. Nauk. SSSR **36** (1942), 91–94. (Russian)
44. _____, *Proof of the regularity theorem*, Uspekhi Mat. Nauk **11** (1956), 191–194. (Russian)
45. _____, *On the theorem about the nilpotent element in a semisimple Lie algebra*, Uspekhi Mat. Nauk **15** (1960), 137–139. (Russian)
46. L. S. Pontryagin, *Homology in compact Lie groups*, Mat. Sb. **6** (1939), 389–422.
47. R. Z. Rozenknop, *Certain questions and applications of the theory of polynomial invariants*, Trudy Moskov. Mat. Obshch. **13** (1965), 246–323; English transl. in Trans. Moscow Math. Soc. **1965** (1967).
48. R. Steinberg, *Finite reflection groups*, Trans. Amer. Math. Soc. **91** (1959), 493–504.
49. J. Tits, *Sous-algèbres des algèbres de Lie semi-simples*, Sémin. Bourbaki (Mai 1955).
50. E. B. Vinberg, *Classification of homogeneous elements of a semisimple graded Lie algebra*, Trudy Seminara po Vekt. Tenz. Analizu **19** (1979), 155–177; English transl., Selecta Math. Societica **6** (1987), 15–35.

51. E. B. Vinberg and A. G. Elashvili, *Classification of trivectors of nine-dimensional space*, Trudy Seminara po Vekt. Tenz. Analizu **18** (1978), 198–233. (Russian)

F. I. KARPELEVICH: MOSCOW INSTITUTE OF TRANSPORT ENGINEERING, OBRAZTSOVA STR. 15, 101475 MOSCOW, RUSSIA

A. L. ONISHCHIK: YAROSLAVL UNIVERSITY, 150000 YAROSLAVL, RUSSIA

E. B. VINBERG: CHAIR OF ALGEBRA, DEPARTMENT OF MATHEMATICS, MOSCOW STATE UNIVERSITY, 119899 MOSCOW, RUSSIA

Matrix Vieta Theorem

DMITRY FUCHS AND ALBERT SCHWARZ

§1. Introduction

Consider a matrix algebraic equation

(1) $$X^n + A_1 X^{n-1} + \cdots + A_n = 0,$$

where the coefficients $X_1, \ldots, X_n$ as well as solutions X are supposed to be square complex matrices of some order k. For a usual algebraic equation of degree n the classical Vieta formulas express the coefficients in terms of the n solutions. However, a matrix equation of degree n generically has $\binom{nk}{k}$ rather than n solutions. (Throughout this article the word *generic* refers to a Zarisky open set.)

We call n solutions $X_1, \ldots, X_n$ of the equation (1) *independent* if they determine the coefficients $A_1, \ldots, A_n$ (a more technical explanation of independence seen in §2). However, the expressions of $A_1, \ldots, A_n$ in terms of $X_1, \ldots, X_n$ are much less elegant than Vieta formulas (see, e.g., formulas (3)–(5) below). Still there are some relations between A's and X's, which are very similar to Vieta.

THEOREM 1.1. *If solutions $X_1, \ldots, X_n$ of the equation* (1) *are independent, then*

(2) $$\operatorname{tr} A_1 = -(\operatorname{tr} X_1 + \cdots + \operatorname{tr} X_n),$$
$$\det A_n = (-1)^{nk} \det X_1 \ldots \det X_n.$$

Theorem 1.1 is proved in §4 (a more direct proof for $n = 2$ is given in §3). In §5 we discuss a generalization of Theorem 1.1 from complex matrix algebras to arbitrary associative unitary rings.

Although Theorem 1.1 is elementary, it is nevertheless connected with some constructions of modern mathematics. For every associative algebra A one can construct a linear space $F(A) = A^2/[A, A]$. This space appeared in [K] as the space of 0-forms on a noncommutative formal manifold. (Such a manifold is determined by a free associative algebra A. For the free associative algebra A with generators $a_1, \ldots, a_n$ the space $F(A)$ is spanned by cyclic words of length ≥ 2 in letters $\vec{1}na$.) The space $F(A)$ appeared also in [GS] and [AW] in relation to other problems. It is easy to understand that the formula for $\operatorname{tr} A_1$ can be interpreted as a nontrivial identity in $F(A)$.

1991 *Mathematics Subject Classification.* Primary 15A24.
This work was partially supported by NSF grant DMS-9201366. Research at MSRI was supported by NSF grant DMS-9022140.

§2. Independent matrices

DEFINITION 2.1. Matrices $X_1, \ldots, X_n$ are called *independent* if the bloc Vandermonde determinant does not vanish,

$$\begin{vmatrix} I & I & \cdots & I \\ X_1 & X_2 & \cdots & X_n \\ \multicolumn{4}{c}{\dotfill} \\ X_1^{n-1} & X_2^{n-1} & \cdots & X_n^{n-1} \end{vmatrix} \neq 0.$$

For $n = 2$ the independence condition means that $\det(X_1 - X_2) \neq 0$. For $n \geq 3$ it does not imply and is not implied by the condition $\det(X_i - X_j) \neq 0$, $1 \leq i < j \leq n$.

It is obvious that a generic equation (1) has n independent solutions (otherwise the above determinant would have vanished for any n solutions of any equation (1)). It is clear also that the matrices $X_1, \ldots, X_n$ are independent if and only is there exist unique $A_1, \ldots, A_n$ such that $X_1, \ldots, X_n$ satisfy the equation (1). In other words, for independent $X_1, \ldots, X_n$ the matrices $A_1, \ldots, A_n$ may be expressed in form of $X_1, \ldots, X_n$. For example, if $n = 2$ then

$$\begin{aligned} A_1 &= -(X_1^2 - X_2^2)(X_1 - X_2)^{-1}, \\ A_2 &= -X_1^2 + (X_1^2 - X_2^2)(X_1 - X_2)^{-1} X_1. \end{aligned} \tag{3}$$

For $n \geq 3$ it is impossible to write a formula that would be valid for all independent matrices; for example, if $n = 3$, then

$$\begin{aligned} A_1 = &-((X_1^3 - X_2^3)(X_1 - X_2)^{-1} - (X_1^3 - X_3^3)(X_1 - X_3)^{-1}) \\ &\times ((X_1^2 - X_2^2)(X_1 - X_2)^{-1} - (X_1^2 - X_3^2)(X_1 - X_3)^{-1})^{-1} \end{aligned} \tag{4}$$

provided that the right-hand side exists (which does not follow from the independence of X_1, X_2, X_3); otherwise the expression will be different.

An expression of $A_1, \ldots, A_n$ as functions of $X_1, \ldots, X_n$, that is valid for generic independent $X_1, \ldots, X_n$, can be given in terms of the Gelfand–Retakh quasideterminants, which are defined as follows.

DEFINITION 2.2. [GR]. Let $A = \{\|a_{ij}\|, i \in I, j \in J\}$ be a square matrix of order $n = \operatorname{card} I = \operatorname{card} J$ with formal noncommuting entries a_{ij}. For $p \in I$, $q \in J$ denote by A^{pq} the submatrix $\{\|a_{ij}\|, i \in I - p, j \in J - q\}$ of A. The formula

$$|A|_{pq} = a_{pq} - \sum_{\substack{i \in I - p \\ j \in J - q}} a_{pj} |A^{pq}|_{ij}^{-1} a_{ip}$$

(which reduces to $|A|_{pq} = a_{pq}$ if $n = 1$) defines inductively n^2 *quasideterminants* $|A|_{pq}$ of the matrix A. (In the commutative case $|A|_{pq} = \pm \det A / \det A_{pq}$.)

Quasideterminants possess some basic properties of determinants; in particular, the following *Kramer rule* holds.

PROPOSITION 2.3. [GR]. *If $(x_1, \ldots, x_n)$ is a solution of system of equations*

$$\sum_{j=1}^{n} a_{ij} x_j = \xi_i, \ (i = 1, \ldots, n),$$

then for any i,

$$x_j = |A|_{ij}^{-1} |A_j(\xi)|_{ij},$$

where $A = \|a_{ij}\|$ and $A_j(\xi)$ is obtained from A be replacing its j-th column by the column $(\xi_1, \ldots, \xi_n)$.

COROLLARY 2.4. *For generic independent $X_1, \ldots, X_n$ and for arbitrary i,*

(5)
$$-A_j = \begin{vmatrix} I & X_1 & \ldots & X_1^{n-1} \\ I & X_2 & \ldots & X_2^{n-1} \\ \cdots & \cdots & \cdots & \cdots \\ I & X_n & \ldots & X_n^{n-1} \end{vmatrix}_{i,n-j}^{-1}$$

$$\times \begin{vmatrix} I & X_1 & \ldots & X_1^{n-j-1} & X_1^n & X_1^{n-j+1} & \ldots & X_1^{n-1} \\ I & X_2 & \ldots & X_2^{n-j-1} & X_2^n & X_2^{n-j+1} & \ldots & X_2^{n-1} \\ \cdots & \cdots & \cdots & \cdots & \cdots & \cdots & \cdots & \cdots \\ I & X_n & \ldots & X_n^{n-j-1} & X_n^n & X_n^{n-j+1} & \ldots & X_n^{n-1} \end{vmatrix}_{i,n-j}.$$

This expression multiplied by $(-1)^j$ (with a minor change of notations and with $i = n$) is called the j-th elementary symmetric function in $X_1, \ldots, X_n$ in [GKLLRT, Section 7.1]. It is proved in [GKLLRT] and is obvious from Corollary 2.4 that for generic $X_1, \ldots, X_n$ it is really symmetric in $X_1, \ldots, X_n$.

§3. The case $n = 2$

Since for any square matrices X, Y we have

$$\begin{vmatrix} I & I \\ X & Y \end{vmatrix} = \det(Y - X),$$

the following result coincides with Theorem 1.1 for $n = 2$.

PROPOSITION 3.1. *Let A, B, X, Y be square matrices of the same order with $X - Y$ being nondegenerate. If*

(6)
$$X^2 + AX + B = 0,$$
$$Y^2 + AY + B = 0,$$

then

$$\operatorname{tr} A = -(\operatorname{tr} X + \operatorname{tr} Y),$$
$$\det B = \det X \det Y.$$

PROOF. 1. Equalities (6) imply

(7)
$$-A = (X^2 - Y^2)(X - Y)^{-1},$$

and since $X^2 - Y^2 = (X+Y)(X-Y) + XY - YX$, we have

$$-A = X + Y + (XY - YX)(X - Y)^{-1}.$$

Therefore,

(8) $\quad -\operatorname{tr} A = \operatorname{tr} X + \operatorname{tr} Y + \operatorname{tr} XY(X - Y)^{-1} - \operatorname{tr} YX(X - Y)^{-1},$

and the relations
$$X(X - Y)^{-1} = Y(X - Y)^{-1} + I,$$
$$(X - Y)^{-1} X = (X - Y)^{-1} Y + I$$

together with the identity $\operatorname{tr} UV = \operatorname{tr} VU$ imply

$$\operatorname{tr} XY(X - Y)^{-1} = \operatorname{tr} Y(X - Y)^{-1} X = \operatorname{tr} Y(X - Y)^{-1} Y + \operatorname{tr} Y,$$
$$\operatorname{tr} YX(X - Y)^{-1} = \operatorname{tr} X(X - Y)^{-1} Y = \operatorname{tr} Y(X - Y)^{-1} Y + \operatorname{tr} Y.$$

Hence $\operatorname{tr} XY(X - Y)^{-1} = \operatorname{tr} YX(X - Y)^{-1}$, and (8) yields

$$-\operatorname{tr} A = \operatorname{tr} X + \operatorname{tr} Y.$$

2. The first of the equalities (6) implies

(9) $\qquad\qquad\qquad B = -(X + A)X,$

and using (7) we get

$$(X + A)(X - Y) = (X - (X^2 - Y^2)(X - Y)^{-1})(X - Y)$$
$$= X^2 - XY - X^2 + Y^2 = Y^2 - XY = (X - Y)(-Y).$$

Therefore,

$$X + A = (X - Y)(-Y)(X - Y)^{-1},$$
$$\det(X + A) = \det(-Y),$$

and by (9)

$$\det B = \det(-(X + A)) \det X = \det Y \det X.$$

§4. The general case

Unlike the above proof for $n = 2$, our proof in the general case does not reduce to direct calculations and uses specifics of the matrix algebra.

LEMMA 4.1. *Let* $A_1, \ldots, A_n, X_1, \ldots, X_n$ *be square matrices of some order* k *and let the eigenvalues of the matrices* $X_1, \ldots, X_n$ *be* kn *pairwise different complex numbers. If*

$$X^n + A_1 X^{n-1} + \cdots + A_n = 0 \tag{10}$$

for $i = 1, \ldots, n$, *then*

$$\operatorname{tr} A_1 = -(\operatorname{tr} X_1 + \cdots + \operatorname{tr} X_n),$$
$$\det A_n = (-1)^{kn} \det X_1 \ldots \det X_n.$$

PROOF. Let λ_{ij}, $j = 1, \ldots, k$, be eigenvalues of X_i and let $X_i v_{ij} = \lambda_{ij} v_{ij}$, $v_{ij} \neq 0$. Then for arbitrary i, j, (10) implies

$$0 = (X_i^n + A_1 X_i^{n-1} + \cdots + A_n) v_{ij} = (\lambda_{ij}^n + A_1 \lambda_{ij}^{n-1} + \cdots + A_n) v_{ij},$$

whence

$$\det(\lambda_{ij}^n + A_1 \lambda_{ij}^{n-1} + \cdots + A_n) = 0.$$

Obviously,

$$P(\lambda) = \det(\lambda^n + A_1 \lambda^{n-1} + \cdots + A_n)$$

is a monic polynomial in λ of degree kn, and since $P(\lambda_{ij}) = 0$ and all λ_{ij} are different, we have

$$\det(\lambda^n + A_1 \lambda^{n-1} + \cdots + A_n) = \prod_{i,j} (\lambda - \lambda_{ij}). \tag{11}$$

Comparing constant terms and the coefficients in the term with λ^{kn-1} of the two sides of (11), we have

$$\det A_n = \prod_{i,j} (-\lambda_{ij}) = (-1)^{kn} \prod_{i,j} \lambda_{ij} = (-1)^{kn} \prod_i \det X_i$$

$$\operatorname{tr} A_1 = \sum_{i,j} (-\lambda_{ij}) = -\sum_i \operatorname{tr} A_i.$$

REMARK 4.2. Comparing other coefficients of polynomials in (11), we may get $kn - 2$ more identities. Most of them involve matrices obtained by combinations of columns of different matrices A_i, but in the two extreme cases (those of λ^p with $p = kn - 2$ and 1) we get the formulas which are worth mentioning:

$$\operatorname{tr} A_2 + \sigma_2(A_1) = \sum_{1 \leq i < j \leq n} (\operatorname{tr} X_i \operatorname{tr} X_j) + \sum_{1 \leq i \leq n} \sigma_2(X_i); \tag{12}$$

$$\det A_{n-1} \cdot \operatorname{tr} A_n A_{n-1}^{-1} = (-1)^{kn-1} \left(\prod \det X_i \right) \left(\sum \operatorname{tr} X_i^{-1} \right), \tag{13}$$

where σ_2 in (12) denotes the second coefficient (the coefficient in the term with λ^{k-2}) of the characteristic polynomial of a matrix, and (13) is valid only if the matrices $A_{n-1}, X_1, \ldots, X_n$ are nondegenerate.

REMARK 4.3. The second formula (2) implies the first formula (2), and many other formulas as well. Put $X_i = Y_i + \lambda I$. Then $Y_1, \ldots, Y_n$ are solutions of the equation

$$(14) \quad (Y+\lambda I)^n + A_1(Y+\lambda I)^{n-1} + \cdots + A_n = Y^n + \cdots + (\lambda^n + A_1 \lambda^{n-1} + \cdots + A_n) = 0.$$

Obviously, these solutions are independent (the coefficients of the equation (14) are expressed in terms $X_1, \ldots, X_n$, and hence in terms $Y_1, \ldots, Y_n$), and the second formula (2) yields

$$(-1)^{nk} \det(X_1 - \lambda I) \ldots \det(X_n - \lambda I) = \det(\lambda^n + A_1 \lambda^{n-1} + \cdots + A_n)$$

(which is the same as (11)), or, equivalently,

$$(15) \qquad \det(I - \mu X_1) \ldots \det(I - \mu X_n) = \det(I + A_1 \mu + \cdots + A_n \mu^n)$$

($\mu = \lambda^{-1}$). Taking the derivatives of the both sides with respect to μ at $\mu = 0$, we have

$$\text{tr}(X_1 + \cdots + X_n) = \text{tr}\, A_1,$$

which is the first formula in (2). Moreover, using the identity

$$\log \det C = \text{tr} \log C,$$

we derive from (15) that

$$\text{tr} \sum_{i=1}^{n} \log(I - \mu X_i) = \text{tr} \log(I + A_1 \mu + \cdots + A_n \mu^n),$$

or

$$\text{tr} \sum_{k=1}^{\infty} (-1)^k \frac{\mu^k}{k} \sum_{i=1}^{n} X_i^k = \text{tr} \sum_{s=0}^{\infty} \sum_{i_1, \ldots, i_s = 1}^{n} \frac{\mu^{i_1 + \cdots + i_s}}{s} A_{i_1} \ldots A_{i_s}.$$

Hence we have the following sequence of formulas:

$$(16) \qquad (-1)^k \sum_{i=1}^{n} \text{tr}\, X_i^k = \sum_{\substack{i_1 + \cdots + i_s = k \\ 1 \le i_u \le n}} \frac{k}{s} \text{tr}(A_{i_1} \ldots A_{i_s})$$

($k = 1, 2, \ldots$). For $k = 1$ the formula (16) is the first formula in (2). For $k = 2, 3, 4, \ldots$ we get infinitely many formulas (which are equivalent to the formulas of Remark 4.2); here are the first three formulas:

$$\sum_{i=1}^{n} X_i^2 = \text{tr}(2A_2 + A_1^2);$$

$$-\sum_{i=1}^{n} X_i^3 = \text{tr}(3A_3 + (3/2)A_1 A_2 + A_1^3);$$

$$\sum_{i=1}^{n} X_i^4 = \text{tr}(4A_4 + 4A_1 A_3 + 2A_2^2 + 4A_1^2 A_2 + A_1^4).$$

PROOF OF THEOREM 1.1. Let U, $V \subset \text{Mat}_k \mathbb{C} \times \cdots \times \text{Mat}_k \mathbb{C}$ (n factors) be respectively the set of all n-tuples $(X_1, \ldots, X_n)$ of independent matrices and the set of all n-tuples $(X_1, \ldots, X_n)$ of matrices whose eigenvalues are kn pairwise different complex numbers. Obviously, both sets are Zarisky open and nonempty, hence $U \cap V$ is dense in U. According to the Lemma, both equalities (2) hold in $U \cap V$. Since both sides of each of the equalities (2) are continuous on U (with respect to $X_1, \ldots, X_n$), equalities (2) hold everywhere in U.

§5. A further generalization

Let R be an associative ring with unity and k be a field. Suppose that we fix either an additive homomorphism $\text{tr}\colon R \to k$ satisfying the condition $\text{tr}\, UV = \text{tr}\, VU$ for any U, $V \in R$, or a ring homomorphism $\det\colon R \to k$ (or both).

PROPOSITION 5.1. *Let A, B, X, $Y \in R$ with $X - Y$ being invertible. If $X^2 + AX + B = 0$, $Y^2 + AY + B = 0$, then $\text{tr}\, A = -(\text{tr}\, X + \text{tr}\, Y)$ and/or $\det B = \det X \det Y$, whichever is defined.*

The proof is the same as in §3.

To generalize to arbitrary rings the general case of Theorem 1.1, we need an explicit expression of A_1, A_n in terms of $X_1, \ldots, X_n$.

THEOREM 5.2. *Let*

$$A_1 = a_1(X_1, \ldots, X_n),$$
$$A_n = a_n(X_1, \ldots, X_n)$$

be expressions of A_1, A_n in terms of $X_1, \ldots, X_n$ involving ring operations and taking inverse elements; we assume that these expressions are valid whenever these inverse elements exist (like (3), (4), (5)). Then for any $X_1, \ldots, X_n \in R$ each of the equalities

$$\text{tr}\, a_1(X_1, \ldots, X_n) = -(\text{tr}\, X_1 + \cdots + \text{tr}\, X_n),$$
$$\det a_n(X_1, \ldots, X_n) = \det(-X_1) \ldots \det(-X_n)$$

holds provided that both sides exist (that is, tr or $\det$ is defined and the inverse elements involved in a_1 or a_n exist in R).

Theorem 5.2 cannot be proved by arguments similar to that of §4. But it is known that identities which hold for complex matrices hold also in arbitrary associative rings with unities (see [A, Section 12.4.3]). Hence Theorem 5.2 follows from Theorem 1.1. The formulas at the end of Remark 4.3 are generalized in a similar way.

Acknowledgements. We are indebted to C. Itzykson and I. Kaplansky for interesting discussions. A. Schwarz is grateful to MSRI and the Isaac Newton Institute for their hospitality.

References

[A]   M. Artin, *Algebra*, Prentice Hall, Englewood Cliffs NJ, 1991.
[AW]   S. L. Adler and Yong-Shi Wu, *Algebraic and geometric aspects of generalized quantum dynamics.*, Preprint hep-th 9405054.

[GR] I. M. Gelfand and V. S. Retakh, *A theory of noncommutative determinants and characteristic functions of graphs.* I, Publ. LACIM, no. 14, Univ. de Québec, Montréal, 1–26.

[GKLLRT] I. M. Gelfand, D. Krob, A. Lascoux, B. Leclerc, V. S. Retakh and J.–Y. Thibon, *Noncommutative symmetric functions*, Preprint hep-th 9407124.

[GS] I. M. Gelfand and M. M. Smirnov, *The algebra of Chern-Simons classes and the Poisson brackets on it*, Preprint hep-th 9404103.

[K] M. Kontsevich, *Formal (non)-commutative symplectic geometry*, Gelfand Mathematical Seminar, 1992, Birkhäuser, Boston, 1993, pp. 173–189.

DEPARTMENT OF MATHEMATICS, UNIVERSITY OF CALIFORNIA, DAVIS, CA 95616, USA

Integral Geometry on Real Quadrics

SIMON GINDIKIN

We consider the problem of the reconstructing sections of a line bundle on a quadric Q in the real projective space P^n from their integrals on hyperplane sections. The dimension of the quadric is $n - 1$, but the manifold of hyperplanes has the dimension n. So the problem is overdeterminate and the inversion formula is not unique. The ideology of considering overdeterminate problems in integral geometry was developed in [1]. The method of the operator κ of Gelfand-Graev-Shapiro gives, in a sense, the universal structure of inversion formulas. Several examples are now known when the operator κ can be constructed. In all these cases, the problem of integral geometry is local (roughly speaking, the inverse operator is differential). We construct here the operator κ for real quadrics. This problem of integral geometry is local for even n, but nonlocal for odd n. Probably, this is the first occasion when the operator κ is constructed for a nonlocal overdeterminate problem of integral geometry. Our method can be easily generalized to other projective varieties but we prefer to focus here on the structure of nonlocal operator κ in simple examples. This integral transformation is already quite interesting and it includes as special cases the Minkowski-Funk transform, the geodesic and horospherical Radon transforms on hyperbolic spaces, etc.

I want to mention that I first heard about integral geometry at the Dynkin seminar. It was Gelfand's lecture in 1959 about the method of horospheres (following the joint results, unpublished at that time, with Graev), from which all the modern activity of integral geometry had started.

1. Basic definition. Let P^n be the (real) projective space with homogeneous coordinates $x = (x_0, x_1, \ldots, x_n)$ and Q the quadric

$$(1) \qquad \Box_\varepsilon(x) = x_0^2 + \varepsilon_1 x_1^2 + \cdots + \varepsilon_n x_n^2 = 0,$$

where $\varepsilon_j = \pm 1$. Of course, we exclude the case $\varepsilon_1 = \cdots = \varepsilon_n = 1$. If $\varepsilon_1 = \cdots = \varepsilon_n = -1$, then in affine coordinates $\{x_0 = 1\}$ we have the sphere. On Q, we consider homogeneous functions (sections of a line bundle):

$$(2) \qquad f(\lambda x) = |\lambda|^{-n+2} f(x), \qquad \lambda \in \mathbb{R} \setminus 0.$$

1991 *Mathematics Subject Classification.* Primary 44A12, 53C65.
Research partially supported by NSF Grant DMS 92 02049.

©1995, American Mathematical Society

We need functions f only on Q, but it is convenient sometimes to extend them to a neighborhood of Q. For $\xi = (\xi_0, \ldots, \xi_n)$ we put

$$\langle \xi, x \rangle = \xi_0 x_0 + \cdots + \xi_n x_n. \tag{3}$$

Let

$$\omega(x) = \det(x, dx, \ldots, dx) = \sum_{j=0}^{n} (-1)^j x_j \bigwedge_{i \neq j} dx_i. \tag{4}$$

Now we can define the basic integral transform:

$$\hat{f}(\xi) = \int_{P^n} f(x) \delta(\langle \xi, x \rangle) \delta(\square_\varepsilon(x)) \omega(x)$$
$$= \int_{Q} f(x) \delta(\langle \xi, x \rangle) \omega(x), \qquad \xi \in \mathbb{R}^{n+1} \setminus \{0\}. \tag{5}$$

We integrate here over any section of $\mathbb{R}^{n+1} \setminus \{0\} \to P^n$, so we integrate f on hyperplane sections of Q. The result will satisfy the condition

$$\hat{f}(\lambda \xi) = |\lambda|^{-1} \hat{f}(\xi), \qquad \lambda \in \mathbb{R} \setminus 0. \tag{6}$$

We can rewrite (5) in terms of inhomogeneous (affine) coordinates ($x_0 = 1$)

$$\hat{f}(\xi) = \int_{Q} f(x) \delta(\xi_0 + \xi_1 x_1 + \cdots + \xi_n x_n) \omega_\varepsilon(x_1, \ldots, x_n), \tag{5'}$$

where

$$\omega_\varepsilon(x_1, \ldots, x_n) = \frac{1}{2} \sum_{j=1}^{n} \varepsilon_j (-1)^{j-1} x_j \bigwedge_{i \neq j} dx_i \tag{7}$$

is the invariant measure on Q relative to the pseudo-orthogonal group for the quadratic form

$$\varepsilon_1 x_1^2 + \cdots + \varepsilon_n x_n^n.$$

Our goal is to describe the image of the operator $f \mapsto \hat{f}$ and to give an explicit construction of the inverse operator (inversion formula). Let us consider the ultrahyperbolic (in this particular case, hyperbolic) differential operator

$$\square_\varepsilon \left(\frac{\partial}{\partial \xi} \right) = \frac{\partial^2}{\partial \xi_0^2} + \varepsilon_1 \frac{\partial^2}{\partial \xi_1^2} + \cdots + \varepsilon_n \frac{\partial^2}{\partial \xi_n^2}. \tag{8}$$

We will prove the following result.

THEOREM. *The image of the operator $f \mapsto \hat{f}$ coincides with the space of the solutions of the equation*

$$\square_\varepsilon \left(\frac{\partial}{\partial \xi}\right) F = 0 \tag{9}$$

satisfying the condition (6).

The necessity of this condition can be obtained by the direct application of the operator (8) to the integrand in (5). We will derive the sufficiency from the construction of the inverse operator.

2. Construction of the operator κ. The basic point in the construction of inverse operators is the construction of the operator κ. Let

$$\omega_j(\xi) = d\xi_j \,\lrcorner\, \omega(\xi) \tag{10}$$

and

$$\kappa_j F[x] = \left[\frac{x_j F(\xi)}{\langle x,\xi\rangle^{n-1}} + \frac{1}{n-2}\frac{F'_j}{\langle x,\xi\rangle^{n-2}}\right]\omega_j(\xi), \quad F'_j = \frac{\partial F(\xi)}{\partial \xi_j}, \tag{11}$$
$$\kappa F[x] = \kappa_0 F[x] + \varepsilon_1 \kappa_1 F[x] + \cdots + \varepsilon_n \kappa_n F[x].$$

REMARK. We can rewrite (11) as follows:

$$\kappa_j F[x] = \sum_{i<j} (-1)^{i+j} \left[\frac{\delta_{ij} F}{\langle x,\xi\rangle^{n-1}} + \frac{1}{n-2}\frac{D_{ij} F}{\langle x,\xi\rangle^{n-2}}\right] \bigwedge_{k\neq i,j} d\xi_k,$$
$$\delta_{ij} = \varepsilon_i x_i \xi_j - \varepsilon_j x_j \xi_i, \tag{11'}$$
$$D_{ij} = \varepsilon_i \xi_j \frac{\partial}{\partial \xi_i} - \varepsilon_j \xi_i \frac{\partial}{\partial \xi_j}.$$

This $(n-1)$-form $\kappa F[x]$ on $\xi \in P^n$ has singularities along the hyperplane $\langle x, \xi \rangle = 0$ in P^n_ξ.

PROPOSITION. *If $x \in Q$ and F is a solution of* (9) *with* (6), *then the form $\kappa F[x]$ is closed at regular points.*

The proof is quite straightforward. Let us compute $d\kappa_0[x]$. If

$$\kappa_0 = G\omega_0, \qquad G(\lambda\xi) = |\lambda|^{-n} G(\xi),$$

then

$$d\kappa_0 = G'_0 d\xi_0 \wedge \omega_0(\xi) + \sum_{j=1}^{w} (-1)^j G'_j \xi_j d\xi_j \bigwedge_{i\neq 0,j} d\xi_i - nG d\xi_1 \wedge \cdots \wedge d\xi_n$$
$$= G'_0(d\xi_0 \wedge \omega_0) + \left(-\sum_{j\neq 0} G'_j \xi_j - nG\right) d\xi_1 \wedge \cdots \wedge d\xi_n$$
$$= G'_0(d\xi_0 \wedge \omega_0) + \xi_0 G'_0 d\xi_1 \wedge \cdots \wedge d\xi_n = G'_0 \omega(\xi)$$

(we used the Euler formula for homogeneous functions). Now, $G = [u'_0 v - u v'_0]$, where
$$u = F, \quad v = \frac{1}{n-2}\langle x, \xi \rangle, \quad x \in Q_j.$$
So
$$G'_0 = [u''_{00} v - u v''_0].$$
Repeating these calculations for all j and summing up the results, we obtain $d\kappa[x] = 0$ so that
$$\Box_\varepsilon \left(\frac{\partial}{\partial z}\right) u = 0 \quad \text{and} \quad \Box_\varepsilon \left(\frac{\partial}{\partial z}\right) v = 0 \quad \text{for } x \in Q.$$

REMARK. It is possible to show that if $d\kappa F[x] = 0$ for all $x \in Q$, then $\Box_\varepsilon \left(\frac{\partial}{\partial x}\right) F = 0$.

We will construct different inversion formulas by integrating $\kappa \hat{f}[x]$ on different cycles. The problem is that $\kappa \hat{f}[x]$ has singularities (on the hyperplane $\langle x, \xi \rangle$) and we need to regularize integrals. We will replace $\langle x, \xi \rangle^{-n+1}$, $\langle x, \xi \rangle^{-n+2}$ in $\kappa \hat{f}[x]$ with $(\langle x, \xi \rangle - i0)^{-n+1}$, $(\langle \kappa, \xi \rangle - i0)^{-n+2}$ correspondingly. Here (see [2])

$$(12) \qquad (t - i0)^{-k} = \lim_{\varepsilon \to 0}(t - i\varepsilon)^{-k} = t^{-k} + \frac{i\pi(-1)^{k-1}}{(k-1)!}\delta^{k-1}(t)$$

in the sense of distributions. We keep the notation $\kappa \hat{f}[\xi]$ for the new form with distributions-valued coefficients. The form $\kappa \hat{f}[x]$ continues to be closed. We need to be careful when we integrate singular closed forms and apply the Stokes theorem. Here we will stay on a primitive level: in all examples in this paper we will integrate over manifolds (often planes) transversal to singular hyperplane and it is possible to check directly that integrals do not change when we deform the cycles. Of course, it would be interesting to develop the formalism for singular differential forms with an eye on possible applications to other nonlocal problems of the integral geometry.

3. Connection with the projective Radon transform. Let us consider examples of cycles. We can assume that $\varepsilon_1 = -1$ and then $u = (1, 1, 0, \ldots, 0) \in Q$. Let Γ_u be the hyperplane

$$(13) \qquad \xi_0 + \xi_1 = 0.$$

Any such $\xi \in \Gamma_u$ define the hyperplane

$$(14) \qquad L_\xi = \{x \in P^n; \ \langle \xi, x \rangle = 0\},$$

which passes through the point u. On Γ_u we will use $\xi_1, \ldots, \xi_n$ as the parameters. Let us compute the restriction of $\kappa \hat{F}[x]$ to Γ_u. We have $\kappa_j \hat{f}|_{\Gamma_u} \equiv 0$ for $j > 1$ and we need to consider only

$$(15)$$
$$\kappa \hat{f}[x]|_{\Gamma_u} = (\kappa_0 \hat{f} + \kappa_1 \hat{f})|_{\Gamma_u}$$
$$= \left[\frac{(x_0 - x_1)\hat{f}(-\xi_1, \xi_1, \ldots \xi_n)}{(\langle x, \xi \rangle - i0)^{n-1}}\right.$$
$$\left. + \frac{1}{(n-2)(\langle x, \xi \rangle - i0)^{n-2}} \frac{\partial \hat{f}(-\xi_1, \xi_1, \xi_2, \ldots, \xi_n)}{\partial \xi_1}\right]\omega_0(\xi).$$

It is essential that in this form we differentiate only along Γ_u, so we can compute $\kappa \hat{f}|_{\Gamma_u}$ in terms of $\hat{f}|_{\Gamma_u}$ only. In integral geometry, such cycles are called *admissible*.

PROPOSITION. *We have*

(16)
$$\int_{\Gamma_u} \kappa \hat{f}[x] = 2 \int_{\Gamma_u} \frac{(x_0 - x_1)\hat{f}(-\xi_1, \xi_1, \ldots, \xi_n)}{(\langle x, \xi \rangle - i0)^{n-1}} \omega_0(\xi) = cf(x),$$

$$\omega_0(\xi) = \sum_{j=1}^{n} (-1)^j \xi_j \bigwedge_{i \neq j, q \leq i \leq n} d\xi_i, \qquad c = \frac{4(2\pi)^{n-1}}{(-1)^{(n-1)/2}(n-2)!}.$$

The first equality results from integration by parts. As for the second one, we can see from (5) that $\hat{f}(-\xi_1, \xi_1, \xi_2, \ldots, \xi_n)$ is the projective Radon transform of $\varphi(y_1, \ldots, y_n)$, where

$$(x_1 - x_0)\varphi(x_1, -x_0, x_2, x_3, \ldots, x_n) = f(x_0, x_1, \ldots, x_n) \quad \text{on } Q,$$
$$\hat{f}(-\xi_1, \xi_1, \xi_2, \ldots, \xi_n) = \hat{\varphi}(\xi_1, \ldots, \xi_n)$$
$$= \int_{P^{n-1}} \varphi(y_1, y_2, \ldots, y_n) \delta(y_1 \xi_1 + \cdots + y_n \xi_n) \omega(y).$$

Now we need only to recall the inversion formula for the projective Radon transform for φ (see [3]).

COROLLARY. *We have*

(17)
$$\int_\gamma \kappa \hat{f}[x] = cf(x)$$

for any cycle γ homologous to Γ_u.

Let us recall again that we need to care about singularities when we deform cycles. Of course the formula (16) will be true for any $u \in Q$ ($\Gamma_u = \{\xi; L_\xi \ni u\}$). If we take $F(\xi)$ satisfying (9) and consider

$$\int_{\Gamma_u} \kappa F[x] = c\varphi(x),$$

where the function φ is independent of $u \in Q$ and $F(\xi) = \hat{\varphi}(\xi)$ (for a fixed $u \in Q$ we use the relation with the projective Radon transform). This proves the theorem.

4. Local inversion formula for even n. The operator $\kappa \hat{f}[x]$ is the sum of local (κ') and nonlocal (κ'') operators:

$$\kappa \hat{f}[x] = \kappa' \hat{f}[x] + \kappa'' \hat{f}[x],$$

(18)
$$\kappa' \hat{f}[x] = \frac{\pi i}{(n-2)!} \sum_{j=0}^{n} (-1)^n [\varepsilon_j F(\xi) \delta^{(n-2)}(\langle x, \xi \rangle)$$
$$- F'_j \delta^{(n-3)}(\langle x, \xi \rangle)] \omega_j(\xi), \qquad \varepsilon_0 = 1,$$

$$\kappa'' F[x] = \sum_{j=0}^{n} \varepsilon_j \left[\frac{x_j F(\xi)}{\langle x, \xi \rangle^{n-1}} + \frac{1}{n-2} \frac{F'_j}{\langle x, \xi \rangle^{n-2}} \right] \omega_j(\xi).$$

in the sense of distributions.

It is natural to consider cycles symmetric with respect to $\xi \mapsto -\xi$, so for the integration we can keep only the even part of κF. For even n it will be the local part κ' of κ and for odd n the nonlocal part κ''. So for even n we can consider the local (differential) operator κ' and $\kappa' \hat{f}[x]$ is a closed differential form on hyperplane $\langle x, \xi \rangle = 0$, so that in this case we do not need to worry anymore about singularities. We obtain local inversion formulas for even n by restricting $\kappa' \hat{f}$ to different cycles.

5. Other examples of inversion formulas. We considered the cycles (hyperplanes) Γ_u for $u \in Q$. Now we consider the situation when $u \notin Q$, i.e., $u = (1, 0, \ldots, 0)$ and Γ_u is the hyperplane of such ξ that $L_\xi \ni u$. Then

(19) $$\xi_0 = 0.$$

We have $\kappa_j \hat{f}[x] \equiv 0$ for $j \geq 1$ and

$$\kappa F[x]|_{\Gamma_u} = \kappa_0 F[x] = \tilde{\kappa}_0 F + \tilde{\tilde{\kappa}}_0 F,$$

$$\tilde{\kappa}_0 F = \frac{x_0 F}{(\langle x, \xi \rangle - i0)^{n-1}} \omega_0(\xi), \quad \tilde{\tilde{\kappa}}_0 F = \frac{1}{n-2} \frac{F_0'(\xi)}{(\langle x, \xi \rangle - i0)^{n-2}} \omega_0(\xi).$$

Here we already have a nonadmissible situation: to compute $\kappa F[x]|_{\Gamma_u}$ we need to know not only $F|_{\Gamma_u}$ but also $F_0'|_{\Gamma_u} = \frac{\partial F}{\partial \xi_0}|_{\Gamma_u}$. The differentiation $\partial/\partial \xi_0$ is transversal to Γ_u.

If f is (affine) even, i.e.,

$$f(x_0, x_1, \ldots, x_n) = f(x_0, -x_1, \ldots, -x_n),$$

then

$$\hat{f}(\xi_0, \xi_1, \ldots, \xi_n) = \hat{f}(\xi_0, -\xi_1, \ldots, -\xi_n),$$

and the integral of $\tilde{\tilde{\kappa}} \hat{f}$ equals zero, so that we can reconstruct $\kappa \hat{f}$ and f from $\hat{f}|_{\Gamma_u}$ alone (in general, we can reconstruct

$$f(x_0, x_1, \ldots, x_n) + f(x_0, -x_1, \ldots, -x_n)).$$

If n is odd, then

$$\int_{\Gamma_u} \tilde{\kappa} \hat{f}[x] = 0$$

and we can reconstruct f from $\hat{f}_0'|_{\Gamma_u}$ only (the first term in the integral vanishes).

In the case of the sphere ($\varepsilon_1 = \cdots = \varepsilon_n = 1$) in affine coordinates $\{x_0 = 1\}$ the sections by $\langle \xi, x \rangle = 0$, $\xi_0 = 0$ are great circles. So in this case the transform

$$f(x) \mapsto \hat{f}(0, \xi_1, \ldots, \xi_n)$$

is the Minkowski-Funk transform. It is known that a function on the sphere can be reconstructed from its Minkowski-Funk transform only if it is even. We remark here that in order to reconstruct a general function, we need to know not only integrals on great circles but also derivatives along parallel sections, which already are not great circles. About the application of this formula, see [4].

Let us remark also that if we consider all plane sections of the sphere, we can reconstruct the function as a limit of integrals over sections (when the radius tends to 0), but we are not interested in this kind of reconstruction.

6. Admissible cycles with tangency condition. Let us take the cone γ_η with vertex $\eta = (0, 0, \ldots, 0, 1)$,

$$\xi_0^2 + \varepsilon_1 \xi_1^2 + \cdots + \varepsilon_{n-1} \xi_{n-1}^2 = 0 \tag{20}$$

as the cycle. The geometrical meaning of this condition is: hyperplanes $L_\xi, \xi \in \gamma_\eta$, intersect the hyperplane $L_\eta = \{\langle \eta, x \rangle = 0\}$ at an $(n-1)$-plane which is tangent to $Q \cap L_\eta$. In other words, $Q \cap L_\xi, \xi \in \gamma_\eta$, and $Q \cap L_\eta$ have a common tangent $(n-1)$-plane. After the stereographic projection the images of the sections $Q \cap L_\xi, \xi \in \gamma_\eta$, are tangent to the image of $Q \cap L_\eta$ and are characterized by this property.

We have

$$\kappa F[x]|_{\gamma_\eta} = \sum_{j=1}^{n-1} \varepsilon_j \tilde{\kappa}_j F[x] + \frac{1}{(n-2)(\langle x, \xi \rangle - i0)^{n-2}} \sum_{i<j<n} D_{ij} F \bigwedge_{k \neq i,j} d\xi_k, \tag{21}$$

$$D_{ij} = \varepsilon_i \xi_j \frac{\partial}{\partial \xi_i} - \varepsilon_j \xi_i \frac{\partial}{\partial \xi_j}.$$

To use this formula, we must require that differential monomials without $d\xi_n$ vanish on γ_η. The differential operators D_{ij} are tangent to γ_η. So the cycle γ_η is admissible: we can reconstruct $\kappa \hat{f}[x]|_{\gamma_\eta}$ and $f(x)$ from $\hat{f}|_{\gamma_\eta}$ alone.

REMARKS. 1. We can replace $\eta = (0, 0, \ldots, 1)$ with any other $\eta \in P_\xi^n$ and the cone γ_η ($\xi \in \gamma_\eta \Leftrightarrow L_\xi \cap L_\eta$ is a tangent plane for $L_\eta \cap Q$) will be an admissible cycle.

2. This construction admits the following generalization. Let γ be given by

$$\xi_0^2 + \varepsilon_1 \xi_1^2 + \cdots + \varepsilon_J \xi_J^2 = 0. \tag{22}$$

The corresponding hyperplanes $L_\xi, \xi \in \gamma$, are characterized by the condition that $L_\xi \cap \{\xi_{J+1} = \cdots = \xi_n = 0\}$ is a tangent plane to $Q \cap \{\xi_{J+1} = \cdots = \xi_n = 0\}$. The cycle γ is also admissible and

$$\kappa F_{[x]}|_\gamma = \sum_{j \leq J} \varepsilon_j \tilde{\kappa}_j F[x] + \frac{1}{(n-2)(\langle x, \xi \rangle - i0)^{n-2}} \sum_{i<j\leq J} D_{ij} F \bigwedge_{k \neq i,j} d\xi_k,$$

where $D_{ij}, i < j \leq J$, are differential operators tangent to γ. In particular, for $J = 1$ we have the example $\Gamma_u, u = (1, 1, 0, \ldots, 0)$. Then $Q \cap \{\xi_2 = \cdots = \xi_n = 0\}$ are the points $(\pm 1, \pm 1, 0, \ldots, 0)$. Of course, we obtain an admissible cycle γ if we consider any $(n-J)$-plane instead of $\xi_{J+1} = \cdots = \xi_n = 0$ and take an analogue of (22).

3. Finally we can take a curve γ and consider $(n-J)$-dimensional submanifold $R \subset Q$ and the set γ_R of $\xi \in P^n$ such that for some $x \in R$ the hyperplane L_ξ contains the tangent plane to R. Then $\kappa F|_{\gamma_R}$ in ξ coincides with $\kappa F|_{\gamma_{\tilde{R} \cap Q}}$, where $\tilde{R}$ is the osculating hyperplane for R at x.

7. Connection with the hyperbolic geometry. Let us consider the affine hyperboloid of two sheets $H \subset \mathbb{R}^n$ ($x_n = 1$):

$$x_0^2 - x_1^2 - \cdots - x_{n-1}^2 = 1.$$

Identifying antipodal points x and $-x$, we obtain a model of the hyperbolic geometry with the transformation group $SO(1, n-1)$. There are two natural $SO(1, n-1)$-invariant families of hyperplanes: hyperplanes passing through the origin and hyperplanes parallel to the asymptotic hyperplanes. Hyperplane sections corresponding to the first family are hyperbolic hyperplanes (they are realized as hyperboloids of two sheets), and those corresponding to the second family are horospheres (they are realized as paraboloids).

We will consider even functions on H. We have the inversion formulae for the geodesic Radon transform (15) and for the horospherical Radon transform (it corresponds to the cone γ_η with $\eta = (0, \ldots, 0, 1)$ (cf. (20))). This interpretation explains why the known inversion formulae for the geodesic and horospherical Radon transforms on the hyperbolic space are very similar to one another and very similar to the inversion formula for the Euclidean Radon transform. They can be interpreted as restrictions of the universal form $\kappa \hat{f}$ to different cycles.

8. Interpretation on the language of differential equations. We started from a (homogeneous) solution of the differential equation $\Box_\varepsilon(\frac{\partial}{\partial \xi})F = 0$. Using Cauchy data on a cycle γ we construct $\kappa F[x]|_\gamma$ and reconstruct an f such that $F = \hat{f}$. By integrating f over hyperplanes L_γ we can find $\hat{f}(\xi) = F(\xi)$ for all $\xi \in P^n$ (not only on γ). As a result, we can reconstruct the solution F from the Cauchy data on γ, i.e., we solve the Cauchy problem. If a cycle γ is admissible, then for the reconstruction of F on P^n we need $F|_\gamma$ only. So we can solve the Goursat problem with data on admissible γ.

9. Generalization to algebraic varieties. Let $P(x_0, \ldots, x_n)$ be a homogeneous polynomial, $\deg P = k$, and $Q \subset P^n$ define by

$$P(x) = 0.$$

Let $f(x)$ be a homogeneous function of degree $-n + k$. Then we can define integrals of f on hyperplane sections of Q:

$$f(\xi) = \int_{P^n} f(x) \delta(P(x)) \delta(\langle \xi, x \rangle) \omega(x).$$

We have

(23) $$\hat{f}(x\xi) = |\lambda|^{-1} \hat{f}(\xi) \qquad P(\frac{\partial}{\partial \xi}) \hat{f}(\xi) = 0.$$

We want to invert $f \mapsto \hat{f}$ by constructing an operator κ on homogeneous solutions of $P(\frac{\partial}{\partial \xi}) F(\xi) = 0$ that transforms the solutions to singular closed $(n-1)$-forms on P^n.

The operator κ must be linear in the space of homogeneous polynomials of the degree k. So it is sufficient to construct κ when P is a monomial:

$$P(x) = x^K, \quad |K| = k.$$

In this case, let us put

$$\kappa_K F[x] = \sum_{j \in K} \left(\frac{x^{K\setminus j} \hat{f}}{(\langle x, \xi \rangle - i0)^{n-1}} \right.$$
$$+ \sum_{K \supset J \ni j} \frac{1}{(n-2)\ldots(n-|J|) \cdot (k-1)\ldots(k-|J|+1)}$$
$$\left. \times \frac{F_{J\setminus i}^{(|J|-1)} \cdot x^{K\setminus J}}{(\langle x, \xi \rangle - i0)^{n-|J|}} \right) \omega_j(\xi)$$

(indexes in J, K can be repeated). If $P(x) = \Sigma a_k x^k$, then $\kappa_P F[x] = \Sigma a_K \kappa_K F[x]$. We can check directly that (23) implies

(24) $$d\kappa_P \hat{f}[x] = 0.$$

References

1. I. Gelfand, M. Graev, and Z. Shapiro, *Integral geometry on k-dimensional planes*, Funct. Anal. Appl. **1** (1967), 14–27.
2. I. Gelfand and G. Shilov, *Generalized functions*, vol. 1, Academic Press, New York, 1964.
3. I. Gelfand, S. Gindikin, and M. Graev, *Integral geometry in affine and projecive spaces*, J. Soviet Math. **18** (1980), 39–167.
4. S. Gindikin, J. Reeds, and L. Shepp, *Spherical tomography and spherical integral geometry*, Lectures in Applied Math., vol. 30, Amer. Math. Soc., Providence, RI, 1994, pp. 83–92.

DEPARTMENT OF MATHEMATICS, RUTGERS UNIVERSITY, NEW BRUNSWICK, NJ 08903, USA
E-mail address: gindikin@math.rutgers.edu

Dynkin Diagrams in Singularity Theory

S. M. GUSEIN-ZADE

The notion of Dynkin diagrams is popular in a number of branches of mathematics. The question if these diagrams should be named after Dynkin is still the object of discussion. Some people insist that the contribution of H. S. M. Coxeter and/or of H. Weyl to the introduction of this notion is more important and hence they should be called Coxeter or Coxeter-Dynkin or Coxeter-Dynkin-Weyl diagrams (graphs). The existence of such debates shows the importance which is attached to this notion by mathematicians. Whatever name for these diagrams is more justified, the contribution of E. B. Dynkin to the introduction and the dissemination of this notion is unquestionable. Here we shall use the term "Dynkin diagram" both for brevity and because this paper appears in the volume dedicated to E. B. Dynkin's anniversary.

Dynkin diagrams were introduced in the theory of Lie groups (algebras) to describe particular systems of elements in lattices endowed with integer quadratic forms (so called root systems). A Dynkin diagram is a graphical representation of the matrix of inner products of these elements. At the moment this notion has two meanings corresponding to two possible expansions of its initial field of application. On one hand, it is used for the standard set of (classical) diagrams (A_k, B_k, ...), which appear (sometimes mysteriously) in a way *a priori* unrelated to a subset of a lattice or to an integer quadratic form. For example, this is how they appear in the theory of representations of quivers, i.e., in the classification of simple diagrams of linear transformations (see, e.g., [6]; there you can find some other examples of problems in which the classical Dynkin diagrams arise). In another context, Dynkin diagrams are used for a description of arbitrary integer quadratic forms on lattices with fixed bases. Generally speaking, corresponding graphs are different from classical Dynkin diagrams. In this way Dynkin diagrams appeared in singularity theory ([2]). The amazing thing that it is precisely the classical Dynkin diagrams that arose from the very beginning of the classification of hypersurface singularities. This fact does not have a complete explanation yet.

§1. Integer quadratic forms on lattices and Dynkin diagrams

Dynkin diagrams as graphical representations of integer quadratic forms can be defined in several (slightly) different ways for several (slightly) different situations. We shall start with the one that is most reasonable for the Singularity Theory.

1991 *Mathematics Subject Classification.* Primary 57R45, 58C27.
Partially supported by ISF Grant M91000.

Let L be a lattice of a finite rank with a fixed ordered basis $\Delta_1, \ldots, \Delta_\mu$ ($\mu = \operatorname{rk} L$) and Q an integer quadratic form on L. Let us suppose that $Q(\Delta_i) = (-2)$ for $i = 1, 2, \ldots, \mu$. We shall denote the corresponding (symmetric) bilinear form by $(\,\cdot\,,\,\cdot\,)$ (in particular $(\Delta_i, \Delta_i) = Q(\Delta_i) = -2$). This bilinear form is even, i.e., for each element $a \in L$ its inner square $Q(a)$ is even. Let D be the following graph:

1) its vertices correspond to the elements $\Delta_1, \ldots, \Delta_\mu$ of the basis;

2) the vertices corresponding to Δ_i and Δ_j are connected by an edge if and only if $(\Delta_i, \Delta_j) \neq 0$ and this edge has the multiplicity (Δ_i, Δ_j) (edges of negative multiplicity are depicted by dashed lines).

DEFINITION 1. The graph D with ordered vertices and with the multiplicities of the edges is called the *Dynkin diagram of the form* Q (in the basis $\Delta_1, \ldots, \Delta_\mu$).

REMARK. This definition coincides with the traditional one only for root systems with roots of the same length, i.e., for root systems A_k, D_k, E_6, E_7, E_8 (multiplying the invariant quadratic form by a non-zero factor allows us to assume that the inner squares of all the roots are equal -2). In other root systems, roots have different lengths and Definition 1 cannot be applied. In the traditional definition of the Dynkin diagram for such a root system the multiplicity of an edge is defined in a somewhat different way. The definition can be found in any book on Lie groups and algebras (e.g. [3]). Here we do not want to overload the description by general details.

FIGURE 1

The Dynkin diagram contains all the information about the pair (even quadratic form on a lattice, basis of the lattice).

EXAMPLE. The Dynkin diagram in Figure 1 corresponds to the quadratic form with the matrix
$$\begin{pmatrix} -2 & 1 & 0 \\ 1 & -2 & 1 \\ 0 & 1 & -2 \end{pmatrix}.$$

To a Dynkin diagram D (or to an even quadratic form Q and a fixed basis $\{\Delta_i\}$ with $Q(\Delta_i) = -2$) there corresponds a number of objects: a group generated by reflections, a distinguished element in it, etc.

DEFINITION 2. The *Picard-Lefschetz transformation* (or operator) $h_i : L \to L$ corresponding to the element Δ_i of the basis is defined by the formula

(1) $$h_i(a) = a + (\Delta_i, a) \cdot \Delta_i \quad (a \in L).$$

It is not difficult to see that the Picard-Lefschetz transformation h_i is the reflection with respect to the hyperplane orthogonal to the vector Δ_i (in the sense of the inner product $(\,\cdot\,,\,\cdot\,)$). Let $\operatorname{Aut} L$ be the group of linear automorphisms of the lattice L.

DEFINITION 3. The subgroup M of the group Aut L generated by the Picard-Lefschetz transformations h_i is called the *monodromy group* (corresponding to the Dynkin diagram D). The superposition $h = h_1 \circ \cdots \circ h_\mu \in M$ of the Picard-Lefschetz transformations h_i (in the indicated order) is called the *Coxeter* (or *Killing–Coxeter*) element of the monodromy group (or *the classical monodromy operator*).

EXAMPLE. The monodromy group corresponding to the Dynkin diagram in Figure 1 (which by definition is the Weyl group of type A_3) is isomorphic to the group S_4 of permutations of four elements. The Coxeter element h is given by the matrix

$$\begin{pmatrix} 0 & 0 & -1 \\ 1 & 0 & -1 \\ 0 & 1 & -1 \end{pmatrix}$$

and is of order 4 (under the above isomorphism it corresponds to the cyclic permutation).

DEFINITION 4. Images of the basic elements Δ_i under the action of elements of the monodromy group M are called *roots* (or, in another terminology, *vanishing cycles*).

§2. The skew-symmetric version of the described objects

Besides the symmetric bilinear form on the lattice also a skew-symmetric one corresponds to a Dynkin diagram D.

DEFINITION 5. *The skew-symmetric form* $[\cdot, \cdot]$ *corresponding to the Dynkin diagram D, is defined by the formula*

$$[\Delta_i, \Delta_j] = (\Delta_i, \Delta_j) \quad \text{for } i < j$$

(let us recall that $\{\Delta_i\}$ is the ordered basis).

Vice versa, a skew-symmetric bilinear form on a lattice with a fixed basis defines a Dynkin diagram and a quadratic form on the lattice.

EXAMPLE. The Dynkin diagram in Figure 1 corresponds to the skew-symmetric form with the matrix

$$\begin{pmatrix} 0 & -1 & 0 \\ 1 & 0 & -1 \\ 0 & 1 & 0 \end{pmatrix}.$$

All notions defined above "natural" skew-symmetric versions. They are defined exactly in the same way, but in the definition of the Picard-Lefschetz transformation the symmetric form $(\cdot, \cdot)$ has to be be substituted by the skew-symmetric one $[\cdot, \cdot]$.

REMARK. In formula (1) we wrote (Δ_i, a) instead of (a, Δ_i) (which is a more usual form of the formula) intentionally, to make it valid in the skew-symmetric case also.

EXAMPLE. The "skew-symmetric" monodromy group corresponding to the Dynkin diagram from Figure 1 is isomorphic to the group of affine unimodular transformations of the lattice $\mathbb{Z}^2$, i.e., of the transformations of the form $x \mapsto Ax+v$, where $x \in \mathbb{Z}^2$, $v \in \mathbb{Z}^2$, $A \in SL(2, \mathbb{Z})$.

Thus to a Dynkin diagram there corresponds two bilinear forms, two monodromy groups, two Coxeter elements, etc. However, the Coxeter elements corresponding to the symmetric and the skew-symmetric versions differ only by the factor (-1).

§3. Change of basis, distinguished bases

It turns out that on the set of bases of the lattice L (and thus on the set of Dynkin diagrams corresponding to one quadratic form Q), there is a natural action of the Artin braid group with μ strands. To bases that can be obtained one from the other by the action of an element of the braid group, there correspond identical monodromy groups and Coxeter elements. Moreover, this braid group action preserves the correspondence between symmetric and skew-symmetric forms.

DEFINITION 6. The *Gabrielov operation* α_m, $1 \leq m \leq \mu - 1$, is the operation of the change of the basis $\{\Delta_i\}$ of the lattice L defined by the formulas

$$\widetilde{\Delta}_i = \Delta_i \quad \text{for } i \neq m, m+1;$$
$$\widetilde{\Delta}_{m+1} = \Delta_m;$$
$$\widetilde{\Delta}_m = h_m(\Delta_{m+1}) = \Delta_{m+1} + (\Delta_m, \Delta_{m+1})\Delta_m.$$

REMARKS. 1) The inverse operation α_m^{-1} (sometimes denoted by β_{m+1}) is defined by the formulas

$$\widetilde{\Delta}_i = \Delta_i \quad \text{for } i \neq m, m+1;$$
$$\widetilde{\Delta}_m = \Delta_{m+1};$$
$$\widetilde{\Delta}_{m+1} = h_{m+1}^{-1}(\Delta_m) = \Delta_m + (\Delta_m, \Delta_{m+1})\Delta_{m+1}.$$

2) The correspondence between symmetric and skew-symmetric forms is preserved because in these formulas one can use the skew-symmetric form $[\cdot, \cdot]$ instead of the symmetric form $(\cdot, \cdot)$ (because they involve the values of the indicated forms only for pairs Δ_i, Δ_j with $i < j$, where they coincide with each other).

It is not difficult to check that the operations α_m satisfy the following relations:

$$\alpha_m \alpha_{m'} = \alpha_{m'} \alpha_m \quad \text{for } |m - m'| \geq 2,$$
$$\alpha_{m+1} \alpha_m \alpha_{m+1} = \alpha_m \alpha_{m+1} \alpha_m \quad \text{for } 1 \leq m < \mu - 1.$$

Thus on the set of bases of the lattice L, we have the action of the group with $\mu - 1$ generators α_m ($m = 1, \ldots, \mu - 1$) and the described relations. This group is exactly the braid group with μ strands.

DEFINITION 7. Bases of the lattice L which can be obtained from the fixed one by the action of the braid group and by subsequent multiplication of some of its elements by -1 are called *distinguished*.

The sets of distinguished bases for the symmetric and the skew-symmetric cases are identical.

§4. Dynkin diagrams corresponding to sets of generators of a lattice

The Dynkin diagram is defined by the matrix $((\Delta_i, \Delta_j))$ of values of the (say symmetric) bilinear form on the elements Δ_i. For this definition it is not necessary to demand that the set $\{\Delta_i\}$ forms a basis of the lattice. In principle it can be any system of elements, say a system of generators. However to a Dynkin diagram (i.e., to the matrix $((\Delta_i, \Delta_j))$) there corresponds a lattice (say L') with a quadratic form for which the set $\{\Delta_i\}$ is a (formal) basis. All the described objects (Picard-Lefschetz transformations, the Coxeter element, the monodromy groups, etc.) can be considered both in the lattice L and in the lattice L'. Instead of the notion of distinguished bases in the lattice L we get the notion of distinguished sets of generators.

In the classical theory of Dynkin diagrams for root systems, such a construction was used to define an extended Dynkin diagram (see, e.g. [3]). For a reduced irreducible root system the extended Dynkin diagram is the diagram that describes the inner products of the elements $\{\Delta_i : i = 0, 1, \ldots, \mu\}$, where $\{\Delta_i : i = 1, \ldots, \mu\}$ is a basis of the root system and Δ_0 is the negative to the maximal root with respect to this basis (i.e., $\Delta_0 = -\sum_{i=1}^{\mu} m_i \Delta_i$ with maximal values of the coefficients m_i).

§5. Dynkin diagrams of singularities

In singularity theory, Dynkin diagrams were introduced for a description of the topology of a local non-singular level set of a complex-analytic function near its isolated critical point ([2]). Here we shall describe the construction in a somewhat more general setting, namely for isolated complete intersection singularities (ICIS).

Let $f = (f_1, \ldots, f_m)$ be a germ of a holomorphic map $(\mathbb{C}^{n+m}, 0) \to (\mathbb{C}^m, 0)$ which defines the (n-dimensional) ICIS $X = (f^{-1}(0), 0) \subset (\mathbb{C}^{n+m}, 0)$ (if $m = 1$, then we have an isolated hypersurface singularity X). For $\varepsilon = (\varepsilon_1, \ldots, \varepsilon_m) \in (\mathbb{C}^m, 0)$, $0 \leq \|\varepsilon\| \ll \delta$, let $F_\varepsilon = f^{-1}(\varepsilon) \cap B_\delta = \{x \in \mathbb{C}^{n+m} : f(x) = \varepsilon, \|x\| \leq \delta\}$ be the local level set of the map f (B_δ is the ball of radius δ with center at the origin in $\mathbb{C}^{n+m}$). For a generic point ε from a neighbourhood of the origin in $\mathbb{C}^m$ the set F_ε is a non-singular complex manifold with the boundary $\partial F_\varepsilon = f^{-1}(\varepsilon) \cap S_\delta$ ($S_\delta = \partial B_\delta$ is the sphere of radius δ). The differentiable type of the manifold F_ε does not depend on the choice of a generic ε. This manifold is called the *Milnor fiber* of the map f (or of the ICIS X). By the theorem of Milnor and Hamm, the Milnor fiber of an n-dimensional ICIS X is homotopy equivalent to the wedge of several n-dimensional spheres. The number μ of these spheres is called the *Milnor number* of the ICIS X. Thus the n-dimensional homology group $H_n(F_\varepsilon; \mathbb{Z})$ of the Milnor fiber (which is called the *vanishing homology group*) is a lattice of rank μ (i.e., it is isomorphic to $\mathbb{Z}^\mu$). In this lattice, there is the intersection form. It is symmetric for n even and skew-symmetric for n odd. However, sometimes it is not suitable for the definition of a Dynkin diagram directly. Thus for $n \equiv 0 \mod 4$, the self-intersection numbers of natural generators (see below) are equal to 2, and not -2. Therefore, this form is used for the definition of a Dynkin diagram (in our version of it) only for $n \equiv 2$ or $1 \mod 4$. For $n \equiv 0$ or $3 \mod 4$ it should be multiplied by -1.

Some special sets of generators of the homology group $H_n(F_\varepsilon; \mathbb{Z})$ can be constructed in the following way. Let us assume that the system of equations $f_1 = 0, \ldots, f_m = 0$ of the ICIS X is generic. Here this means that the first $(m - 1)$ equations define the ICIS $X^* = \{f_1 = \cdots = f_{m-1} = 0\} \subset (\mathbb{C}^{n+m}, 0)$ (of dimension $n + 1$) with the Milnor number μ_1 as small as possible. Let us denote

$$f' = (f_1, \ldots, f_{m-1}) : (\mathbb{C}^{n+m}, 0) \to (\mathbb{C}^{m-1}, 0), \qquad \varepsilon' = (\varepsilon_1, \ldots, \varepsilon_{m-1}).$$

Let

$$F_{\varepsilon'}^* = f^{-1}(\varepsilon') \cap B_\delta = \{x \in \mathbb{C}^{n+m} : f(x) = \varepsilon', \|x\| \leq \delta\}$$

be the Milnor fiber of the ICIS X^*.

REMARK. If X is an isolated hypersurface singularity (i.e., if $m = 1$), then the corresponding system of equations is always generic and $X^* = \mathbb{C}^{n+1}$.

Let $\widetilde{f}_i$ be perturbations of the functions f_i ($i = 1, \ldots, m$) such that the complex manifold $S_\mathbb{C} = \{x \in \mathbb{C}^{n+m} : \widetilde{f}_1 = \cdots = \widetilde{f}_{m-1} = 0\}$ is nonsingular and the restriction $\widetilde{f}_m|_{S_\mathbb{C}}$ of the function $\widetilde{f}_m$ is Morse, i.e., it has only nondegenerate critical points $p_1, \ldots, p_\nu$ with different critical values $z_1, \ldots, z_\nu$ ($z_i = \widetilde{f}(p_i)$). It can be shown that the number ν of critical points is equal to the sum $\mu + \mu_1$ of the Milnor numbers of the ICIS X and X^*. Let z_0 be a noncritical value of the function $\widetilde{f}_m|_{S_\mathbb{C}}$ with $|z_0| > |z_i|$. If the perturbations $\widetilde{f}_i$ are small enough (i.e., close to the non-perturbed functions f_i), then the level manifold $V_{z_0} = \{\widetilde{f}_m|_{S_\mathbb{C}} = z_0\}$ of the function $\widetilde{f}_m|_{S_\mathbb{C}}$ is diffeomorphic to the Milnor fiber F_ε of the ICIS X. Let $\{u_i\}$ be a system of disjoint and nonselfintersecting paths connecting the critical values z_i with a noncritical value z_0 inside the circle $\{|z| \leq |z_0|\}$ ($i = 1, \ldots, \nu$; $u_i(0) = z_i$, $u_i(1) = z_0$, $u_i(t) \leq |z_0|$). Let us suppose that the critical values z_j (hence the critical points p_j) are numbered in the same order as the paths u_j enter the noncritical value z_0, counting clockwise beginning from the boundary of the circle $\{|z| \leq |z_0|\}$. By the Morse Lemma, in a neighbourhood of the critical point p_i in $S_\mathbb{C}$ there exists a system of local coordinates $x_1, \ldots, x_{n+1}$ such that

$$\widetilde{f}_m|_{S_\mathbb{C}}(x) = \sum_{i=1}^{n+1} x_i^2.$$

For small enough κ the level manifold $\{\widetilde{f}_m|_{S_\mathbb{C}}(x) = u_i(\kappa)\}$ contains an n-dimensional sphere $\sqrt{u_i(\kappa) - z_i} \cdot S_1^n$, where $S_1^n = \{x \in \mathbb{R}^{n+1} \subset \mathbb{C}^{n+1} : \sum x_i^2 = 1\}$ is the standard sphere of radius 1 in the space $\mathbb{R}^{n+1}$. Lifting the homotopy along the path u_i, we get a sphere in the level manifold $V_{z_0} = \{\widetilde{f}_m|_{S_\mathbb{C}} = z_0\}$, which represents an element Δ_i in the vanishing homology group $H_n(V_{z_0}; \mathbb{Z}) \cong H_n(F_\varepsilon; \mathbb{Z})$ (defined up to multiplication by ± 1). It can be shown that $\{\Delta_i\}$ is a system of generators of the vanishing homology group. A system of generators obtained in such a way is called *distinguished*. A change of the system of paths $\{u_i\}$ results in a change of the system of generators $\{\Delta_i\}$, which can be obtained by a sequence of Gabrielov operations.

§6. Dynkin diagrams and deformations of singularities

The main reason for the appearance of Dynkin diagrams in singularity theory is that they can be used to describe some properties of singularities and especially of their deformations. The most natural properties that can be expressed in such terms are adjacencies and decompositions of singularities.

Let f and g be germs of holomorphic functions $(\mathbb{C}^{n+1}, 0) \to (\mathbb{C}, 0)$ with an isolated critical point at the origin. We say that the singularity g is of the same *type* as the singularity f (or that the singularities f and g are *µ-equivalent*) if there exists an analytic family f_t ($t \in [0, 1]$) with $f_0 = f$, $f_1 = g$, and $\mu(f_t) = \text{const}$. If the singularities f and g are µ-equivalent, then they have identical sets of Dynkin diagrams.

DEFINITION 8. A singularity f is said to be *adjacent* to a singularity g if there exists a deformation f_t of the singularity f such that for all sufficiently small $t \neq 0$ the function f_t has a critical point of the type g in a neighbourhood of the origin (in this case $\mu(f) \geq \mu(g)$).

Let $X_1, \ldots, X_s$ be a set of types of singularities.

DEFINITION 9. The singularity f is said to have a *decomposition of the type* $(X_1, \ldots, X_s)$ if there exists a deformation f_t of the singularity f such that for all sufficiently small $t \neq 0$ the function f_t in a neighbourhood of the origin has critical points of the types $(X_1, \ldots, X_s)$ and no other critical points (in this case the Milnor number $\mu(f)$ of the singularity f is equal to the sum $\mu(X_1) + \cdots + \mu(X_s)$ of Milnor numbers of the singularities X_i).

The relationship between these properties of singularities and Dynkin diagrams is given by the following statements.

STATEMENT 1. *If the singularity f is adjacent to the singularity g, then there exist distinguished bases (in the corresponding vanishing homology groups) such that the Dynkin diagram of the singularity f contains the Dynkin diagram of the singularity g.*

Let D be a Dynkin diagram. Suppose that the set of its vertices is represented as the union of s disjoint subsets. Let D_i ($i = 1, \ldots, s$) be the Dynkin diagram that consists of the vertices of the ith subset with the edges from the diagram D between them. In this case we say that the Dynkin diagram D has a *decomposition* into Dynkin diagrams $(D_1, \ldots, D_s)$.

STATEMENT 2. *If the singularity f has a decomposition of type $(X_1, \ldots, X_s)$, then there exist Dynkin diagrams of the singularities f and X_i such that the Dynkin diagram of the singularity f has a decomposition into the Dynkin diagrams of the singularities $X_1, \ldots, X_s$.*

One should be careful with the formulation of converse statements here. There exists µ-equivalent singularities with different sets of singularities to which they are adjacent. Straightforward converse statements are valid for the so-called simple singularities, which have Dynkin diagrams coinciding with the classical ones ([7]). However, their applications require the knowledge of the set of all Dynkin diagrams of a singularity. For simple singularities of types A_k and D_k, these sets are known. The problem to describe them for an arbitrary singularity seems to be transcendental. It is reasonable to try to deal only with some Dynkin diagrams, preferably with those that can be considered as the simplest or canonical. For simple singularities

A_k, D_k, E_6, E_7, E_8 it is clear which Dynkin diagrams should be considered as the canonical one. These are standard Dynkin diagrams of the corresponding type (with an arbitrary numbering of vertices). There exists a number of characterizations of such diagrams. For instance, a Dynkin diagram of a singularity is a tree if and only if the singularity is simple and the Dynkin diagram coincides with the classical one ([1]). Simple singularities are adjacent if and only if the canonical Dynkin diagram of one of them contains the canonical Dynkin diagram of the other. For unimodal singularities the candidates for canonical Dynkin diagrams are those constructed by Gabrielov [4]. However, it is not known in which sense they can be considered as canonical ones or what the term "canonical" should mean in general.

§7. Curves and Dynkin diagrams

There exists a method of associating a Dynkin diagram (and hence a lattice with a basis and a quadratic form) to a curve with only simple (double) self-intersections on a surface (e.g., in the plane). On one hand, this construction gives a method for calculating Dynkin diagrams of some (isolated complete intersection) singularities. On the other hand, it provides some invariants of such curves.

Let S be a smooth real surface (without a boundary) and L a curve in S that is nonsingular everywhere except several simple (double) self-intersection points p_j. Let us suppose that the set of components of the complement to the curve L can be (and is) divided into two parts ("positive" and "negative") in such a way that each two components with a common side belong to different parts. This can be always done if the surface S is the sphere or the plane. Let U_k^2 (respectively U_i^0) be the bounded components of the complement to the curve L from the positive (respectively negative) set. To each bounded component U_k^2 (respectively U_i^0) and to each self-intersection point p_j of the curve L we associate a formal generator Δ_k^2 (respectively Δ_i^0, Δ_j^1). Let L be the lattice generated by (formal) elements Δ_k^2, Δ_j^1, and Δ_i^0. We assume that the generators are regarded as elements of a distinguished basis of the lattice L in the indicated order (i.e., Δ_k^2, Δ_j^1, Δ_i^0); the order of generators with the same upper index is not essential. Let $n_{21}(k, j)$ and $n_{10}(j, i)$ be the number of vertices of the corresponding component of the complement (U_k^2 or U_i^0) that coincide with the self-intersection point p_j of the curve L. Let $n_{20}(k, i)$ be the number of common sides of the components U_k^2 and U_i^0 that are segments (sides which are circles are not taken into account). It is not difficult to verify that the integers $n_{21}(k, j)$, $n_{10}(j, i)$, and $n_{20}(k, i)$ satisfy the condition

$$n_{20}(k, i) = \frac{1}{2} \sum_j n_{21}(k, j) n_{10}(j, i).$$

This formula means that
$$\begin{pmatrix} -\operatorname{id} & n_{21} & -n_{20} \\ 0 & \operatorname{id} & -n_{10} \\ 0 & 0 & -\operatorname{id} \end{pmatrix}$$

is the matrix of an involution.

DEFINITION 11: The skew-symmetric bilinear form corresponding to the curve $L \subset S$ is defined by the following table of inner products of generators:

$$[\Delta_j^1 \circ \Delta_i^0] = n_{10}(j, i),$$
$$[\Delta_k^2 \circ \Delta_j^1] = n_{21}(k, j),$$
$$[\Delta_k^2 \circ \Delta_i^0] = -n_{20}(k, i).$$

The intersection numbers of cycles with the same upper index are equal to zero.

From this definition the descriptions of the corresponding Dynkin diagram and of the corresponding quadratic form are clear.

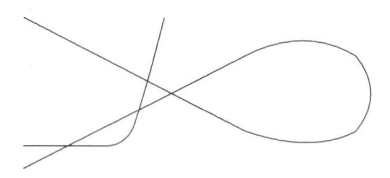

FIGURE 2

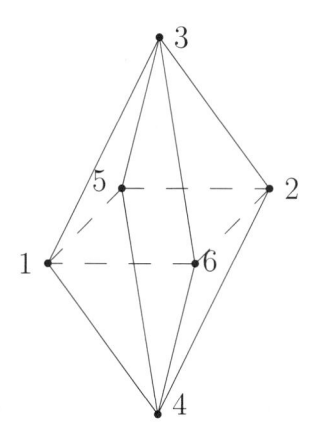

FIGURE 3

EXAMPLE. The Dynkin diagram of the curve in the plane $\mathbb{R}^2$ from Figure 2 is the standard Dynkin diagram E_7. The Dynkin diagram in Figure 3 is the Dynkin diagram of the union of two big circles in the 2-sphere.

Curves in real surfaces and corresponding Dynkin diagrams arise from some isolated complete intersection singularities in the following way. Let us suppose that the ICIS X is one-dimensional (i.e., $n = 1$), is defined by real equations $f_1, \ldots, f_m$, and there exist real perturbations $\widetilde{f}_i$ of the functions f_i, such that

1) the complex surface $S_{\mathbb{C}} = \{x \in \mathbb{C}^{m+1} : \widetilde{f}_1 = \cdots = \widetilde{f}_{m-1} = 0\}$ is nonsingular;

2) all critical points of the restriction of the function $\widetilde{f}_m$ to the surface $S_{\mathbb{C}}$ are nondegenerate and real (i.e., they lie in the real surface $S = \{x \in \mathbb{R}^{m+1} : \widetilde{f}_1 = \cdots = \widetilde{f}_{m-1} = 0\}$);

3) values of the function $\widetilde{f}_m$ in all its saddle points in the surface S are equal to zero.

Let us consider the real curve $L = \{\widetilde{f}_m = 0\}$ in the real surface S. A component of the complement to L is called *positive* (respectively *negative*) if the function $\widetilde{f}_m$ is positive (respectively negative) on it.

THEOREM [5]. *In the described case, the Dynkin diagram of the ICIS X coincides with the Dynkin diagram of the curve L.*

EXAMPLE. Let us consider the intersection of two quadrics in $\mathbb{C}^3$, which can be defined by the equations $f_1 = x_1^2 + x_2^2 + x_3^2 = 0$, $f_2 = x_1 x_2 = 0$. It is not difficult to see that the functions $\widetilde{f}_1 = x_1^2 + x_2^2 + x_3^2 - \varepsilon^2$ ($\varepsilon > 0$), $\widetilde{f}_2 = x_1 x_2$ satisfy the required conditions. The equation $\{\widetilde{f}_1 = 0\}$ defines the standard sphere in the space $\mathbb{R}^3$. The curve $L = \{\widetilde{f}_2 = 0\} \subset S$ is the union of two big circles. Thus a Dynkin diagram of the intersection of two quadrics in $\mathbb{C}^3$ can be seen in Figure 3.

References

1. N. A'Campo, *Le groupe de monodromie du deploiement des singularités isolées de courbes planes* II, Proc. of the Intern Congress of Math. (Vancouver 1974), vol. 1, 1975, pp. 395–404.
2. V. I. Arnold, *Normal forms of functions near degenerate critical points, the Weyl groups A_k, D_k, E_k and Lagrangian singularities*, Functional Anal. Appl. **6** (1972), 254–272.
3. N. Bourbaki, *Lie groups and Lie algebras*, Hermann, Paris, 1971.
4. A. M. Gabrielov, *Dynkin diagrams of unimodal singularities*, Functional Anal. Appl. **8** (1974), 192–196.
5. S. M. Gusein-Zade, *Method of real morsifications for complete intersection singularities*, C. R. Acad. Sci. Paris Sér. 1 **318** (1994), 391–396.
6. M. Hazewinkel, W. Hesselink, D. Siersma, and F. D. Veldkamp, *The ubiquity of Coxeter-Dynkin diagrams (an introduction to the A-D-E problem)*, Nieuw Arch. Wisk. **25** (1977), 257–307.
7. O. V. Lyashko, *Decomposition of simple singularities of functions*, Functional Anal. Appl. **10** (1976), 122–127.

Translated by THE AUTHOR

Variations on the Triangular Theme

A. A. KIRILLOV

Dedicated to E. B. Dynkin on his seventieth birthday

Contents

§0. Introduction
§1. Coadjoint orbits
§2. Adjoint orbits
§3. On $\mathbb{R}$-$\mathbb{H}$ groups
§4. Algebraic structures in sets of orbits.

§0. Introduction

The set of triangular matrices is one of the most fundamental objects in mathematics (as well as the set of natural numbers or the group of permutations). It can be endowed by several structures and the interaction between these structures is very interesting and leads to many remarkable discoveries.

Some aspects of this program are discussed in this article in more detailed form, the others are only briefly described. Due to the lack of time, place, or competence several interesting questions are only mentioned and the corresponding reference is given.

0.1. In the first section we consider the classification problem for coadjoint orbits of the group $G_n(k)$ of upper triangular matrices with elements from a field k. For real, complex, or p-adic field this is equivalent to the description of the unitary dual of this group (i.e. the set of equivalence classes of unitary irreducible representations).

For finite fields the relation of orbits to representations becomes more delicate and the classification problem turns out to be related with interesting combinatorial objects (such as q-analog of the so-called Euler-Bernoulli triangle).

It is also a temptation to consider a triangular matrix over a finite field as a state of some peculiar lattice model and apply to it the intuition and machinery of statistical physics. The very preliminary results in this direction are discussed in 1.5.

1991 *Mathematics Subject Classification.* Primary 15A57, 22E25.
Partly supported by the International Science Foundation.

©1995, American Mathematical Society

0.2. In the second section of the paper we consider the properties of adjoint orbits of $G_n(k)$. Though they look completely different compared with coadjoint orbits, we show that the total number of orbits is the same. This is an euristic argument supporting the conjecture relating coadjoint orbits with irreducible (complex) representations. The detailed form of the conjecture gives a simple explicit formula for an irreducible character in terms of associated orbit.

The next question discussed in this section is the splitting of the set of strictly triangular matrices according to their "partition type". We give here the elementary proof and some generalization of the famous result by Springer that relates this splitting to representation theory of the symmetric group.

0.3. Section 3 is devoted to some questions of the representation theory of finite and compact Lie groups connected with ζ-like expressions

$$\sum_{\lambda \in \widehat{G}} d(\lambda)^k ,$$

where $d(\lambda)$ is the dimension of an irreducible representation of class $\lambda \in \widehat{G}$. The reason to include this section is that one can use the sums of this type in order to estimate the number of irreducible representations and their distribution by dimensions.

0.4. In §4 we introduce some algebraic structures in the union of the sets of triangular matrices of all orders and in the corresponding sets of adjoint and coadjoint orbits and partition types. These structures are connected with the possibility of "fusing" two triangular matrices into one bigger matrix. This leads immediately to the construction of infinite dimensional graded associative algebras with involution $\mathcal{T}_q, \mathcal{A}_q, \mathcal{B}_q, \mathcal{P}_q$. But it seems that a much deeper understanding of the situation can be obtained if we could construct Tannakian categories in the sense [DM] for which these algebras would be the Grothendieck rings.

§1. Coadjoint orbits

1.1. The method of orbits relates the set $\widehat{G}$ of (classes of equivalence of) unitary irreducible representations of a Lie group G to the set $\mathfrak{g}^*/G$ of G-orbits in the space $\mathfrak{g}^*$, dual to the Lie algebra $\mathfrak{g} = \text{Lie}(X)$. This relation is especially simple for connected and simply connected nilpotent Lie groups. In this case $\widehat{G}$ and $\mathfrak{g}^*/G$ are naturally homeomorphic.[1]

However, the structure of the set $\mathfrak{g}^*/G$ for nilpotent G is much more complicated than for semisimple groups (where the coadjoint action is equivalent to the adjoint one). Even for the important case of the group of real strictly upper triangular matrices, this structure is not completely understood despite the numerous attempts (see [An, H, K0]).

Let us accept the algebraic geometry point of view and consider this group as the group of real points of the algebraic group $G_n \subset GL_n$ defined by

(1.1.1) $$g_{ii} = 1, \quad g_{ij} = 0 \quad \text{for } i > j.$$

---
[1] The general references for this and other facts related to the orbit method are [K0], [K1], and [K4].

We denote by $G_n(k)$ the group of k-points of G_n for any field k. The corresponding Lie algebra $\mathfrak{g}_n(k)$ consists of all matrices $X = \|x_{ij}\| \in \operatorname{Mat}_n(k)$ satisfying

$$\tag{1.1.2} x_{ij} = 0 \quad \text{for } i \geq j.$$

The dual k-linear space $\mathfrak{g}_n^*(k)$ can be naturally identified with the set of lower triangular matrices $F = \|f_{ij}\| \in \operatorname{Mat}_n(k)$ with

$$\tag{1.1.3} f_{ij} = 0 \quad \text{for } i \leq j.$$

The value of a functional $F \in \mathfrak{g}_n^*(k)$ on an element $X \in \mathfrak{g}_n(k)$ is given by

$$\tag{1.1.4} \langle F, X \rangle = \sum_{i<j} f_{ji} x_{ij} = \operatorname{tr}(FX).$$

The coadjoint action K of $G_n(k)$ on $\mathfrak{g}_n^*(k)$ has the form

$$K(g)F = (gFg^{-1})_{\text{low}},$$

where the index "low" means that we keep the lower triangular part of the matrix and put zeroes on the main diagonal and above. Indeed,

$$\langle K(g)F, X \rangle = \langle F, \operatorname{Ad} g^{-1} \cdot X \rangle = \operatorname{tr}(Fg^{-1}Xg) = \operatorname{tr}(gFg^{-1}X) = \operatorname{tr}((gFg^{-1})_{\text{low}} X).$$

1.2. We are interested in the structure of the set $\mathcal{O}_n^*(k) = \mathfrak{g}_n^*(k)/G_n(k)$ of $G_n(k)$-orbits in $\mathfrak{g}_n^*(k)$. Unfortunately, this is not (the set of k-points of) a manifold, because these orbits have different dimensions.

Recall the following simple but important fact [K0].

LEMMA 1.2.1. *The dimension of the orbit Ω_F passing through the point F is equal to the rank of the bilinear form B_F on $\mathfrak{g}_n(k)$ defined by*

$$\tag{1.2.1} B_F(X, Y) = \langle F, [X, Y] \rangle.$$

PROOF. The direct computation shows that the Lie algebra $\operatorname{stab}(F)$ of the stabilizer $\operatorname{Stab}(F)$ of F in $G_n(k)$ coincides with the kernel of B_F. Hence, $\dim \Omega_F = \dim T_F \Omega_F = \dim \mathfrak{g} - \dim \ker B_F = \operatorname{rk} B_F$.

Let now $\mathfrak{g}_{n,m}^*(k)$ be the set of points $F \in \mathfrak{g}_n^*(k)$ for which $\operatorname{rk} B_F = 2m$. (The rank of B_F is always even since B_F is skew-symmetric).

LEMMA 1.2.2. *The set $\mathfrak{g}_{n,m}^*(k)$ is the set of k-points of a quasi-affine algebraic manifold $\mathfrak{g}_{n,m}^* \subset \mathfrak{g}_n^*$.*

PROOF. Let $X_{n,m} = \{F \in \mathfrak{g}_n \mid \operatorname{rk} B_F \leq 2m\}$. Then $X_{n,m}$ is an affine algebraic manifold defined by the condition that all minors of order $> 2m$ of the matrix representing B_F vanish. Hence, $\mathfrak{g}_{n,m}^*(k) = X_{n,m} \setminus X_{n,m-1}$ is a quasi-affine algebraic manifold. One can check that m here ranges from 0 to $[(n-1)^2/2]$.

THEOREM 1.2.3. *We have*

$$\mathcal{O}_n^*(k) = \coprod_{m=0}^{[(n-1)^2/2]} \mathcal{O}_{n,m}^*(k), \tag{1.2.2}$$

where $\mathcal{O}_{n,m}^*$ are quasi-affine algebraic manifolds.

SKETCH OF THE PROOF. Let $\mathcal{O}_{n,m}^*(k)$ be the set of all $G_n(k)$-orbits in $\mathfrak{g}_{n,m}^*(k)$. For any $\Omega \in \mathcal{O}_{n,m}^*(k)$ we choose a point $F \in \Omega$ and a linear subspace $L \subset \mathfrak{g}_n^*(k)$ transversal to Ω, i.e. such that

$$\mathfrak{g}_n^*(k) = L \oplus T_F(\Omega). \tag{1.2.3}$$

The set of those $F \in \mathfrak{g}_{n,m}^*(k)$ for which (1.2.3) holds is a Zariski open subset and its intersection with $L + F$ can be taken as a local chart on $\mathfrak{g}_{n,m}^*(k)$ at the point Ω.

For future use we prove the following result.

THEOREM 1.2.4. *Every 2m-dimensional coadjoint orbit in $\mathfrak{g}_n^*(\mathbb{F}_q)$ consists of q^{2m} points.*

REMARK. Let $k = \mathbb{R}$; it is shown in [K0] that for any nilpotent Lie group the coadjoint orbits are affine algebraic manifolds isomorphic to $\mathbb{A}^{2m}$. It seems that the same is true for $G_n(\mathbb{F}_q)$, but I cannot prove it now.

LEMMA 1.2.5. *The map $g \mapsto g - 1$ establishes the bijection between the stabilizer $\mathrm{Stab}(F) \in G_n$ and its Lie algebra $\mathrm{stab}(F) \in \mathfrak{g}_n$.*

PROOF OF THE LEMMA. Let $\mathfrak{b}_n^+$ be the space of all non-strictly upper triangular matrices of order n. Then $g \in \mathrm{Stab}(F)$ is equivalent to $gFg^{-1} \equiv F \mod \mathfrak{b}_n^+$ or, using the relation $g \cdot \mathfrak{b}_n^+ = \mathfrak{b}_n^+$, to $gF \equiv Fg \mod \mathfrak{b}_n^+$. The last is equivalent to $g - 1 \in \mathrm{stab}(F) = \ker B_F$.

PROOF OF THE THEOREM. According to the lemma just proved,

$$\#\mathrm{Stab}(F) = \#\mathrm{stab}(F) = q^{\dim \mathrm{stab}(F)}.$$

Now,

$$\#\Omega = \frac{\#G_n}{\#\mathrm{Stab}(F)} = q^{\dim G_n - \dim \mathrm{stab}(F)} = q^{\dim \Omega}.$$

EXAMPLE 1. $\mathfrak{g}_3^*$ consists of matrices $F = \begin{pmatrix} 0 & 0 & 0 \\ x & 0 & 0 \\ z & y & 0 \end{pmatrix}$; $\mathfrak{g}_{3,0}^*$ is given by the equation $z = 0$ and $\mathfrak{g}_{3,1}^*$ by the inequality $z \neq 0$.

Hence, $\mathcal{O}_{3,0}^* \simeq \mathbb{A}^2$, $\mathcal{O}_{3,1}^* \simeq \mathbb{A}^1 \backslash \mathbb{A}^0 \simeq GL_1$.

EXAMPLE 2. $\mathfrak{g}_4^*$ consists of matrices $F = \begin{pmatrix} 0 & 0 & 0 & 0 \\ u & 0 & 0 & 0 \\ x & v & 0 & 0 \\ z & y & w & 0 \end{pmatrix}$; $\mathfrak{g}_{4,0}^*$ is given by $x = y = z = 0$, $\mathfrak{g}_{4,1}^*$ by $z = 0$, $(x, y) \neq (0, 0)$ and $\mathfrak{g}_{4,2}^*$ by $z \neq 0$.

Hence, $\mathcal{O}_{4,0}^* \simeq \mathbb{A}^3$, $\mathcal{O}_{4,1}^* \simeq \mathbb{A}^3 \backslash \mathbb{A}^1$, $\mathcal{O}_{4,2}^* \simeq \mathbb{A}^2 \backslash \mathbb{A}^1$.

1.3. We note that all $\mathcal{O}_{n,m}^*$ are defined over $\mathbb{Z}$, i.e., are given by the algebraic equations and inequalities with integer coefficients. For such manifolds M the important information can be obtained by considering the number $M(q)$ of points of M over a finite field $\mathbb{F}_q$ of q elements. For instance, for "nice" smooth projective manifolds M the following formula holds:

$$(1.3.1) \qquad M(q) = \sum_{k=0}^{\dim M} b_{2k} q^k, \qquad b_{2k} = \dim H^{2k}(M(\mathbb{C}), \mathbb{R}).$$

The ultimate reason for this is that M can be represented as the union of affine cells of different dimensions. Each cell of dimension k contributes q^k in $M(q)$ and 1 in b_{2k}.

For "nice" quasi-affine algebraic manifolds the quantity $M(q)$ is still a polynomial in q with integer coefficient (which could be negative).

EXAMPLE 1. $M = G_{n,k}$, the Grassmann manifold of k-planes in n-space. Here

$$(1.3.2) \qquad G_{n,k}(q) = \begin{bmatrix} n \\ k \end{bmatrix}_q = \frac{[n]_q}{[k]_q [n-k]_q}$$

is the Gaussian q-binomial coefficient and $[n]_q := 1 + q + q^2 + \cdots + q^{n-1}$.

EXAMPLE 2. $M = GL_n$, the general linear group of order n

$$(1.3.3) \qquad GL_n(q) = \prod_{k=0}^{n-1} (q^n - q^k).$$

EXAMPLE 3. More generally, if $M_{a,b}^c$ is the manifold of $(a \times b)$-matrices of rank c, we have

$$(1.3.4) \qquad M_{a,b}^c(q) = \begin{bmatrix} a \\ c \end{bmatrix}_q \cdot \begin{bmatrix} b \\ c \end{bmatrix}_q \cdot GL_c(q).$$

Sometimes it is more convenient to deal with Laurent polynomials in $q^{\frac{1}{2}}$ and $q^{-\frac{1}{2}}$. Let us define the *balancing map*

$$k[q] \ni P \to \widehat{P} \in k[q^{1/2}, q^{-1/2}]$$

by

$$(1.3.5) \qquad \widehat{P}(q) = P(q) q^{-(M+m)/2},$$

where $M = \deg P$ and $m = \deg q^M P(1/q)$ are the maximal and minimal degrees of nonzero monomials in P.

This map has the following useful properties:

(1.3.6) a) $\widehat{\overline{P \cdot Q}} = \widehat{\overline{P}}(q) \cdot \widehat{\overline{Q}}(q),$

b) $\widehat{\overline{q}} = 1,$

which allow to compute it easily.[2]

To illustrate the use of the balancing procedure we note that there are no simple expression for the sum of q-binomial coefficients, but for corresponding balanced quantities we have the nice formula

$$(1.3.7) \qquad \sum_{k=0}^{n} \widehat{\overline{\begin{bmatrix} n \\ k \end{bmatrix}}}_q = \prod_{s=1}^{n} (1 + q^{(n+1-2s)/2}).$$

1.4. In our case the quantities $\mathcal{O}^*_{n,m}(q)$ and, consequently, $\mathcal{O}^*_n(q)$ turn out to be polynomials in q with integer coefficients. The explicit computation of these polynomials is a rather difficult combinatorial problem. Still more mysterious is their behavior when $n \to \infty$. Even the asymptotics of the sequence $\{\mathcal{O}^*_n(q)\}$ for given q is unknown. However, the behavior of $\mathcal{O}^*_{n,m}(q)$ for given q and m is relatively simple.

THEOREM 1.4.1. *For given q and m we have*

$$(1.4.1) \qquad \mathcal{O}^*_{n,m}(q) = q^n L_m(n),$$

where L_m is a polynomial of degree m with coefficients from $\mathbb{Z}[q, q^{-1}]$.

PROOF. The proof of this theorem is based on the observation that for fixed q, m and $n \to \infty$ the set $\mathfrak{g}^*_{n,m}(\mathbb{F}_q)$ has a *cluster structure*. This means that nonzero elements of matrices $F \in \mathfrak{g}^*_{n,m}(\mathbb{F}_q)$ are not numerous and gather into groups called *clusters*. To be more precise, we introduce some notations and definitions.

Let $\mathfrak{b}^-_n(\mathbb{F}_q)$ be the set of all non-strictly lower triangular matrices in $\mathrm{Mat}_n(\mathbb{F}_q)$. To each $F \in \mathfrak{g}^*_{n+1}(\mathbb{F}_q)$ we associate the matrix $\overline{F} \in \mathfrak{b}^-_n(\mathbb{F}_q)$ which is obtained from F by deleting of the first row and the last column. A simple but important observation is that rk B_F depend only on the lower part of $\overline{F}$ (i.e., does not depend on the diagonal part of $\overline{F}$).

An element $\overline{F} \in \mathfrak{b}^-_n(\mathbb{F}_q)$ is called *decomposable*, if it has a block-diagonal form:

$$(1.4.2) \qquad \overline{F} = \overline{F}_1 \oplus \overline{F}_2 \oplus \cdots \oplus \overline{F}_r, \qquad r > 1,$$

where $\overline{F}_i \in \mathfrak{b}^-_{n_i}(\mathbb{F}_q)$ and $\sum_{i=1}^r n_i = n$.

---

[2] We remark that the quantity $\widehat{\overline{\mathbb{P}^{n-1}}}(q) = \dfrac{q^{n/2} - q^{-n/2}}{q^{1/2} - q^{-1/2}} = \widehat{\overline{[n]}}_q$ is often considered as "the quantum analog" of an integer n, and the quantity $\widehat{\overline{\mathcal{F}^n}}(q)$, where $\mathcal{F}^n$ is the (full) flag manifold, as a quantum analog of $n!$.

LEMMA 1.4.2. *Under the above conditions we have*

(1.4.3) $$\operatorname{rk} B_F = \sum_{i=1}^{r} \operatorname{rk} B_{F_i}.$$

PROOF. Let $\{X_{i,j}\}$, $1 \le i < j \le n+1$, be the standard basis in $\mathfrak{g}_{n+1}(\mathbb{F}_q)$ and $\{X_{i,j}^*\}$ be the dual basis in $\mathfrak{g}_{n+1}^*(\mathbb{F}_q)$. Then

$$B_F = \sum_{i<k<j} f_{j,i} \{X_{i,k}^*\} \wedge \{X_{k,j}^*\}.$$

Now it is easy to see that 2-forms B_{F_i} that correspond to different summands in (1.4.3) depend on disjoint sets of $\{X_{i,j}^*\}$'s. This makes the lemma evident.

Define the *cluster* (or *elementary particle*) of *size n* and *mass* (or *energy*) m as an indecomposable element $\overline{F} \in \mathfrak{b}_n^-(\mathbb{F}_q)$ associated to an element $F \in \mathfrak{g}_{n+1,m}^*(\mathbb{F}_q)$.

It follows from this definition that for any $F \in \mathfrak{g}_{n+1,m}^*(\mathbb{F}_q)$ the associated $\overline{F} \in \mathfrak{b}_n^-(\mathbb{F}_q)$ splits into a certain number of clusters of sizes $n_1, \ldots, n_r$ and masses $m_1, \ldots, m_r$, such that

$$\sum_{i=1}^{r} n_i = n, \qquad \sum_{i=1}^{r} m_i = m.$$

It is also clear that for any q there are only finitely many clusters of given mass m. Let us denote by $C_{\nu,\mu}(q)$ the number of clusters of size ν and mass μ. Then the standard combinatorial argument proves

PROPOSITION 1.4.4. *We have*

$$\sum_{n,m} \mathfrak{g}_{n+1,m}^*(q) t^m x^n = \left(1 - \sum_{\nu,\mu} C_{\nu,\mu}(q) t^\mu x^\nu\right)^{-1}.$$

COROLLARY 1.4.5. *We have*

$$\mathcal{O}_{n+1,m}(q) = q^{-2m} \left\{ \text{the coefficient at } t^m x^n \text{ in } \left(1 - \sum_{\nu,\mu} C_{\nu,\mu}(q) t^\mu x^\nu\right)^{-1} \right\}.$$

We now derive the statement of the theorem from this corollary. We observe that $C_{1,0} = q$: the corresponding clusters ("photons") are just matrices of order 1 with a coefficient in $\mathbb{F}_q$.

Moreover, $C_{n,0} = 0$ for $n > 1$, because all particles of size $n > 1$ have a nonzero mass (the corresponding 2-form B_F is nonzero).

From this we conclude that the right-hand side of (1.4.4) can be written in the form

$$[1 - qx - t\Phi(q, x, t)]^{-1}$$

for some $\Phi \in \mathbb{Z}[q][[x, t]]$. It follows that the right-hand side of (1.4.5) is equal to the coefficient of x^n in

$$\frac{\Psi(q, x, t)}{[1 - qx - t\Phi(q, x, t)]^{m+1}}\bigg|_{t=0} = \frac{\Psi(q, x, 0)}{(1 - qx)^{m+1}}$$

for some $\Psi \in \mathbb{Z}[q][[x, t]]$, and this coefficient has evidently the form (1.4.1).

As an illustration, we give here the description of clusters for small m and the corresponding explicit formulae for $\mathcal{O}_{n,m}$.

$m = 0$. It is clear that $\mathcal{O}_{n,0}^*(q) = q^{n-1}$, $n \geq 1$. One can also formally deduce it from the above relation $C_{n,0} = q \cdot \delta(n)$.

$m = 1$. One can check that $C_{2,1} = q^2(q-1)$, $C_{3,1} = q^3(q-1)^2$, and $C_{n,1} = 0$ for $n > 3$. The corresponding $\overline{F}$'s look as follows:

$$\begin{pmatrix} x & 0 \\ \lambda & y \end{pmatrix} \in \mathfrak{b}_2^- \quad \text{and} \quad \begin{pmatrix} x & 0 & 0 \\ \lambda & y & 0 \\ 0 & \mu & z \end{pmatrix} \in \mathfrak{b}_3^-,$$

where $\lambda, \mu \in \mathbb{F}_q^\times$, $x, y, z \in \mathbb{F}_q$. It follows that $\mathcal{O}_{n,1}^*(q) = q^{n-3}(q-1)(nq-3q+1)$, $n \geq 3$.

$m = 2$. Clusters of mass 2 are:

$$\begin{pmatrix} x & 0 & 0 & 0 \\ \lambda & y & 0 & 0 \\ 0 & \mu & z & 0 \\ 0 & 0 & \nu & u \end{pmatrix}, \begin{pmatrix} x & 0 & 0 & 0 & 0 \\ \lambda & y & 0 & 0 & 0 \\ 0 & \mu & z & 0 & 0 \\ 0 & 0 & \nu & u & 0 \\ 0 & 0 & 0 & \rho & v \end{pmatrix}, \begin{pmatrix} x & 0 & 0 & 0 & 0 \\ p & y & 0 & 0 & 0 \\ 0 & q & z & 0 & 0 \\ 0 & \lambda & r & u & 0 \\ 0 & 0 & 0 & s & v \end{pmatrix},$$

where all latin letters run through $\mathbb{F}_q$ and greek letters run through $\mathbb{F}_q^\times$. It follows that $C_{4,2}(q) = q^4(q-1)^3$, $C_{5,2}(q) = q^9(q-1) + q^5(q-1)^4$, $C_{n,2} = 0$ for $n > 5$.

1.5. The information contained in the double sequence $\{\mathcal{O}_{n,m}^*(q)\}$ can be encoded in the ordinary sequence of polynomials

$$(1.5.1) \qquad \mathcal{O}_n^*(q, t) = \sum_{m \geq 0} \mathcal{O}_{n,m}^*(q) t^m$$

or in the generating series

$$(1.5.2) \qquad \mathcal{O}^*(q, t, z) = \sum_{n \geq 0} \mathcal{O}_n^*(q, t) z^n.$$

At the moment I have almost nothing to say about the last series except the very plausible conjecture that it is a formal power series in z with the zero radius of convergence for all $q > 1$ and $t \geq 1$. On the contrary, the polynomials $\mathcal{O}_n^*(q, t)$ have several interesting interpretations.

Firstly, Theorem 1.2.4 implies that after the substitution $t \to \mu^2 q^2$ the values of these polynomials can be expressed in the form which reminds partition functions in the quantum statistical physics:

$$(1.5.3) \qquad \mathcal{O}_n^*(q, \mu^2 q^2) = \sum_{F \in \mathfrak{g}^*(\mathbb{F}_q)} \mu^{\operatorname{rk} B_F}.$$

In other words, we consider $\mathfrak{g}^*(\mathbb{F}_q)$ as a peculiar lattice model where rk B_F plays the role of the energy function and $-\log \mu$ is the analog of the inverse temperature $\beta = 1/kT$.

The approach suggested in [K3] gives the expression of this partition function for the special cases when β is a nonnegative integer.

THEOREM 1.5.1. *Let $r > 0$ be an integer. Then*

$$\mathcal{O}_n^*(q, q^{2-2r}) = q^{-rn(n-1)} \sum_{X,Y,F} \theta(\langle F, [X, Y] \rangle),$$

where X, Y run through $\mathfrak{g}_n(\mathbb{F}_{q^r})$, F runs through $\mathfrak{g}_n^*(\mathbb{F}_q)$ naturally embedded in $\mathfrak{g}_n^*(\mathbb{F}_{q^r})$ and θ is any nontrivial additive character of $\mathbb{F}_{q^r}$.

The proof of this theorem is based on the following simple facts. Let $V_i, i = 1, 2$, be finite dimensional vector spaces over $\mathbb{F}_q$ and $B: V_1 \times V_2 \to \mathbb{F}_q$ be a bilinear form. For any integer $r \geq 1$ we can consider $V_i^{(r)} = V_i \otimes_{\mathbb{F}_q} \mathbb{F}_{q^r}$ and extend B by $\mathbb{F}_{q^r}$-linearity to a bilinear form $B^{(r)}: V_1^{(r)} \times V_2^{(r)} \to \mathbb{F}_{q^r}$.

LEMMA 1.5.2. *For any nontrivial additive character θ of $\mathbb{F}_{q^r}$ we have*

$$(1.5.4) \qquad \sum_{v_i \in V_i^{(r)}} \theta(B(v_1, v_2)) = q^{r(\dim V_1 + \dim V_2 - \operatorname{rk} B)}.$$

PROOF OF THE LEMMA. By a suitable choice of bases in V_1 and V_2 we can reduce B to the form

$$B(v_1, v_2) = \sum_{j=1}^{\operatorname{rk} B} v_1^j v_2^j,$$

where v_i^j, $1 \leq j \leq \dim V_i$, are coordinates of $v_i \in V_i$. Then we use the formula

$$\sum_{x \in \mathbb{F}_{q^r}} \theta(xy) = q^r \delta(y),$$

where $\delta(y) = 0$ if $y \neq 0$, $\delta(y) = 1$ if $y = 0$, and get the result.

Now, applying the lemma to the form B_F, we obtain

$$\sum_{X,Y,F} \theta(\langle F, [X, Y] \rangle) = \sum_F q^{r[n(n-1) - \operatorname{rk} B_F]}.$$

The contribution of those F for which rk $B_F = 2m$ is equal to $q^{2m} \cdot \mathcal{O}_{n,m}(q)$ times $q^{r[n(n-1)-2m]}$. This gives the left-hand side of the desired equality.

Another interpretation of the polynomials $\mathcal{O}_n^*(q, t)$ arises if we accept Conjecture 2.2.1 (see 2.2 below). Namely

$$(1.5.5) \qquad \mathcal{O}_n^*(q, q^{-s}) = \sum_{\lambda \in \widehat{G_n(\mathbb{F}_q)}} d(\lambda)^{-s},$$

where $\widehat{G_n(\mathbb{F}_q)}$ is the set of (equivalence classes of) unitary irreducible representations of $G_n(\mathbb{F}_q)$ and $d(\lambda)$ is the dimension of a representation from the class λ. These ζ-like sums seem to have interesting properties and will be discussed in §3.

We reproduce now another result of [K4], which is the definition of a more sophisticate partition function associated with a Lie algebra $\mathfrak{g}_n$.

Namely, we put

$$\Phi_n(Z) = \sum_{X,Y,F} \theta(\langle F, [X, Y] - Z\rangle), \quad (1.5.6)$$

where, as before, X, Y run through $\mathfrak{g}_n(\mathbb{F}_q)$, F runs through $\mathfrak{g}_n^*(\mathbb{F}_q)$.

Let us introduce also for any Lie algebra $\mathfrak{g}$ the algebraic manifold consisting of pairs of commuting elements in $\mathfrak{g}$:

$$CP_\mathfrak{g} = \{(X, Y) \in \mathfrak{g} \times \mathfrak{g} \mid [X, Y] = 0\}.$$

THEOREM 1.5.3. *We have*

$$\Phi_n(0) = q^{n(n-1)/2} \cdot CP_{\mathfrak{g}_n}(q).$$

The proof is completely analogous to the proof of Theorem 1.5.1 and is also based on Lemma 1.5.2.

REMARK. This theorem is also true for any Lie algebra over $\mathbb{F}_q$. On the contrary, the equality $\Phi_n(0) = q^{n(n-1)} \cdot \mathcal{O}_n^*(q)$ should be replaced for general Lie algebra $\mathfrak{g}$ by

$$\Phi_n(0) = q^{2\dim \mathfrak{g}} \cdot \sum_{\Omega \in \mathfrak{g}^*/G} \mathrm{ind}(\Omega),$$

where $\mathrm{ind}(\Omega) = q^{-\dim \Omega} \cdot \#\Omega$. If $\Omega \simeq \mathbb{A}^{2m}$, we have $\mathrm{ind}(\Omega) = 1$.

Peculiar lattice models became rather popular in the last time. We indicate to the interested reader the paper [MS] where some of these models are discussed.

1.6. The numerical computations for small n in the simplest case $q = 2$ lead to the remarkable observation: the sequence $\{\mathcal{O}_n^*(2)\}$ is very close to the sequence $\{K_n\}$ of so called *Euler-Bernoulli numbers*, defined by the generating function

$$\sum_{n \geq 0} K_n \frac{x^n}{n!} = \frac{1 + \sin x}{\cos x}. \quad (1.6.1)$$

Namely, we have (see [K2, K3])
(1.6.2)

n	0	1	2	3	4	5	6	7	8	9
$\mathcal{O}_2^*(n)$	?	1	2	5	16	61	275	1430	8506	57205
K_{n+1}	1	1	2	5	16	61	272	1385	7936	50521

Unfortunately, it is too difficult to compute directly next terms of the second row. (The first six entries were computed "manually" by my former student A. Mikhailov about 20 years ago, the next three entries are taken from [G].)

On the contrary, for computing the sequence $\{K_n\}$ there exists a very simple and efficient procedure ([A1, A2, KB]). Consider the triangular table, the so called *Euler-Bernoulli triangle*:

(1.6.3)
$$\begin{array}{c} a_{00} \\ a_{10} \leftarrow a_{01} \\ a_{02} \rightarrow a_{11} \rightarrow a_{20} \\ a_{30} \leftarrow a_{21} \leftarrow a_{12} \leftarrow a_{03} \\ a_{04} \rightarrow a_{13} \rightarrow a_{22} \rightarrow a_{31} \rightarrow a_{40} \end{array}$$

which is filled up shuttlewise according to the rules:
1) $a_{0,n} = \delta(n)$;
2) $a_{k,l} = a_{k-1,l+1} + a_{l,k-1}$, $k \geq 1$, $l \geq 0$.

The numerical values of the first six rows are:

(1.6.4)
$$\begin{array}{c} 1 \\ 1 \leftarrow 0 \\ 0 \rightarrow 1 \rightarrow 1 \\ 2 \leftarrow 2 \leftarrow 1 \leftarrow 0 \\ 0 \rightarrow 2 \rightarrow 4 \rightarrow 5 \rightarrow 5 \\ 16 \leftarrow 16 \leftarrow 14 \leftarrow 10 \leftarrow 5 \leftarrow 0. \end{array}$$

In other words, the number at the end of an arrow is equal to the sum of the number at the tail of the arrow and the number above the arrow.

The provocative conjecture arises that there exists a q-analog of the Euler-Bernoulli triangle whose side elements count the numbers of coadjoint orbits in $\mathfrak{g}_n^*(\mathbb{F}_q)$.

Namely, let us keep the rule 1) and substitute the rule 2) by the two more sophisticate ones:
2') $a_{k,l} = t^{-1} a_{k-1,l+1} + (q-1) t^{l-1} a_{l,k-1}$ for $k \geq 1$, $l \geq 1$;
2'') $a_{n,0} = (q-1)^{-1} a_{n-1,1}$.

We obtain in this way a triangle table whose elements are polynomials in q and t. The first few rows are:

(1.6.5)
$$\begin{array}{c} 1 \\ 1 \leftarrow 0 \\ 0 \rightarrow q-1 \rightarrow 1 \\ q \leftarrow q(q-1) \leftarrow (q-1)t \leftarrow 0 \\ 0 \rightarrow q(q-1)t^2 \rightarrow q^2(q-1)t \rightarrow q^2(q-1) + (q-1)^2 t \rightarrow q^2 + (q-1)t. \end{array}$$

It turns out that $a_{n,0}(q,t) = \mathcal{O}_n^*(q,t)$ for $n \leq 5$.

It is natural to conjecture that the other quantities $a_{k,l}(q,t)$ also have an interpretation in terms of orbits. One easily checks inductively that 2') and 2'') imply the equality

(1.6.6) $$a_{n+1,0}(q,t) = \sum_{k+l=n} a_{k,l}(q,t).$$

So the orbit interpretation of $a_{k,l}(q,t)$ should be connected with some stratification of the set $\mathcal{O}_n^*(q)$. Namely, $a_{k,l}(q,t) = \sum_m a_{k,l}^m(q) t^m$, where $a_{k,l}^m(q)$ is the number of $2m$-dimensional orbits in the k-th stratum of $\mathfrak{g}_{n,m}^*(\mathbb{F}_q)$. The appropriate stratification can be defined as follows. Let $X_{k,l} \subset \mathfrak{g}_{k+l}^*(\mathbb{F}_q)$ be defined by the following conditions on the coefficients $\{f_{i,j}\}$ of $F \in \mathfrak{g}_{k+l}^*(\mathbb{F}_q)$:

(1.6.7) $\quad f_{n,1} = \cdots = f_{n,k-1} = 0, \quad f_{n,k} \neq 0; \quad n = k + l.$

Let $x_{k,l}^m(q)$ denote the number of $2m$-dimensional orbits in $X_{k,l}$ and put $x_{k,l}(q,t) = \sum_m x_{k,l}^m(q) t^m$. In [K3] we have established that $x_{k,l}(q,t) = a_{k,l}(q,t)$ for $k + l \leq 6$ with only one exception: $x_{4,2}(q,t) - a_{4,2}(q,t) = (q-1)^3 t^2 (t-q) \cdot (t - q^2) \neq 0$.

REMARK 1. The latter formula suggests that the generalized Euler-Bernoulli triangle may have special properties when $t = q$ or $t = q^2$. Indeed, for $t = q^2$ it looks like

$$
\begin{array}{ccccccccccccc}
& & & & & & 1 & & & & & & \\
& & & & & 1 & \leftarrow & 0 & & & & & \\
& & & & 0 & \rightarrow & q-1 & \rightarrow & 1 & & & & \\
& & & q & \leftarrow & q^2-q & \leftarrow & q^3-q^2 & \leftarrow & 0 & & & \\
& & 0 & \rightarrow & q^6-q^5 & \rightarrow & q^5-q^4 & \rightarrow & q^4-q^3 & \rightarrow & q^3 & & \\
q^6 & \leftarrow & q^7-q^6 & \leftarrow & q^8-q^7 & \leftarrow & q^9-q^8 & \leftarrow & q^{10}-q^9 & \leftarrow & 0. & &
\end{array}
$$

The case $t = q$ is still under investigation.

REMARK 2. Still more intriguing is the following coincidence: for $q = 3$, $t = 1$, the sequence $\{x_{n,0}\}$ have the same first entries 1, 1, 1, 3, 11, 57, 361, ... as the sequence $\{b_n\}$ of Springer numbers of type D_n (see [A2, Sh]).

All this gives rise to many interesting but rather difficult problems. We formulate here only three of them.

1. Could one give the explicit formula for $a_{k,l}(q,t)$ or for an appropriate generating function?

We observe that in the case $q = 2$, $t = 1$ our generalized Euler-Bernoulli triangle becomes the usual one and the following formula holds:

(1.6.8) $\quad A(u,v) := \sum_{k,l} a_{k,l}(2,1) \frac{u^k v^l}{k! \, l!} = \frac{\cos v + \sin u}{\cos(u+v)}.$

It can be derived from the differential equation and the initial condition satisfied by $A(u,v)$

(1.6.9) $\quad (\partial_u - \partial_v) A(u,v) = A(v,u), \quad A(0,v) = 1,$

which follow from the rules 2'), 2'').

2. What is the asymptotic behavior of $a_n(q) := \sum_{k+l=n} a_{k,l}(q,1)$ as $n \to \infty$?

For $q = 2$, (1.6.1) implies that $a_n(2) \approx (2/\pi)^n n!$, but in general case we have no results.

3. Euler and Bernoulli numbers are related to the ζ-like sums:

$$(-1)^{n-1} B_{2n} \frac{2^{2n}-1}{2(2n)!} \pi^{2n} = \sum_{k=0}^{\infty} \frac{1}{(2k+1)^{2n}},$$

$$E_n \frac{1}{2(2n)!} \left(\frac{\pi}{2}\right)^{2n+1} = \sum_{k=0}^{\infty} \frac{(-1)^k}{(2k+1)^{2n+1}}.$$

Recently in [BCK] the very ingenious geometric interpretation of these relations has been found in terms of volumes of some polytopes. Namely,

$$\frac{K_{n-1}}{2(n-1)!} = \left(\frac{2}{\pi}\right)^n \sum_{k=0}^{\infty} \frac{(-1)^{nk}}{(2k+1)^n} = \text{vol}(\Delta_n),$$

where Δ_n is the convex polytope in $\mathbb{R}^n$ defined by the conditions:

$$x_i \geq 0, \quad x_i + x_{i+1} \leq 1, \quad i \in \mathbb{N} \bmod n.$$

Could it be possible to find the analogous interpretation for all entries of the Euler-Bernoulli triangle and its q-analogue?

§2. Adjoint orbits

2.1. Adjoint orbits, i.e., G-orbits in $\mathfrak{g} = \text{Lie}(G)$ are in a certain sense the linear approximations to the conjugacy classes in G. In our case this approximation has especially nice properties.

THEOREM 2.1.1. *The map*

(2.1.1) $$\mathfrak{g}_n(\mathbb{F}_q) \ni X \to 1 + X \in G_n(\mathbb{F}_q)$$

defines the bijection between adjoint orbits in $\mathfrak{g}_n(\mathbb{F}_q)$ and conjugacy classes in $G_n(\mathbb{F}_q)$.

PROOF. The relation $X = gXg^{-1}$ is equivalent to $(1 + X) = g(1 + Y)g^{-1}$.

There is also an analog of Lemma 1.2.5 which relates the centralizers of $X \in \mathfrak{g}_n(\mathbb{F}_q)$ in $\mathfrak{g}_n(\mathbb{F}_q)$ and in $G_n(\mathbb{F}_q)$.

LEMMA 2.1.2. *The map* (2.1.1) *defines the bijection*

$$Z_{\mathfrak{g}_n(\mathbb{F}_q)}(X) \to Z_{G_n(\mathbb{F}_q)}(1 + X).$$

COROLLARY. *A conjugacy class C in $G_n(\mathbb{F}_q)$ contains $q^{\dim C}$ points.*

Thus, the map (2.1.1) has the usual properties of the exponential map $X \to e^X$, which is usually defined only over topological fields of zero characteristics.

The set $\mathcal{O}_n(\mathbb{F}_q) = \mathfrak{g}_n(\mathbb{F}_q)/G_n(\mathbb{F}_q)$ of adjoint orbits as well as $\mathcal{O}_n^*(\mathbb{F}_q)$ admits a stratification by algebraic manifolds (according to the dimension of an orbit), but this stratification looks completely different (compare examples below with examples in 1.2).

EXAMPLE 1. $\mathfrak{g}_3(\mathbb{F}_q)$ consists of matrices $X = \begin{pmatrix} 0 & \alpha & \gamma \\ 0 & 0 & \beta \\ 0 & 0 & 0 \end{pmatrix}$; $\mathfrak{g}_{3,0}^*$ is defined by $\alpha = \beta = 0$ and $\mathfrak{g}_{3,1}$ by $(\alpha, \beta) \neq (0, 0)$.
So $\mathcal{O}_{3,0} = \mathbb{A}^1$, $\mathcal{O}_{3,1} = \mathbb{A}^2 \setminus \mathbb{A}^0$.

EXAMPLE 2. $\mathfrak{g}_4(\mathbb{F}_q)$ consists of matrices $X = \begin{pmatrix} 0 & \lambda & \alpha & \gamma \\ 0 & 0 & \mu & \beta \\ 0 & 0 & 0 & \nu \\ 0 & 0 & 0 & 0 \end{pmatrix}$; $\mathfrak{g}_{4,0}(\mathbb{F}_q)$ is defined by $\alpha = \beta = \lambda = \mu = \nu = 0$, $\mathfrak{g}_{4,1}(\mathbb{F}_q)$ by $\lambda = \mu = \nu = 0$, $(\alpha, \beta) \neq (0, 0)$, $\mathfrak{g}_{4,2}(\mathbb{F}_q)$ by $\mu = 0$, $(\lambda, \nu) \neq (0, 0)$ or $\lambda = \nu = 0$, $\mu \neq 0$ and $\mathfrak{g}_{4,3}(\mathbb{F}_q)$ by $\mu \neq 0$, $(\lambda, \nu) \neq (0, 0)$.

So $\mathcal{O}_{4,0} = \mathbb{A}^1$, $\mathcal{O}_{4,1} = \mathbb{A}^2 \backslash \mathbb{A}^0$, $\mathcal{O}_{4,2} = \mathbb{A}^3 \backslash \mathbb{A}^1 \sqcup \mathbb{A}^2 \backslash \mathbb{A}^1$, $\mathcal{O}_{4,3} = (\mathbb{A}^1 \backslash \mathbb{A}^0) \times (\mathbb{A}^2 \backslash \mathbb{A}^0)$.

Nevertheless, we have

THEOREM 2.1.2. *The total number of adjoint and coadjoint orbits for $G_n(\mathbb{F}_q)$ are the same:*
$$\#\mathcal{O}_n(\mathbb{F}_q) = \#\mathcal{O}_n^*(\mathbb{F}_q).$$

PROOF. It follows from the following two observations.

PROPOSITION 2.1.3. *Let G be a finite group acting on a finite set X. Then the number of G-orbits in X is equal to the multiplicity of the trivial representation in the natural (permutational) representation T of G in the space $\mathbb{C}[X]$ of complex-valued functions on X.*

PROOF. Evident.

PROPOSITION 2.1.4. *The natural representations of the group $G_n(\mathbb{F}_q)$ in $\mathbb{C}[\mathfrak{g}_n(\mathbb{F}_q)]$ and $\mathbb{C}[\mathfrak{g}_n^*(\mathbb{F}_q)]$ are equivalent.*

PROOF. Choose a nontrivial additive character θ of the field $\mathbb{F}_q$[3] and define the Fourier transform
$$\mathcal{F} : \mathbb{C}[\mathfrak{g}_n(\mathbb{F}_q)] \to \mathbb{C}[\mathfrak{g}_n^*(\mathbb{F}_q)]$$
by

(2.1.3) $$\mathcal{F}\varphi(F) = \sum_{X \in \mathfrak{g}_n(\mathbb{F}_q)} \varphi(X)\theta(\langle F, X \rangle).$$

It is the standard Fourier transform corresponding to dual abelian groups $\mathfrak{g}_n(\mathbb{F}_q)$ and $\mathfrak{g}_n^*(\mathbb{F}_q)$. The isomorphism between $\mathfrak{g}_n^*(\mathbb{F}_q)$ and $\widehat{\mathfrak{g}_n(\mathbb{F}_q)}$ is given by $F \to \theta(\langle F, \cdot \rangle)$.

It is well known (and easy to check) that $\mathcal{F}$ is a $G_n(\mathbb{F}_q)$-equivariant bijection.

---

[3] It is convenient to take θ given by
$$\theta(\lambda) = \exp\frac{2\pi i}{p} \cdot \operatorname{tr}(\lambda),$$
where p is the characteristic of $\mathbb{F}_q$, $q = p^l$, and $\operatorname{tr} : \mathbb{F}_q \to \mathbb{F}_p$ is the trace map from $\mathbb{F}_q$ to $\mathbb{F}_p$ arising from a realization of $\mathbb{F}_q$ as a subalgebra in $\operatorname{Mat}_l(\mathbb{F}_p)$.

2.2. Theorems 2.1.1 and 2.1.2 imply that the number of coadjoint orbits $\mathcal{O}_n^*(q)$ is equal to the number of (equivalence classes of) irreducible complex representations of $G_n(\mathbb{F}_q)$.

This makes plausible the conjecture that the machinery of the orbit method is still valid for the groups $G_n(\mathbb{F}_q)$. The strongest form of this conjecture can be formulated as follows.

CONJECTURE 2.2.1. *There is a bijection* $\Omega \to \pi_\Omega$ *of* $\mathcal{O}_n^*(\mathbb{F}_q)$ *onto* $\widehat{G_n(\mathbb{F}_q)}$ *such that the character* χ_Ω *of* π_Ω *is given by the formula*

$$(2.2.1) \qquad \chi_\Omega(1+X) = q^{-\frac{1}{2}\dim \Omega} \cdot \sum_{F \in \Omega} \theta(\langle F, X \rangle),$$

where θ *is a fixed nontrivial additive character of* $\mathbb{F}_q$.

In particular, putting $X = 0$ in (2.2.1), we obtain that $\dim \pi_\Omega = q^{\frac{1}{2}\dim \Omega}$. Hence, the dimensions of irreducible representations of $G_n(\mathbb{F}_q)$ should be powers of q and not only powers of p as for arbitrary p-group. (Cf. papers [Ka] and [Gu] where analogous results were obtained as affirmative answers to the questions of Springer and Thompson respectively).

One possible way to prove the conjecture is to use the technique of abelian hypergroups of conjugacy classes developed in [W1, W2, DW]. Another approach is discussed in 4.3.

2.3. The set $\mathcal{O}_n(\mathbb{F}_q)$ admits one more very interesting stratification in addition to $\{\mathcal{O}_{n,m}(\mathbb{F}_q)\}$. Namely, for $X \in \mathfrak{g}_n(\mathbb{F}_q)$ define the *partition type* of X as a partition $\lambda = \{\lambda_1 \geq \lambda_2 \geq \cdots \geq \lambda_r\}$ of n such that X is equivalent, as a nilpotent linear operator in $\mathbb{F}_q^n$, to the direct sum $J_{\lambda_1} \oplus J_{\lambda_2} \oplus \cdots \oplus J_{\lambda_r}$, where J_k is a Jordan block of size k with zero eigenvalue. It is clear that the partition type of X depend only on the adjoint orbit which contains X. Let $\mathcal{P}_n$ be the set of all partitions of n and $\mathcal{P} = \coprod_{n \geq 0} \mathcal{P}_n$. We denote by $\mathfrak{g}_{n,\lambda}(\mathbb{F}_q)$ the set of all $X \in \mathfrak{g}_n(\mathbb{F}_q)$ of given type $\lambda \in \mathcal{P}_n$ and by $\mathcal{O}_{n,\lambda}(\mathbb{F}_q)$ the set of adjoint orbits in $\mathfrak{g}_{n,\lambda}(\mathbb{F}_q)$.

The remarkable discovery of Springer ([Sp1, Sp2], cf. also [St]) is that $\mathfrak{g}_{n,\lambda}$ is an algebraic manifold whose irreducible components all have the same dimension $n(\lambda)$ and the number $d(\lambda)$ of these components is equal to the dimension of the irreducible representation π_λ of S_n associated to the partition λ. Although this fact belongs essentially to the basic linear algebra, the original proof is rather complicated and uses the technique of intersection homology and derived categories.

We give here a simple proof of this result which uses only elementary facts from linear algebra and from representation theory of the symmetric group.[4] It seems that this approach can be modified to get a similar result for the nilpotent radicals of all classical semisimple Lie algebras (I checked it for the split form of symplectic algebras with q odd) and I hope that it still works for the exceptional Lie algebra G_2. The original proof by Springer has the advantage that it embraces all semisimple Lie algebras in a uniform way (under some restrictions on the characteristic of the base field).

---

[4]As I found out after the first draft of this paper was written, a similar result is contained in [Spa], where it is formulated in other terms: for the unipotent matrix u of the given partition type λ the author considers the fixed point set X_u in the flag manifold. The algebraic manifolds $\mathfrak{g}_{n,\lambda}$ and X_u are different but closely related, so they have the same structure of connected components.

LEMMA 2.3.1. $\mathfrak{g}_{n,\lambda}(\mathbb{F}_q)$ is an algebraic manifold defined over $\mathbb{Z}$.

PROOF. For $X \in \mathfrak{g}_n(k)$ we denote by $v_l(X)$ the dimension of $\ker X^l$. It is clear that

$$(2.3.1) \qquad 0 = v_0(X) \leq v_1(X) \leq \cdots \leq v_n(X) = n.$$

In particular, for $X = J_n$ we have $v_l(X) = l$, $0 \leq l \leq n$.

Since $\ker \oplus_i X_i = \oplus_i \ker X_i$, we easily get that for $X \in \mathfrak{g}_{n,\lambda}(k)$ the following equalities hold:

$$(2.3.2) \qquad v_l(X) = \sum_i \min(l, \lambda_i) = \sum_k \alpha_k \cdot \min(l, k),$$

where α_k denote the number of λ_i's which are equal to k.[5] It follows that

$$(2.3.3) \qquad v_l(X) - v_{l-1}(X) = \sum_{k \geq l} \alpha_k \quad \text{and} \quad \alpha_l = 2v_l - v_{l-1} - v_{l+1}.$$

Hence, the numbers $\{\alpha_l\}$, which define the partition λ, can be expressed in terms of $\{v_l(X)\}$ and the latter have obviously the algebraic level sets defined over $\mathbb{Z}$.

Now we consider $\mathfrak{g}_n(k)$ as a Lie subalgebra of $\mathfrak{g}_{n+1}(k)$ and write the general element of the latter algebra in the block form

$$(2.3.4) \qquad \mathcal{X} = \begin{pmatrix} X & x \\ 0 & 0 \end{pmatrix}.$$

We say that partition $\lambda \in \mathcal{P}_n$ and $\Lambda \in \mathcal{P}_{n+1}$ are *adjacent* if there exists an $\mathcal{X} \in \mathfrak{g}_{n+1,\Lambda}(k)$ of the form (2.3.4) for which $X \in \mathfrak{g}_{n,\lambda}(k)$. Formally this definition depends on the field k, but we shall see in a moment that it really does not.

It will be convenient to associate with a partition $\lambda \in \mathcal{P}_n$ the corresponding Young diagram $D(\lambda)$ consisting of n boxes arranged in r rows so that the s-th row contains λ_s boxes.

We say that a diagram D is *j-adjacent* to D' if D is obtained from D' by deleting one box from j-th column.

THEOREM 2.3.2. (i) *The partitions $\lambda \in \mathcal{P}_n$ and $\Lambda \in \mathcal{P}_{n+1}$ are adjacent iff $D(\lambda)$ is j-adjacent to $D(\Lambda)$ for some j, $1 \leq j \leq \lambda_1 + 1$.*

(ii) *Let $D(\lambda)$ be j-adjacent to $D(\Lambda)$. Then for any $X \in \mathfrak{g}_{n,\lambda}(k)$ the elements $\mathcal{X}$ of the form (2.3.4), which belong to $\mathfrak{g}_{n+1,\Lambda}(k)$, form an irreducible quasi-affine manifold of the type $\mathbb{A}^{M(\lambda,\Lambda)} \setminus \mathbb{A}^{m(\lambda,\Lambda)}$, where*

$$(2.3.5) \quad M(\lambda, \Lambda) = n - \sum_{l \geq j} \alpha_l, \quad m(\lambda, \Lambda) = \begin{cases} n - \sum_{l \geq j-1} \alpha_l & \text{for } j > 1, \\ -1 & \text{for } j = 1, \end{cases}$$

and we agree that $\mathbb{A}^{-1} = \emptyset$.

PROOF. It is based on the following result.

---
[5] Sometimes the partition $\lambda \in \mathcal{P}_n$ is written in the form $1^{\alpha_1} 2^{\alpha_2} \ldots n^{\alpha_n}$.

LEMMA 2.3.3. *Let $\mathcal{X}$ be a matrix of the form (2.3.4) with $X \in \mathfrak{g}_{n,\lambda}(k)$. Then*
(i) $\mathcal{X} \in \mathfrak{g}_{n+1,\Lambda}(k)$, *where $D(\lambda)$ is j-adjacent to $D(\Lambda)$ if and only if x satisfies the conditions*

(2.3.6) $\quad$ 1) $\quad X^{l-1}x \in \operatorname{Im} X^l \quad$ for $l \geq j$,

$\qquad\qquad$ 2) $\quad X^{j-2}x \notin \operatorname{Im} X^{j-1} \quad$ (*this condition is empty for $j = 1$*).

(ii) *Any vector x satisfies (2.3.6) for some j, $1 \leq j \leq \lambda_1 + 1$.*

PROOF. Let $\lambda = 1^{\alpha_1} 2^{\alpha_2} \ldots n^{\alpha_n}$ and $\Lambda = 1^{A_1} 2^{A_2} \ldots (n+1)^{A_{n+1}}$. Then $D(\lambda)$ is j-adjacent to $D(\lambda)$ if and only if $A_j = \alpha_j + 1$, $A_{j-1} = \alpha_{j-1} - 1$ and $A_k = \alpha_k$ for $k \neq j, j-1$. From (2.3.2) we conclude that $D(\lambda)$ is j-adjacent to $D(\lambda)$ if and only if

$$v_l(\mathcal{X}) - v_l(X) = \begin{cases} 1 \text{ for } l \geq j, \\ 0 \text{ for } l < j. \end{cases}$$

Now, $\operatorname{rk}(\mathcal{X}^l) = n + 1 - v_l(\mathcal{X})$ and $\operatorname{rk}(X^l) = n - v_l(X)$. Hence,

(2.3.7) $\qquad\qquad \operatorname{rk}(\mathcal{X}^l) - \operatorname{rk}(X^l) = \begin{cases} 0 \text{ for } l \geq j, \\ 1 \text{ for } l < j. \end{cases}$

But $\mathcal{X}^l = \begin{pmatrix} X^l & X^{l-1}x \\ 0 & 0 \end{pmatrix}$ and, according to Cramer's rule, (2.3.7) is equivalent to (2.3.6). We have proved the first part of the lemma. The second part follows if we observe that the first condition (2.3.6) defines the increasing family of subspaces $\{V_j\}$ in $\mathbb{F}_q^n$ with $V_1 = \ker X$ and $V_{\lambda_1+1} = \mathbb{F}_q^n$.

PROOF OF THE SPRINGER THEOREM. We recall some definitions from the partition theory. To each partition $\lambda \in \mathcal{P}_n$ there corresponds the *dual partition* $\lambda' \in \mathcal{P}_n$ such that the diagram $D(\lambda')$ is obtained from $D(\lambda)$ by the transposition. Moreover, the quantity $c(\lambda)$ is defined by

(2.3.8) $\qquad\qquad c(\lambda) = \sum_{s=1}^{r} \frac{\lambda'_s(\lambda'_s - 1)}{2} = \frac{1}{2} \sum_{k,l} \min(k, l) \alpha_k \alpha_l.$

It is the important characteristic of λ which occurs in many situations.[6]

Let now $\lambda \in \mathcal{P}_n$ is j-adjacent to $\Lambda \in \mathcal{P}_{n+1}$. Then $c(\Lambda) - c(\lambda) = \lambda'_j$ because $\Lambda'_k = \lambda'_k$ for $k \neq j$ and $\Lambda'_j = \lambda'_j + 1$. So we can rewrite the expression for $M(\lambda, \Lambda)$ from (2.3.5) in the form

(2.3.9) $\qquad\qquad M(\lambda, \Lambda) = n + c(\lambda) - c(\Lambda).$

Assume now that the Springer Theorem is true for all $\mathfrak{g}_{n,\lambda}$, $\lambda \in \mathcal{P}_n$, with additional condition:

(2.3.10) $\qquad \operatorname{codim} \mathfrak{g}_{n,\lambda} = c(\lambda) \qquad$ (or $\dim \mathfrak{g}_{n,\lambda} = \dfrac{n^2 - \sum_{s=1}^{r}(\lambda'_s)^2}{2}$),

---

[6]E.g., $c(\lambda)$ is the minimal degree of polynomials in $x_1, x_2, \ldots, x_n$ which are transformed according to the representation π_λ of S_n.

and consider $\Lambda \in \mathcal{P}_{n+1}$. The manifold $\mathfrak{g}_{n+1,\Lambda}$ is the union of submanifolds $\mathfrak{g}_{n+1,\lambda}^{\Lambda}$ consisting of those $\mathcal{X}$ of the form (2.3.4) for which $X \in \mathfrak{g}_{n,\lambda}$. Here λ runs through all partition from $\mathcal{P}_n$ adjacent to Λ. According to Theorem 2.3.2 (ii), $\mathfrak{g}_{n+1,\lambda}^{\Lambda}$ is a manifold of dimension $d(\lambda) + M(\lambda, \Lambda)$. Using the induction hypothesis and (2.3.9) we can write

$$\operatorname{codim} \mathfrak{g}_{n+1,\lambda}^{\Lambda} = \frac{n(n+1)}{2} - \left(\frac{n(n-1)}{2} - c(\lambda) + n + c(\lambda) - c(\Lambda) \right) = c(\Lambda).$$

Hence, all components of $\mathfrak{g}_{n+1,\Lambda}$ have the same codimension $c(\Lambda)$. The number of these component is equal to the sum $\sum d(\lambda)$ extended to all $\lambda \in \mathcal{P}_n$ adjacent to Λ. But the sum above is equal to $d(\Lambda) = \dim \pi_\Lambda$ according to the well known branching rule for the representations of symmetric groups. This completes the induction step and proves the theorem.

§3. On ℝ-ℍ groups

3.1. Let G be a finite group with the property
(ℝ-ℍ) All irreducible characters of G are real-valued.

THEOREM 3.1.1. *This property is equivalent to the following properties:*
(ℝ-ℍ$_1$) *Any $g \in G$ is conjugate to g^{-1}.*
(ℝ-ℍ$_2$) *The function $Q(g) = \#\{x \in G \mid x^2 = g\}$ is convolutory invertible, i.e., there exists a function $k(g)$ such that $Q * k = \delta$.*
(ℝ-ℍ$_3$) *The anti-involution A of $\mathbb{C}[G]$ defined by $Af(g) = f(g^{-1})$ commutes with the convolutions with the elements of the center $Z[G]$ of $\mathbb{C}[G]$.*
(ℝ-ℍ$_4$) *A can be written in the form*

$$(3.1.1) \qquad A = \sum_{g_1, g_2} K(g_1, g_2) L_{g_1} R_{g_2},$$

where L_g (resp. R_g) is the operator of the left (resp. right) regular representation of G.

PROOF. The equivalence of (ℝ-ℍ$_1$) and (ℝ-ℍ) follows from the fact that the irreducible characters are independent and separate the conjugacy classes. To prove the other statements we shall use the Fourier transform on G. This is the map $f \mapsto \widehat{f}$ from $\mathbb{C}[G]$ to the space $\mathbb{C}[\widehat{G}]$ of matrix-valued functions on the set $\widehat{G}$ of (equivalence classes of) irreducible complex representations of G given by

$$(3.1.2) \qquad \widehat{f}(\lambda) = \sum_{g \in G} f(g) T_\lambda(g),$$

where T_λ is a representation from the class $\lambda \in \widehat{G}$. The inverse transform is given by

$$(3.1.3) \qquad f(g) = |G|^{-1} \sum_{\lambda \in \widehat{G}} d(\lambda) \operatorname{tr}(\widehat{f}(\lambda) T_\lambda^*(g)),$$

where $d(\lambda)$ is the dimension of T_λ. The main property of this transform is

$$(3.1.4) \qquad \widehat{(f_1 * f_2)}(\lambda) = \widehat{f_1}(\lambda)\widehat{f_2}(\lambda),$$

i.e., the convolution of functions goes to the multiplication of their Fourier transforms.

Note also that the Fourier transform is a unitary operator from $L^2(G)$ to $L^2(\widehat{G})$ if the norms on these spaces are given by:

$$(3.1.5) \qquad \|f\|^2_{L^2(G)} = \sum_{g \in G} |f(g)|^2, \qquad \|\tilde{f}\|_{L^2(\widehat{G})} = \sum_{\lambda \in \widehat{G}} \frac{d(\lambda)}{|G|} \operatorname{tr}[\tilde{f}(\lambda)\tilde{f}(\lambda)^*].$$

Now consider the map $\widehat{A} : \mathbb{C}[\widehat{G}] \to \mathbb{C}[\widehat{G}] : \widehat{f} \mapsto \widehat{Af}$. It is an anti-involution of $\mathbb{C}[\widehat{G}]$. Moreover, if $\operatorname{supp} \phi = \{\lambda\}$, then $\operatorname{supp} A\phi = \{\bar{\lambda}\}$, where $\bar{\lambda}$ denotes the class of the representation contragredient (= complex conjugate) to λ. (This follows, for example, from the equality $A\chi_\lambda = \overline{\chi_\lambda} = \chi_{\bar{\lambda}}$ for the irreducible character χ_λ of the class λ.) Hence, ($\mathbb{R}$-H) and ($\mathbb{R}$-H$_3$) are equivalent, because the convolution with an element of $Z[G]$ goes under Fourier transform to the multiplication by a scalar function on $\widehat{G}$.

Further, we recall

SCHUR CRITERION. *The representation T_λ belongs to the complex, real, or quaternionic type if and only if the quantity*

$$(3.1.6) \qquad \operatorname{ind}(\lambda) := |G|^{-1} \sum_{g \in G} \chi_\lambda(g^2)$$

is equal to 0, 1, or -1 respectively.

This criterion implies that the function Q from ($\mathbb{R}$-H$_2$) is equal to $\sum_{\lambda \in \widehat{G}} \operatorname{ind}(\lambda)\chi_\lambda$ and its Fourier transform is $\widehat{Q}(\lambda) = \operatorname{ind}(\lambda)|G|d(\lambda)^{-1}\mathbf{1}_{V_\lambda}$, where $\mathbf{1}_{V_\lambda}$ is the identity operator in the representation space V_λ of T_λ. This immediately gives the equivalence of ($\mathbb{R}$-H) and ($\mathbb{R}$-H$_2$).

Finally, the equivalence of ($\mathbb{R}$-H) and ($\mathbb{R}$-H$_4$) follows from the fact that the algebras L and R of operators on $\mathbb{C}[G]$ generated by the operators $L(g)$ and $R(g)$, respectively, satisfy

$$(3.1.7) \qquad L \vee R = (L \cap R)' = Z[G]',$$

where $L \vee R$ is the algebra generated by L and R, and A' denotes the von Neumann dual to the algebra A, i.e., the set of all operators commuting with every operator from A.

We will use the term "$\mathbb{R}$-H group" for a group satisfying the equivalent conditions ($\mathbb{R}$-H)–($\mathbb{R}$-H$_4$) of Theorem 1. If all irreducible representations of group G are of real type (i.e., $\operatorname{ind}(\lambda) = 1$ for all $\lambda \in \widehat{G}$) then we will call it an "$\mathbb{R}$ group".

EXAMPLES OF $\mathbb{R}$-$\mathbb{H}$ GROUPS.

(a) All dihedral groups D_n of order $2n$ are $\mathbb{R}$-$\mathbb{H}$ groups.
(b) For any n the symmetric group S_n of order $n!$ is an $\mathbb{R}$-$\mathbb{H}$ group.
(c) The group $G = G_n(\mathbb{F}_2)$, consisting of all (strictly) upper triangular $(n \times n)$-matrices with entries from the field $\mathbb{F}_2$, is an $\mathbb{R}$-$\mathbb{H}$ group.[7]
(d) the quaternionic group of order 8, consisting of ± 1, $\pm i$, $\pm j$, $\pm k$ (the standard quaternionic units) is an $\mathbb{R}$-$\mathbb{H}$ group.

Note that the groups from the examples a)–c) are in fact $\mathbb{R}$ groups, while the last one has also an irreducible representation of the quaternionic type.

3.2. Consider now in more detail the representation of the anti-ivolution A in the form (3.1.1). This means that there exists a function $K(g_1, g_2)$ such that for any f we have $f(g^{-1}) = \sum_{g_1,g_2} K(g_1, g_2) f(g_1^{-1} g g_2)$. Or, taking $f = \delta_h(g) = \delta(g^{-1}h)$, where $\delta(x) = 1$ if $x = 1$, $\delta(x) = 0$ if $x \neq 1$, we get

(3.2.1)
$$\delta(gh) \sum_{g_1,g_2} K(g_1, g_2) \delta(g_2^{-1} g^{-1} g_1 h) = \sum_{g_1} K(g_1, g^{-1} g_1 h) = \sum_{x} K(gx, xh).$$

Such a function K, if it exists, is not unique, because the map $\mathbb{C}[G \times G] \to \operatorname{End} \mathbb{C}[G] : K \mapsto \sum_{g_1,g_2} K(g_1, g_2) L_{g_1} R_{g_2}$ factors through $\mathbb{C}[G] \otimes_{Z[G]} \mathbb{C}[G]$.

Let us suppose that $K(g_1, g_2) = k(g_1 g_2)$. Then (3.2.1) becomes

(3.2.2)
$$\sum_{x} k(gx^2 h) = \delta(gh),$$

which yields that k is the central function satisfying $k * Q = \delta$.

Conversely, if ($\mathbb{R}$-$\mathbb{H}_2$) is true, then

$$\sum_{g_1,g_2} k(g_1 g_2) f(g_1^{-1} g g_2) = \sum_{g_1,x} k(g_1 g^{-1} x) f(x)$$
$$= \sum_{x,y} k(y^2 gx) f(x) = \sum_{x} \delta(gx) f(x) = f(g^{-1}).$$

Thus, $\sum_{g_1,g_2} k(g_1 g_2) L_{g_1} R_{g_2} = A$.

It turns out that in terms of convolution powers of k and Q, one can easily express the sums of powers of $d(\lambda)$. Namely, we have

THEOREM 3.2.1. *The following equalities hold for any $\mathbb{R}$-$\mathbb{H}$ group G:*

(3.2.3) $$\sum_{\lambda \in \widehat{G}} \operatorname{ind}(\lambda)^m d(\lambda)^{m+2} = |G|^{m+1} k^{*m}(1), \quad m \geq 0;$$

(3.2.4) $$\sum_{\lambda \in \widehat{G}} \operatorname{ind}(\lambda)^m d(\lambda)^{2-m} = |G|^{1-m} Q^{*m}(1), \quad m \geq 0.$$

---

[7]This can be deduced from a result by Gutkin [Gu].

In particular,

$$\sum_{\lambda \in \widehat{G}} d(\lambda)^2 = |G| \quad \text{(the Burnside identity)} \quad (m = 0 \text{ in } (3.2.3) \text{ or } (3.2.4)),$$

$$\sum_{\lambda \in \widehat{G}} \mathrm{ind}(\lambda) d(\lambda) = Q(1) = \text{number of involutions in } G \quad (m = 1 \text{ in } (3.2.4)),$$

$$\sum_{\lambda \in \widehat{G}} d(\lambda)^0 = |\widehat{G}| = |G|^{-1} Q * Q(1) = |G|^{-1} \|Q\|_{L^2(G)}^2 \quad (m = 2 \text{ in } (3.2.4)).$$

The functions Q^{*m} and k^{*m} have simple expressions in terms of irreducible characters:

(3.2.5)
$$Q^{*m}(x) = \sum_{\lambda \in \widehat{G}} \mathrm{ind}^m(\lambda) \frac{|G|^{m-1}}{d(\lambda)^{m-1}} \chi_\lambda(x),$$

$$k^{*m}(x) = \sum_{\lambda \in \widehat{G}} \mathrm{ind}^m(\lambda) \frac{d(\lambda)^{m+1}}{|G|^{m+1}} \chi_\lambda(x).$$

Their Fourier transforms look as follows:

(3.2.6)
$$(Q^{*m})\widehat{\ }(\lambda) = \widehat{Q}(\lambda)^m = \left(\frac{\mathrm{ind}(\lambda)|G|}{d(\lambda)} \right)^m \mathbf{1}_{V_\lambda},$$

$$(k^{*m})\widehat{\ }(\lambda) = \widehat{k}(\lambda)^m = \left(\frac{\mathrm{ind}(\lambda) d(\lambda)}{|G|} \right)^m \mathbf{1}_{V_\lambda}.$$

3.3. Generalization to compact groups. Let now G be a compact group with the Haar measure dg normalized so that the total mass of G is equal to 1. We assume that there exists an integrable function $q(x)$ satisfying the equality

(3.3.1)
$$\int_G f(g^2) \, dg = \int_G f(g) q(g) \, dg \quad \text{for all } f \in L^1(G, dg),$$

which is certainly the case if G is a Lie group. Symbolically we can write $d\sqrt{x} = q(x) dx$. For G semisimple the function $q(x)$ can be computed explicitly via H. Weyl's integration formula.

PROPOSITION 1. *For $t \in T$ (a maximal torus in G),*

(3.3.2)
$$q(t) = \left\langle \prod_{\alpha > 0} |t^{\alpha/4} + t^{-\alpha/4}|^{-2} \right\rangle.$$

Here α runs through the set of positive roots of (G, T), t^α denotes $e^{\langle \alpha, \log t \rangle}$, where $\log : T \to \mathfrak{t} = \mathrm{Lie}(T)$ is the inverse to the exponential map $\exp : \mathfrak{t} \to T$, and $\langle \phi \rangle$ denotes the mean value of a multivalued function ϕ.

PROOF. It is clear that the function q satisfying (3.3.1) must be central. Then, assuming that f is also central and applying Weyl's formula we rewrite (3.3.1) in the form

$$\int_T f(t^2) |j(t)|^2 \, dt = \int_T f(t) q(t) |j(t)|^2 \, dt \quad \text{for all } f \in L^1(T, |j(t)|^2 dt).$$

Now, it is easy to verify that $\int_T \varphi(t)\,dt = \int_T \langle \varphi(\sqrt{t}) \rangle \, dt$ for any $\varphi \in L^1(T, dt)$. (Indeed, the right-hand side defines a translation invariant normalized measure on T which is unique.) It follows that

$$q(t) = \frac{\langle |j(t^{1/2})|^2 \rangle}{|j(t)|^2} = \left\langle \prod_{\alpha > 0} \left| \frac{t^{\alpha/4} - t^{-\alpha/4}}{t^{\alpha/2} - t^{-\alpha/2}} \right|^2 \right\rangle = \left\langle \prod_{\alpha > 0} |t^{\alpha/4} + t^{-\alpha/4}|^{-2} \right\rangle.$$

A simple compact group G belongs to the $\mathbb{R}$-$\mathbb{H}$ type iff the corresponding Weyl group contains -1 (i.e., G is not of type A_n, $n \geq 2$, or D_{2n+1}, $n \geq 1$, or E_6).

Some of the results of §§1, 2 still hold for these compact groups. Namely, formulas (3.2.5) and (3.2.6) define certain distributions on G and for m big enough Q^{*m} is a regular distribution, so the formula (3.2.4) makes sense. This gives an interesting integral representation for ζ-like sums involving dimensions of irreducible representations:

(3.3.3) $$\sum_{\lambda \in \hat{G}} d(\lambda)^{-m} (\mathrm{ind}(\lambda))^m = \int_{G^{m+1}} q(y_1^2 \ldots y_{m+1}^2)\, dy_1 \ldots dy_{m+1}.$$

In particular,

(3.3.4) $$\zeta(2m) = \int_{SU(2)^{2m+1}} \frac{dy_1 \ldots dy_{2m+1}}{2 + \mathrm{tr}(y_1^2 \ldots y_{2m+1}^2)}.$$

§4. Algebraic structures in sets of orbits

4.1. We are interested in the asymptotic behavior of $\mathfrak{g}_n$, $\mathfrak{g}_n^*$, $\mathcal{O}_n$ and $\mathcal{O}_n^*$ when $n \to \infty$. For this it is useful to endow the sets $\mathfrak{g}_\infty = \coprod_{n \geq 0} \mathfrak{g}_n$, $\mathfrak{g}_\infty^* = \coprod_{n \geq 0} \mathfrak{g}_n^*$, $\mathcal{O}_\infty = \coprod_{n \geq 0} \mathcal{O}_n$, and $\mathcal{O}_\infty^* = \coprod_{n \geq 0} \mathcal{O}_n^*$ (or, rather, the set of functions on these sets) by some algebraic structures. This idea is suggested, from one side, by the "fusion rules" in the conformal field theory and, from the other side, by the geometric approach to quantum groups and enveloping algebras proposed in [BLM, Gi1, Gi2]. Of course, our situation is different, but I believe that the resulting algebras are also interesting in many aspects.

Consider first $\mathfrak{g}_\infty$. For any field k let $\mathcal{T}_q(k) = \bigoplus_{n \geq 0} \mathcal{T}_q^{(n)}(k)$ denote the k-linear $\mathbb{Z}_+$-graded space whose n-th component is spanned by the elements e_X, $X \in \mathfrak{g}_n(\mathbb{F}_q)$. We can think about elements of $\mathcal{T}_q(k)$ as on k-valued functions on $\mathfrak{g}_\infty$ with finite supports.

We define the composition law in $\mathcal{T}_q(k)$ by

(4.1.1) $$e_{X_1} \cdot e_{X_2} = \sum e_X, \qquad X_i \in \mathfrak{g}_{n_i}(\mathbb{F}_q), \quad i = 1, 2,$$

where the sum in the right-hand side is taken over all $X \in \mathfrak{g}_{n_1+n_2}(\mathbb{F}_q)$ of the form

(4.1.2) $$X = \begin{pmatrix} X_1 & X_3 \\ 0 & X_2 \end{pmatrix}.$$

For any $X \in \mathfrak{g}_n(\mathbb{F}_q)$ we denote by $\check{X}$ the matrix obtained from X by transposition with respect to the second diagonal:

(4.1.3) $$\check{X}_{i,j} = X_{n+1-j, n+1-i}.$$

The algebraic properties of $\mathcal{T}_q(k)$ are described in

PROPOSITION 4.1.4. $\mathcal{T}_q(k)$ is a $\mathbb{Z}_+$-graded associative noncommutative k-algebra with the anti-involution $\sigma : e_X \to e_{\check{X}}$.

PROOF. A straightforward verification.

The analogous construction can be applied to $\mathfrak{g}_\infty^*$ and leads to another $\mathbb{Z}_+$-graded algebra $\mathcal{T}_q^*(k) = \bigoplus_{n \geq 0} \mathcal{T}_q^{*(n)}(k)$ with the basis $\{e_F^*\}$, $F \in \mathfrak{g}_\infty^*(\mathbb{F}_q)$ and the involution $\sigma^* : e_F^* \mapsto e_{\check{F}}^*$.

Since $\dim \mathcal{T}_q(n)(k) = \dim \mathcal{T}_q^{*(n)}(k)$, the Poincaré series for both algebras are the same:

$$(4.1.4) \qquad P_{\mathcal{T}_q(k)}(t) = P_{\mathcal{T}_q^*(k)}(t) = \sum_{n \geq 0} q^{n(n-1)/2} t^n.$$

The comparison of (4.1.4) with the classical θ-series suggest to unite $\mathcal{T}_q(k)$ and $\mathcal{T}_q^*(k)$ in one algebraic structure, but it is not yet clear how to do it.

In the rest of this section we assume that the field k is of zero characteristic and contains q-th roots of unity.

Then one can establish the natural duality between $\mathcal{T}_q(n)(k)$ and $\mathcal{T}_q^{*(n)}(k)$:

$$(4.1.5) \qquad (e_F, e_X) = \theta(\langle F, X \rangle),$$

and thus identify $\mathcal{T}_q^{*(n)}(k)$ with $\mathcal{T}_q^{(n)*}(k)$.

4.2. The algebras $\mathcal{T}_q(k)$ and $\mathcal{T}_q^*(k)$ are too big in a sense and we are essentially interested in smaller algebras $\mathcal{A}_q(k)$ and $\mathcal{B}_q(k)$ which we are going to define now.

Let the group $G_n(\mathbb{F}_q)$ act linearly on $\mathcal{T}_q^{(n)}(k)$ by

$$(4.2.1) \qquad g \cdot e_X = e_{gXg^{-1}}.$$

We denote by $\mathcal{A}_q(k)$ the subspace of $G_n(\mathbb{F}_q)$-invariant elements in $\mathcal{T}_q^{(n)}(k)$. It is clear that it is spanned by the elements

$$(4.2.2) \qquad e_O = \sum_{X \in O} e_X, \qquad O \in \mathcal{O}_n(\mathbb{F}_q).$$

LEMMA 4.2.1. $\mathcal{A}_q(k) := \bigoplus_{n \geq 0} \mathcal{A}_q^{(n)}(k)$ is a $\mathbb{Z}_+$-graded σ-invariant subalgebra of $\mathcal{T}_q(k)$.

PROOF. Fix $O_i \in \mathcal{O}_{n_i}$, $i = 1, 2$, and denote by $S(O_1, O_2)$ the set of $Y \in \mathfrak{g}_{n_1+n_2}(\mathbb{F}_q)$ for which the coefficient of e_Y in $e_{O_1} \cdot e_{O_2}$ is positive. (From (4.1.1) it follows that this coefficient can only be 0 or 1.) It is clear that $S(O_1, O_2)$ is a $G_{n_1+n_2}(\mathbb{F}_q)$-invariant subset in $\mathfrak{g}_{n_1+n_2}(\mathbb{F}_q)$, hence

$$(4.2.3) \qquad e_{O_1} \cdot e_{O_2} = \sum_{O \subset S(O_1, O_2)} e_O.$$

This proves the lemma and introduces also the natural $\mathbb{Z}$-structure in $\mathcal{A}_q(k)$.

Consider now the the linear action of $G_n(\mathbb{F}_q)$ in $\mathcal{T}_q^{*(n)}(k)$ defined by

$$(4.2.4) \qquad g \cdot e_F^* = e_{K(g)F}^*,$$

and denote by $\mathcal{B}_q^{(n)}(k)$ the subspace of $G_n(\mathbb{F}_q)$-invariant elements in $\mathcal{T}_q^{*(n)}(k)$. It is spanned by the elements

$$(4.2.5) \qquad e_\Omega^* = \sum_{F \in \Omega} e_F^*, \qquad \Omega \in \mathcal{O}_n^*(\mathbb{F}_q).$$

However, the space $\mathcal{B}_q(k) := \oplus_{n \geq 0} \mathcal{B}_q^{(n)}(k)$ is not a subalgebra in $\mathcal{T}_q^*(k)$. But we can nevertheless define a composition law $\diamond$ in $\mathcal{B}_q(k)$ as follows. Let P_n denote the projection of $\mathcal{T}_q^{*(n)}(k)$ on $\mathcal{B}_q^{(n)}(k)$:

$$P_n e_F = \frac{1}{\#G_n(\mathbb{F}_q)} \sum_{g \in G_n(\mathbb{F}_q)} e_{K(g)F}^* = \frac{1}{\#\Omega} e_\Omega^*,$$

where Ω is the coadjoint orbit of F and put

$$(4.2.6) \qquad e_{\Omega_1}^* \diamond e_{\Omega_2}^* = P_{n_1+n_2}(e_{\Omega_1}^* \cdot e_{\Omega_2}^*) \quad \text{for } \Omega_i \in \mathcal{O}_{n_i}^*, \; i = 1, 2.$$

LEMMA 4.2.2. $(\mathcal{B}_q(k), \diamond)$ is a $\mathbb{Z}_+$-graded associative algebra with the involution induced by σ^*.

PROOF. The only nonevident part of the statement is the associativity law. It follows from the useful relations

$$(e_{\Omega_1}^* \diamond e_{\Omega_2}^*) \diamond e_{\Omega_3}^* = P_{n_1+n_2+n_3}(e_{\Omega_1}^* \cdot e_{\Omega_2}^* \cdot e_{\Omega_3}^*) = e_{\Omega_1}^* \diamond (e_{\Omega_2}^* \diamond e_{\Omega_3}^*),$$

which can be easily verified.

4.3. There is another algebra connected with the family of groups $\{G_n(\mathbb{F}_q)\}$. Namely, let $\mathcal{R}_q^{(n)}(k)$ be the k-linear span of elements $\lambda \in \widehat{G_n(\mathbb{F}_q)}$ and $\mathcal{R}_q(k) = \oplus_{n \geq 0} \mathcal{R}_q^{(n)}(k)$. To any finite dimensional complex representation π of $G_n(\mathbb{F}_q)$ we associate the element $[\pi] = \sum_{\lambda \in \widehat{G_n(\mathbb{F}_q)}} m_\lambda(\pi) \cdot \lambda$ of $\mathcal{R}_q^{(n)}(k)$, where $m_\lambda(\pi)$ is the multiplicity of λ in π.

Now, for $\lambda_i \in \widehat{G_{n_i}(\mathbb{F}_q)}$, $i = 1, 2$, we choose representations π_i of classes λ_i and define

$$(4.3.1) \qquad \lambda_1 \circ \lambda_2 = \left[\mathrm{Ind}_{G_{n_1}(\mathbb{F}_q) \times G_{n_2}(\mathbb{F}_q)}^{G_{n_1+n_2}(\mathbb{F}_q)} (\pi_1 \times \pi_2) \right].$$

LEMMA 4.3.1. *The algebra $(\mathcal{R}_q(k), \circ)$ is a $\mathbb{Z}_+$-graded associative noncommutative algebra with the anti-involution σ induced by the family $\{\sigma_n\}$ of anti-involutions of groups $\{G_n(\mathbb{F}_q)\}$.*

PROOF. Follows from the principle of induction by stage (See, e.g., [K1] or [K4]).

The following statement can be considered as an argument in the favor of Conjecture 2.2.1.

PROPOSITION 4.3.4. *Conjecture 2.2.1 implies that the graded algebras with anti-involutions* $\mathcal{B}_q(k)$ *and* $\mathcal{R}_q(k)$ *are isomorphic under the map*

(4.3.2) $$[\pi_\Omega] \mapsto q^{-\frac{1}{2}\dim \Omega} \cdot e_\Omega^*.$$

PROOF. First of all, we represent the map (4.3.2) as the composition of two natural maps. The first map from $\mathcal{R}_q(k)$ to $\mathcal{A}_q(k)$ is taking the character of a representation in "exponential" coordinates. It sends $\lambda \in \widehat{G_n(\mathbb{F}_q)}$ to the function $X \mapsto \chi_\lambda(1+X)$ from $\mathcal{A}_q^{(n)}(k)$.

The second map is the Fourier transform from $\mathcal{T}_q(k)$ to $\mathcal{T}_q^*(k)$ which maps $\mathcal{A}_q(k)$ to $\mathcal{B}_q(k)$.

So, to $\lambda \in \mathcal{R}_q^{(n)}(k)$ we associate the element

$$\sum_{X,F} \operatorname{tr} \pi_\lambda(1+X)\overline{\theta(\langle F, X \rangle)} e_F^* \in \mathcal{B}_q^{(n)}(k).$$

We apply now the Frobenius formula for the character of an induced representation:

$$\chi_{\operatorname{Ind}_H^G \pi}(g) = \sum_{x \in G/H} \check{\chi}_\pi(x^{-1}gx)$$

where $\check{\chi}_\pi$ is the extension of χ_π from H to G by the zero valued function outside H.

In our case $H = G_{n_1}(\mathbb{F}_q) \times G_{n_2}(\mathbb{F}_q)$, $G = G_{n_1+n_2}(\mathbb{F}_q)$ and one can identify G/H with the subgroup of matrices of the form $\begin{pmatrix} 1 & A \\ 0 & 1 \end{pmatrix}$. We have

$$\begin{pmatrix} 1 & -A \\ 0 & 1 \end{pmatrix}\begin{pmatrix} 1+X_1 & X_3 \\ 0 & 1+X_2 \end{pmatrix}\begin{pmatrix} 1 & A \\ 0 & 1 \end{pmatrix} = \begin{pmatrix} 1+X_1 & X_3+X_1A-AX_2 \\ 0 & 1+X_2 \end{pmatrix}$$

and

$$\check{\chi}_{\pi_1 \times \pi_2}\begin{pmatrix} 1+Y_1 & Y_3 \\ 0 & 1+Y_2 \end{pmatrix} = \chi_{\pi_1}(1+Y_1)\chi_{\pi_2}(1+Y_2)\delta(Y_3).$$

Now for $(n_1 \times n_2)$-matrix Y with elements in $\mathbb{F}_q$ we can use the relation

$$\delta(Y) = q^{-n_1 n_2} \sum_F \theta(\operatorname{tr}(FY))$$

where F runs over all $(n_2 \times n_1)$-matrices over $\mathbb{F}_q$.

So, using (2.2.1) we obtain

(4.3.3)
$$\chi_{\pi_1 \circ \pi_2}\begin{pmatrix} 1+X_1 & X_3 \\ 0 & 1+X_2 \end{pmatrix} = q^{-n_1 n_2 - \frac{1}{2}(\dim \Omega_1 + \dim \Omega_2)}$$
$$\times \sum_{F_1 \in \Omega_1, F_2 \in \Omega_2, F_3, A} \theta(\operatorname{tr}(F_1 X_1) + \operatorname{tr}(F_2 X_2) + \operatorname{tr}(F_3(X_3 + X_1 A - AX_2))).$$

On the other hand, using the relation

$$K\left(\begin{pmatrix} 1 & -A \\ 0 & 1 \end{pmatrix}\right)\begin{pmatrix} F_1 & 0 \\ F_3 & F_2 \end{pmatrix} = \begin{pmatrix} F_1+AF_3 & 0 \\ F_3 & F_2-F_3A \end{pmatrix},$$

we obtain:

$$e^*_{\Omega_1} \diamond e^*_{\Omega_2} = P_{n_1+n_2}(e^*_{\Omega_1} \cdot e^*_{\Omega_2})$$
(4.3.4)
$$= q^{-n_1 n_2} \sum_{F_1 \in \Omega_1, F_2 \in \Omega_2, F_3, A} e\begin{pmatrix} F_1 + AF_3 & 0 \\ F_3 & F_2 - F_3 A \end{pmatrix}^*.$$

The comparison of (4.3.3) and (4.3.4) proves that (4.3.2) is an algebra homomorphism. It evidently commutes with the anti-involution (since $\operatorname{tr} \check{F}\check{X} = \operatorname{tr}(XF)^{\vee} = \operatorname{tr} XF = \operatorname{tr} FX$) and preserves the grading.

It would be interesting to find the independent proof of the proposition, not using the conjecture. One possible way to do so is to find a dual (in the sense of Weyl–Howe) group $\mathfrak{G}_q$ such that the representation algebra $\mathcal{R}_{\mathfrak{G}_q}(k)$ for this group[8] would be isomorphic to $\mathcal{R}_q(k)$.

Recall that if we take the family of symmetric groups $\{S_n\}$ instead of $\{G_n(\mathbb{F}_q)\}$ and restrict ourself by the representation of S_n corresponding to the Young diagrams with $\leq N$ rows (resp. $\leq N$ columns), then the dual group would be GL_N acting on the tensor superalgebra $T(V)$, where V is an N-dimensional vector space consisting of even (resp. odd) elements.

For the unrestricted family of representations of $\{S_n\}$ one should consider the group GL_∞ and certain category of "tame" representations of this group (see, e.g., [O]).

4.4. Splitting of Lie algebras $\mathfrak{g}_n(\mathbb{F}_q)$ on partition types leads to a remarkable commutative algebra $\mathcal{P}_q(k) = \oplus_{n \geq 0} \mathcal{P}^n_q(k)$. As a linear space $\mathcal{P}^n_q(k)$ is spanned by the elements

(4.4.1)
$$e_\lambda = \sum_{X \in \mathfrak{g}^*_{n,\lambda}(\mathbb{F}_q)} e_X.$$

Note that $\mathcal{P}_q(k)$ is a subspace, but *not a subalgebra* in $\mathcal{A}_q(k)$. We can proceed in the same way as in 4.2, where we defined the algebra $\mathcal{B}_q(k)$. For $|\lambda_i| = n_i$, $i = 1, 2$, put

(4.4.2)
$$e_{\lambda_1} \diamond e_{\lambda_2} := Q_{n_1+n_2}(e_{\lambda_1} \cdot e_{\lambda_2}),$$

where Q_n is a projection of $\mathcal{T}^{(n)}_q(k)$ on $\mathcal{P}^{(n)}_q(k)$ given by

$$Q_n e_X = \frac{1}{N_\lambda(q)} e_\lambda \quad \text{if } X \in \mathfrak{g}_{n,\lambda}; \qquad N_\lambda(q) = \#\mathfrak{g}_{n,\lambda}.$$

THEOREM 4.4.1. *$\mathcal{P}_q(k)$ is an associative commutative graded algebra with trivially acting anti-involution σ.*

PROOF. Fix partition types λ_i, $i = 1, 2$, and consider the elements $X_i \in \mathfrak{g}_{n_i, \lambda_i}$. The set $S_\lambda(X_1, X_2)$ of all $(n_1 \times n_2)$-matrices X_3 for which the matrix $\begin{pmatrix} X_1 & X_3 \\ 0 & X_2 \end{pmatrix}$ belongs to $\mathfrak{g}_{n_1+n_2, \lambda}$, has the properties

(4.4.3) $\quad S_\lambda(g_1 X_1 g_1^{-1}, g_2 X_2 g_2^{-1}) = g_1 S_\lambda(X_1, X_2) g_2^{-1}, \qquad g_i \in GL_{n_i}, \; i = 1, 2$

---

[8] I.e., the algebra spanned over k by the classes of equivalence of irreducible representations of the group with composition law induced by the tensor multiplication of representations.

and
(4.4.4) $$S_\lambda(\check{X}_1, \check{X}_2) = S_\lambda(X_2, X_1)^{\vee}.$$

From (4.4.3) we conclude that the cardinality of $S_\lambda(X_1, X_2)$ depends only on λ_1, λ_2 and λ. Denote it by $C^\lambda_{\lambda_1, \lambda_2}(q)$.[9]

From (4.4.2) we obtain
$$e_{\lambda_1} \diamond e_{\lambda_2} = \sum_\lambda \frac{N_{\lambda_1}(q) N_{\lambda_2}(q)}{N_\lambda(q)} C^\lambda_{\lambda_1, \lambda_2}(q) e_\lambda.$$

It is natural to introduce the new basis $\left\{E_\lambda = \frac{1}{N_\lambda(q)} e_\lambda\right\}$ in $\mathcal{P}_q(k)$. Then the multiplication law will be

(4.4.5) $$E_{\lambda_1} \diamond E_{\lambda_2} = \sum_\lambda C^\lambda_{\lambda_1, \lambda_2}(q) E_\lambda.$$

The associativity of this law follows from the following result.

LEMMA 4.4.2. *Fix matrices* $X_i \in \mathfrak{g}_{n_i, \lambda_i}$, $i = 1, 2, 3$. *Both quantities*
$$\sum_\mu C^\mu_{\lambda_1, \lambda_2} \cdot C^\Lambda_{\mu, \lambda_3} \quad \text{and} \quad \sum_\mu C^\Lambda_{\lambda_1, \mu} \cdot C^\mu_{\lambda_2, \lambda_3}$$
are equal to the number of matrices of the form $\begin{pmatrix} X_1 & X_{12} & X_{13} \\ 0 & X_2 & X_{23} \\ 0 & 0 & X_3 \end{pmatrix}$, *which belong to* $\mathfrak{g}_{n_1 + n_2 + n_3, \Lambda}(\mathbb{F}_q)$.

The proof of the lemma is left for the reader.

To prove the commutativity of the operation $\diamond$, we simply remark that the antiinvolution σ preserves $\mathfrak{g}_{n, \lambda}$ and, hence, acts trivially on $\mathcal{P}_q(k)$. So
$$e_{\lambda_1} \diamond e_{\lambda_2} = (e_{\lambda_1} \diamond e_{\lambda_2})^{\vee} = \check{e}_{\lambda_2} \diamond \check{e}_{\lambda_1} = e_{\lambda_2} \diamond e_{\lambda_1}.$$

LEMMA 4.4.3. (i) *The quantities* $C^\lambda_{\lambda_1, \lambda_2}(q)$ *are polynomials in q with integer coefficients.*

(ii) *The degrees of these polynomials satisfy*

(4.4.6) $\deg C^\lambda_{\lambda_1, \lambda_2} = \deg N_\lambda - \deg N_{\lambda_1} - \deg N_{\lambda_2} = |\lambda_1| \cdot |\lambda_2| - c(\lambda) + c(\lambda_1) + c(\lambda_2).$

PROOF. The first statement is the generalization of Theorem 2.3.2 (ii) and can be proved by the same arguments. The second statement can be obtained by the direct computation analogous to the proof of Lemma 2.3.3.

We can now construct a new algebra $\mathcal{P}_\mathbb{Z}$ over $\mathbb{Z}[q]$ with generators E_λ, $\lambda \in \mathcal{P}$, and relations (4.4.5).[10]

This algebra is related to $\mathcal{P}_q(k)$ as follows:

(4.4.7) $$\mathcal{P}_q(k) = \mathcal{P}_\mathbb{Z} \otimes_{\mathbb{Z}[q]} k;$$

here we consider k as a $\mathbb{Z}[q]$-module, where the generator $q \in \mathbb{Z}[q]$ acts as the multiplication by the integer q.

---
[9] One can express this quantity in terms of Hall polynomials (see the next footnote).

[10] This algebra is in fact one of realizations of the Hall algebra from [M], but we shall not discuss here this phenomenon.

4.5. The algebra $\mathcal{P}_\mathbb{Z}$ has several interesting contractions. Recall that a *contraction* of an algebra $\mathcal{A}$ is an algebra $\mathcal{A}_0$ with a basis $\{a_i\}_{i \in I}$ such that there exists a one-parameter family of bases $\{a_i(t)\}_{i \in I}$ in $\mathcal{A}$ and

$$\lim_{t \to 0} c_{i,j}^k(t) = c_{i,j}^k \tag{4.5.1}$$

where $c_{i,j}^k(t)$ (resp. $c_{i,j}^k$) are the structure constants of the algebra $\mathcal{A}$ (resp. $\mathcal{A}_0$) with respect to the basis $\{a_i(t)\}_{i \in I}$ (resp. $\{a_i\}_{i \in I}$), i.e.,

$$a_i(t) \cdot a_j(t) = \sum_k c_{i,j}^k(t) a_k(t), \quad a_i \cdot a_j = \sum_k c_{i,j}^k a_k. \tag{4.5.2}$$

This definition varies according to which bases are considered (e.g. homogeneous with respect to some grading) and what kind of dependence on t is allowed for $a_i(t)$ (e.g. continuous, smooth, analytic or formal).

EXAMPLE. Let $\mathcal{E}_\lambda(q) = q^{\deg N_\lambda} E_\lambda$ and put $q = t^{-1}$. Then

$$c_{\lambda_1, \lambda_2}^\lambda(t) = C_{\lambda_1, \lambda_2}^\lambda(q) q^{\deg N_{\lambda_1} + \deg N_{\lambda_2} - \deg N_\lambda}\Big|_{q = t^{-1}} \in \mathbb{Z}[t].$$

It turns out that the contracted algebra with the structure constants $c_{\lambda_1, \lambda_2}^\lambda(0)$ is isomorphic to the well-known graded algebra $\mathbb{Z}[x_1, x_2, \ldots, x_n, \ldots]$ of polynomials in a countable number of homogeneous generators x_i, $\deg x_i = i$, with the basis consisting of Schur polynomials (see e.g., [Z])

$$S_\lambda = \det \| x_{\lambda + j - i} \|.$$

So, the multiplication in $\mathcal{P}_\mathbb{Z}$ gives us a deformation of the multiplication law for Schur polynomials. E.g., the equality

$$S_{2,1} \cdot S_{2,1} = S_{4,2} + S_{4,1^2} + S_{3,3} + 2S_{3,2,1} + S_{3,1^3} + S_{2^2, 1^2}$$

is deformed to

$$\mathcal{E}_{2,1} \cdot \mathcal{E}_{2,1} = (q-1)^2 q^7 \mathcal{E}_{4,2} + (q-1) q^7 \mathcal{E}_{4,1^2} + (q-1)^2 q^6 \mathcal{E}_{3,3}$$
$$+ (q-1) q^5 (2q+1) \mathcal{E}_{3,2,1} + (q-1) q^4 \mathcal{E}_{3,1^3} + q^4 \mathcal{E}_{2^2, 1^2}.$$

Another interesting algebra arises if we take the quotient of $\mathcal{P}_\mathbb{Z}$ over the ideal generated by $\mathcal{E}_\lambda(q)$ with $\lambda_1 \geq 3$.

The corresponding subset in $\mathfrak{g}_n(\mathbb{F}_q)$ consists of matrices X satisfying $X^2 = 0$. For $q = 2^l$ this is equivalent to $(1 + X)^2 = 1$. So, the dimension of the n-th component of our quotient algebra is equal to the number I_n of involutions in $G_n(\mathbb{F}_2^l)$. It seems that the asymptotics of the sequence $\{I_n\}$ can be computed more easily than the asymptotics of $\{O_n(2)\}$. At the same time both sequences are related (see 3.2) and one can obtain an estimation for the latter sequence in terms of the former one.

More precisely, if we denote by I_n^r the number of involutions $s \in G_n(\mathbb{F}_2)$ with $\text{rk}(s - 1) = r$, then $I_n^r = N_{1^{n-2r} 2^r}(2)$.

The leading term of the polynomial $N_{1^{n-2r}2^r}(q)$ is equal to $\binom{n}{r}\frac{n-2r+1}{n-r+1}q^{r(n-r)}$. E.g., for $n = 2r$ (resp. $2r + 1$), i.e., in the case of involutions s with maximal rank of $s - 1$, the leading coefficient is the so called *Catalan number* c_r (resp. c_{r+1}).

In order to compute this leading coefficient (and, in particular, the Catalan numbers) one can use the skew analog of the Pascal triangle $\{a_{k,l}\}$ satisfying the same relations

$$a_{k,l} = a_{k-1,l} + a_{k,l-1}$$

as the classical one, but the different initial conditions:

$$a_{0,0} = 0, \quad a_{1,0} = 1, \quad a_{0,1} = -1.$$

Of course, the entries of this triangle are simply the differences of two binomial coefficients. The first lines of this triangle look like

$$
\begin{array}{ccccccccc}
 & & & & 0 & & & & \\
 & & & -1 & & 1 & & & \\
 & & -1 & & 0 & & 1 & & \\
 & -1 & & -1 & & 1 & & 1 & \\
-1 & & -2 & & 0 & & 2 & & 1 \\
-1 & -3 & & -2 & & 2 & & 3 & 1 \\
-1 & -4 & -5 & & 0 & & 5 & 4 & 1 \\
-1 & -5 & -9 & -5 & & 5 & 9 & 5 & 1.
\end{array}
$$

The q-version of it is defined by the rule

(4.5.3) $$a_{k,l} = (q^k - q^{l-1})a_{k,l-1} + q^l a_{k-1,l},$$

and the initial conditions $a_{0,0}(q) = 0$, $a_{1,0}(q) = 1$.

Here $a_{k,l}(q) = 0$ automatically for $k < l$ and $a_{k,l}(q) = N_{1^{k-l}2^l}(q)$ for $k \geq l$. It would be very interesting to find an explicit formula for $a_{k,l}(q)$ and the asymptotic behavior of $I_n = \sum_{k+l=n} a_{k,l}(2)$ when $n \to \infty$. Some results in this direction are obtained in [K5] where the connection of $a_{k,l}(q)$ with q-Hermite polynomials was established.

4.6. As is well known, the graded algebra $\mathbb{Z}[x_1, x_2, \ldots, x_n, \ldots]$ of polynomials in countable number of homogeneous generators x_i, $\deg x_i = i$, can be interpreted as the representation algebra for GL_∞ or as the algebra constructed from the set $\coprod_{n\geq 0} \widehat{S_n}$ via the so called "circle multiplication"(cf. [Z]).

Namely, to every finite dimensional polynomial representation π of GL_∞ one can associate the symmetric polynomial $\mathrm{ch}_\pi(\lambda_1, \lambda_2, \ldots) = \mathrm{tr}\,\pi(D(\lambda))$ where $D(\lambda)$ is an infinite matrix with eigenvalues $\lambda_1, \lambda_2, \ldots$. Let x_i denote the i-th elementary symmetric function of $\lambda_1, \lambda_2, \ldots$. Then $\mathrm{ch}_\pi(\lambda_1, \lambda_2, \ldots)$ defines an element of

$\mathbb{Z}[x_1, x_2, \ldots, x_n, \ldots]$. This gives the desired interpretation of the latter algebra because of evident relations

$$\mathrm{ch}_{\pi_1 \oplus \pi_2} = \mathrm{ch}_{\pi_1} + \mathrm{ch}_{\pi_2}, \qquad \mathrm{ch}_{\pi_1 \otimes \pi_2} = \mathrm{ch}_{\pi_1} \times \mathrm{ch}_{\pi_2}.$$

The Weyl duality between irreducible polynomial representations of GL_∞ of degree n and irreducible representations of S_n allows us to identify the representation algebra for GL_∞ with the $\mathbb{Z}$-span of $\coprod_{n \geq 0} \widehat{S_n}$. The multiplication law in the latter ring has the form

$$\rho_1 \circ \rho_2 = \mathrm{Ind}_{S_{n_1} \times S_{n_2}}^{S_{n_1+n_2}} (\rho_1 \times \rho_2)$$

for $\rho_i \in \widehat{S_{n_i}}$, $i = 1, 2$ and is called the "circle multiplication".

The passage from $\mathbb{Z}[x_1, x_2, \ldots, x_n, \ldots]$ to the deformed algebra over $\mathbb{Z}[q]$ (the Hall algebra) corresponds to the replacement of GL_∞ by the corresponding quantum group.

I hope that the algebras $\mathcal{B}_q(k)$ and $\mathcal{R}_q(k)$ also can be interpreted in a similar way so that the role of the sequence $\{S_n\}$ is played by $\{G_n(\mathbb{F}_q)\}$. The very intriguing problem is to find the appropriate analog of GL_∞.

References

[An] A. M. André Carlos, *Irreducible characters of the unitriangular group and coadjoint orbits*, Dissertation, Math. Institute, University of Warwick, May 1992.

[A1] V. I. Arnold, *Bernoulli-Euler updown numbers associated with function singulatities, their combinatorics and arithmetics*, Duke Math. J. **63** (1991), no. 2, 537–555.

[A2] _____, *Snake calculus and combinatorics of Bernoulli, Euler and Springer numbers for Coxeter groups*, Russian Math. Surveys **47** (1992), no. 1, 3–45.

[BKC] F. Beukers, J. A. C. Kolk, and E. Calabi, *Sums of generalized harmonic series and volumes*, Nieuw Archief voor Wiskunde **11** (1993), no. 3, 217–224.

[BLM] A. A. Beilinson, G. Lusztig, and R. D. MacPherson, *A geometric setting for the quantum deformation of GL_n*, Duke Math. J. **61** (1990), no. 2, 655–677.

[DM] P. Deligne and J. S. Milne, *Tannakian categories*, Lecture Notes in Math., vol. 900, Springer-Verlag, Berlin, Heidelberg, and New York, 1982, pp. 101–228.

[DW] A. H. Dooley and N. J. Wildberger, *Harmonic analysis and the global exponential map for compact Lie groups*, Functional Anal. Appl. **27** (1993), no. 1, 25–32.

[Gi1] V. A. Ginzburg, *G-modules, Springer's representations and bivariant Chern classes*, Adv. in Math. **61** (1986), no. 1, 1–48.

[Gi2] _____, *Lagrangian construction of the enveloping algebra $U(sl_n)$*, C. R. Acad. Sci. Paris Sér. I **312** (1991), 907–912.

[G] P. M. Gudivok et al., *Conjugacy classes in the unitriangular group*, Kybernetika (Kiev) (1990), no. 1, 40–48. (Russian)

[Gu] E. A. Gutkin, *Representations of algebraic nilpotent groups over a self-dual field*, Funktsional. Anal. i Prilozhen. **7** (1973), no. 4, 80; English transl. in Functional Anal. Appl. **7** (1973).

[H] W. H. Hesselink, *A classification of the nilpotent triangular matrices*, Compositio Math. **55** (1985), 89–133.

[Ka] D. Kazhdan, *Proof of Springer hypothesis*, Israel J. Math. **28** (1977), no. 4, 272–284.

[K0] A. A. Kirillov, *Unitary representations of nilpotent Lie groups*, Russian Math. Surveys **17** (1962), no. 4, 57–101.

[K1] _____, *Elements of the Theory of Representations*, vol. 220, Springer-Verlag, Berlin, Heidelberg, and New York, 1975.

[K2] _____, *On the combinatorics of coadjoint orbits*, Funktsional. Anal. i Prilozhen. **27** (1993), no. 1, 73–75; English transl. in Functional Anal. Appl. **27** (1993).

[K3] _____, *Combinatorics of coadjoint orbits*, Proceedings of the Sophus Lie Memorial Conference, Oslo (1992) (O. A. Laudal and B. Jahren, eds.), Scandinavian Univ. Press, 1994.

[K4] _____, *Introduction to the theory of representations and noncommutative harmonic analysis*, Encyclopaedia of Mathematical Sciences, vol. 22, Springer-Verlag, 1994, pp. 1–162.

[K5] _____, *On the number of solutions to the equation $X^2 = 0$ in triangular matrices over a finite field*, Funktsional. Anal. i Prilozhen. **29** (1995), no. 1, 82–89; English transl. in Functional Anal. Appl. **29** (1995).

[KB] D. E. Knuth and T. J. Buckholtz, *Computation of tangent, Euler and Bernoulli numbers*, Math. Comp. **21** (1992), no. 1, 3–45.

[M] I. G. Macdonald, *Symmetric functions and Hall polynomials*, Oxford Univ. Press, Oxford, 1979.

[MS] P. Martin and H. Saleur, *On an algebraic approach to higher dimensional statistical mechanics*, Comm. Math. Phys. **158** (1993), 155–190.

[O] G. I. Olshanskii, *Unitary representations of infinite dimensional pairs (G, K) and the formalism of R. Howe.*, Representations of Lie groups and related topics. Adv. Stud. Contemp. Math., vol. 7, Gordon and Breach, London, 1990, pp. 269–463.

[Sh] D. Shanks, *Generalized Euler and Bernoulli numbers*, Math. Computations **21** (1967), 639–694.

[Spa] N. Spaltenstein, *The fixed point set of a unipotent transformation on the flag manifold*, Indagationes Math. **38** (1976), 452–456.

[Sp1] T. A. Springer, *The unipotent variety of a semi-simple Lie group*, Proc. Bombay Colloq. Algebraic Geometry, 1968, pp. 373–391.

[Sp2] _____, *Green functions of finite groups and representations of Weyl groups*, Invent. Math. **36** (1976), 173–207.

[St] R. Steinberg, *On the desingularization of the nilpotent variety*, Invent. Math. **36** (1976), 209–224.

[W1] N. J. Wildberger, *On a relationship between adjoint orbits and conjugacy classes of a Lie group*, Canad. Math. Bull. **33** (1990), no. 3, 297–304.

[W2] _____, *Hypergroups and harmonic analysis*, In: Proc. Centre Math. Anal., vol. 29, ANU, 1992, pp. 238–253.

[Z] A. V. Zelevinskii, *Representations of finite classical groups*, Lecture Notes in Math., vol. 869, Springer-Verlag, Berlin and Heidelberg, 1981.

DEPARTMENT OF MATHEMATICS, THE UNIVERSITY OF PENNSYLVANIA, PHILADELPHIA, PA 19104-6395
E-mail address: kirillov@math.upenn.edu

Vector Fields and Deformations of Isotropic Super-Grassmannians of Maximal Type

A. L. ONISHCHIK AND A. A. SEROV

ABSTRACT. We determine the holomorphic vector fields and the deformations of the isotropic super-Grassmannians of maximal type $I°Gr_{2r|2s,\,r|s}$ associated with the complex orthosymplectic Lie superalgebras.

§1. Preliminaries

In [2, 6, 7] the holomorphic vector fields and the deformations of complex super-Grassmannians were studied. It was proved, in particular, that for a wide class of super-Grassmannians, all holomorphic vector fields are induced by linear transformations and the tangent sheaf 1-cohomology vanishes. Here we want to apply the same methods in order to get similar results for isotropic super-Grassmannians of maximal type associated with orthosymplectic Lie superalgebras. It turns out that the super-Grassmannian of maximal type associated with the Lie superalgebra $\mathfrak{osp}_{2r-1|2s}(\mathbb{C})$ is isomorphic to a connected component of the Grassmannian associated with $\mathfrak{osp}_{2r|2s}(\mathbb{C})$ (this is well known in the classical situation), and so we shall study only the latter case.

Let us denote by $IGr_{2r|2s,\,r|s}$ the isotropic super-Grassmannian of maximal type associated with the classical Lie superalgebra $\mathfrak{osp}_{2r|2s}(\mathbb{C})$ (see [4]). Its reduction is the product of two isotropic complex Grassmannians $IGr_{2r,\,r}^s \times IGr_{2s,\,s}^a$, where the first factor is the Grassmannian of isotropic r-planes in the vector space $\mathbb{C}^{2r}$ endowed with a nondegenerate symmetric bilinear form, while the second one is that of isotropic s-planes in $\mathbb{C}^{2s}$ endowed with a nondegenerate skew-symmetric bilinear form. The supermanifold $IGr_{2r|2s,\,r|s}$ admits a natural transitive action of the orthosymplectic Lie supergroup $OSp_{2r|2s}(\mathbb{C})$, inducing on its reduction the standard transitive action of the Lie group $O_{2r}(\mathbb{C}) \times Sp_{2s}(\mathbb{C})$.

Let $(e_1, \ldots, e_{2r})$, $(f_1, \ldots, f_{2s})$ be the standard bases of $\mathbb{C}^{2r}$, $\mathbb{C}^{2s}$ respectively. We suppose that the orthosymplectic Lie supergroup leaves invariant the bilinear

---

1991 *Mathematics Subject Classification*. Primary 58A50, 17C70.

Key words and phrases. Supermanifold, Lie superalgebra, isotropic super-Grassmannian.

The work partially supported by Centre for Advanced Study at The Norwegian Academy of Science and Letters and International Sophus Lie Centre (the first author), and by the International Science Foundation (both authors).

©1995, American Mathematical Society

form in $\mathbb{C}^{2r|2s}$ given in the basis $(e_1, \ldots, e_{2r}, f_1, \ldots, f_{2s})$ by the matrix

$$\begin{pmatrix} 0 & 1_r & 0 & 0 \\ 1_r & 0 & 0 & 0 \\ 0 & 0 & 0 & 1_s \\ 0 & 0 & -1_s & 0 \end{pmatrix}.$$

We denote by o the graded isotropy subspace of maximal dimension

$$o = \langle e_{r+1}, \ldots, e_{2r}, f_{s+1}, \ldots, f_{2s} \rangle$$

of $\mathbb{C}^{2r|2s}$. It is well known that the manifold $\mathrm{IGr}_{2r,r}^s$ has two connected components, while $\mathrm{IGr}_{2s,s}^a$ is connected. We choose the connected component

$$M = \mathrm{I}^\circ \mathrm{Gr}_{2r,r}^s \times \mathrm{IGr}_{2s,s}^a$$

of $\mathrm{IGr}_{2r,r}^s \times \mathrm{IGr}_{2s,s}^a$, containing the point o, and denote by $\mathrm{I}^\circ\mathrm{Gr}_{2r|2s,r|s}$ the corresponding connected component of $\mathrm{IGr}_{2r|2s,r|s}$. Sometimes we shall denote this supermanifold by $(M, \mathcal{O})$, where $\mathcal{O}$ is its structure sheaf.

The natural action of the Lie supergroup $\mathrm{OSp}_{2r|2s}(\mathbb{C})$ induces the transitive action of its identity component $\mathrm{SOSp}_{2r|2s}(\mathbb{C})$ on $(M, \mathcal{O})$. The reduction of the latter supergroup is $G = G_0 \times G_1$, where $G_0 = \mathrm{SO}_{2r}(\mathbb{C})$, $G_1 = \mathrm{Sp}_{2s}(\mathbb{C})$. Let P denote the stabilizer G_o of the point $o \in M$ in G; we have $P = P_0 \times P_1$, where $P_0 \subset G_0$, $P_1 \subset G_1$. The subgroup $R = R_0 \times R_1$, where $R_0 \simeq \mathrm{GL}_r(\mathbb{C})$, $R_1 \simeq \mathrm{GL}_s(\mathbb{C})$, leaves invariant the subspaces

$$\langle e_1, \ldots, e_r \rangle, \ \langle e_{r+1}, \ldots, e_{2r} \rangle, \ \langle f_1, \ldots, f_s \rangle, \ \langle f_{s+1}, \ldots, f_{2s} \rangle,$$

and is the reductive part of P. The matrices from R are of the form

$$\begin{pmatrix} A & 0 & 0 & 0 \\ 0 & (A^t)^{-1} & 0 & 0 \\ 0 & 0 & B & 0 \\ 0 & 0 & 0 & (B^t)^{-1} \end{pmatrix},$$

where $A \in \mathrm{GL}_r(\mathbb{C})$, $B \in \mathrm{GL}_s(\mathbb{C})$, while those from P have the form

$$\begin{pmatrix} A & 0 & 0 & 0 \\ U & (A^t)^{-1} & 0 & 0 \\ 0 & 0 & B & 0 \\ 0 & 0 & V & (B^t)^{-1} \end{pmatrix}.$$

The tangent Lie algebras and Lie superalgebras of Lie groups and Lie supergroups will be denoted by the corresponding lower case fraktur (gothic) letters. We have

$$\mathfrak{g} = \mathfrak{g}_0 \oplus \mathfrak{g}_1, \quad \mathfrak{g}_0 = \mathfrak{so}_{2r}(\mathbb{C}), \quad \mathfrak{g}_1 = \mathfrak{sp}_{2s}(\mathbb{C}).$$

The Lie algebra $\mathfrak{p}$ of P admits the semi-direct decomposition $\mathfrak{p} = \mathfrak{r} + \mathfrak{n}$, where $\mathfrak{n}$ is the nil-radical of $\mathfrak{p}$. We have $\mathfrak{n} = \mathfrak{n}_0 \oplus \mathfrak{n}_1$, where $\mathfrak{n}_0 \subset \mathfrak{g}_0$, $\mathfrak{n}_1 \subset \mathfrak{g}_1$ consist of the matrices

$$(1) \qquad u = \begin{pmatrix} 0 & 0 \\ U & 0 \end{pmatrix}, \quad v = \begin{pmatrix} 0 & 0 \\ V & 0 \end{pmatrix},$$

U and V being a skew-symmetric $r \times r$ and a symmetric $s \times s$-matrix respectively. The subalgebra $\mathfrak{n}$ is commutative.

We shall use the standard coordinate system on $\mathrm{IGr}_{2r|r,\,2s|s}$ in a neighborhood of o introduced in [4, Chapter 5, §6], changing the notation slightly. More precisely, we transpose the coordinate matrix, which will have the form

$$(2) \qquad Z = \begin{pmatrix} X & \Xi \\ 1_r & 0 \\ -\Xi^t & Y \\ 0 & 1_s \end{pmatrix},$$

where $X = (x_{\alpha\beta})$ and $Y = (y_{ij})$ are an $r \times r$-matrix and an $s \times s$-matrix of even coordinates, $X^t = -X$, $Y^t = -Y$, and $\Xi = (\xi_{\alpha s})$ is an $r \times s$-matrix of odd ones. At the point o we have $x_{\alpha\beta} = y_{ij} = 0$. The natural action of $\mathrm{OSp}_{2r|2s}(\mathbb{C})$ on $\mathrm{IGr}_{2r|2s,\,r|s}$ is given by matrix multiplication from the left.

Let ρ_0, ρ_1 be the standard representations of $\mathrm{GL}_r(\mathbb{C})$, $\mathrm{GL}_s(\mathbb{C})$ and σ_0, σ_1 their adjoint representations in the corresponding derived algebras $\mathfrak{sl}_p(\mathbb{C})$, $p = r, s$. The trivial 1-dimensional representation of any group will be denoted by 1. In what follows, we shall omit for simplicity the trivial factors 1 in the notation for the representations.

As in [6], we use the theory of homogeneous vector bundles. Let $E = E_\psi$ be a finite-dimensional P-module determined by a holomorphic linear representation ψ of P. We denote by $\mathbf{E} = \mathbf{E}_\psi$ the corresponding homogeneous vector bundle over M and by $\mathcal{E} = \mathcal{E}_\psi$ the sheaf of its holomorphic sections. As is well known, the tangent sheaf Θ on M is isomorphic to $\mathcal{E}_\tau$, where the isotropy representation τ of P is completely reducible and satisfies the condition

$$(3) \qquad \tau|R = \bigwedge^2 \rho_0 + S^2 \rho_1.$$

The supermanifold $(M, \mathcal{O})$ is, in general, nonsplit. As usual, we associate with it the split supermanifold $(M, \mathrm{gr}\,\mathcal{O})$. Its structure sheaf is the graded sheaf associated with the filtration

$$(4) \qquad \mathcal{O} = \mathcal{J}^0 \supset \mathcal{J}^1 \supset \mathcal{J}^2 \supset \dots,$$

where $\mathcal{J} = (\mathcal{O}_{\bar{1}})$. We have $\mathrm{gr}\,\mathcal{O} \simeq \bigwedge \mathcal{E}$, where $\mathcal{E} = \mathcal{J}/\mathcal{J}^2$. The holomorphic vector bundle $\mathbf{E}$ over M associated with $\mathcal{E}$ has the fibers $\mathbf{E}_x = \mathcal{J}_x/m_x\mathcal{J}_x$, $x \in M$, where m_x is the maximal ideal of $\mathcal{O}_x$.

Clearly, the action of $\mathrm{OSp}_{2r|2s}(\mathbb{C})$ on the super-Grassmannian induces actions of G on the sheaves $\mathcal{O}$, $\mathcal{J}$, $\mathcal{E}$ and on the vector bundle $\mathbf{E}$, covering the standard action of G on M. Thus, $\mathbf{E}$ is a homogeneous vector bundle over M.

PROPOSITION 1. *We have* $\mathrm{gr}\,\mathcal{O} \simeq \bigwedge \mathcal{E}_\varphi$, *where φ is the irreducible representation of P such that $\varphi|R = \rho_0^* \otimes \rho_1^*$.*

PROOF. Clearly, $\mathcal{J}/\mathcal{J}^2 = \mathcal{E}_\varphi$, where φ is the representation of P induced in the fiber $\mathbf{E}_o = \mathcal{J}_o/m_o\mathcal{J}_o$. To calculate it, we use the coordinate matrix (2). The action of P on $(M, \mathcal{O})$ is expressed in the coordinates as follows:

$$\tilde{Z} = \begin{pmatrix} A & 0 & 0 & 0 \\ U & (A^t)^{-1} & 0 & 0 \\ 0 & 0 & B & 0 \\ 0 & 0 & V & (B^t)^{-1} \end{pmatrix} \begin{pmatrix} X & \Xi \\ 1_r & 0 \\ -\Xi^t & Y \\ 0 & 1_s \end{pmatrix} \qquad (5)$$

$$= \begin{pmatrix} AX & A\Xi \\ (A^t)^{-1} + UX & U\Xi \\ -B\Xi^t & BY \\ -V\Xi^t & (B^t)^{-1} + VY \end{pmatrix}.$$

We must reduce the result to the form (2) by multiplying from the right by the matrix $\begin{pmatrix} (A^t)^{-1} + UX & U\Xi \\ -V\Xi^t & (B^t)^{-1} + VY \end{pmatrix}^{-1}$. We may set $X = 0$, $Y = 0$ which simplifies the calculation. Then

$$\begin{pmatrix} (A^t)^{-1} & U\Xi \\ -V\Xi^t & (B^t)^{-1} \end{pmatrix}^{-1} \equiv \begin{pmatrix} A^t & -A^t U\Xi B^t \\ B^t V\Xi^t A^t & B^t \end{pmatrix}$$

modulo $\mathcal{J}_o^2$. Hence,

$$\tilde{Z} \equiv \begin{pmatrix} 0 & A\Xi B^t \\ 1_r & 0 \\ -B\Xi^t A^t & 0 \\ 0 & 1_s \end{pmatrix}$$

modulo $m_o \mathcal{J}_o^2$. Since the entries of Ξ determine a basis of $\mathbf{E}_o$, this implies our assertion.

Our goal is to calculate the 0- and 1-cohomology of the tangent sheaf $\mathcal{T} = \mathcal{D}er\mathcal{O}$ of $\mathrm{IGr}_{2r|2s,r|s}$. As in [6], we consider first the $\mathbb{Z}$-graded sheaf $\widetilde{\mathcal{T}} = \mathcal{D}er\,\mathrm{gr}\,\mathcal{O}$. It is known (see [4]) that for any $q \geq -1$ there exists a natural exact sequence of sheaves

$$0 \to \mathcal{T}_{(q+1)} \to \mathcal{T}_{(q)} \to \widetilde{\mathcal{T}}_q \to 0, \qquad (6)$$

where $\mathcal{T}_{(q)}$ are the subsheaves of $\mathcal{T}$ forming a filtration of this sheaf and are defined by

$$\mathcal{T}_{(-1)} = \mathcal{T},$$
$$\mathcal{T}_{(q)} = \{\delta \in \mathcal{T} \mid \delta \mathcal{O} \subset \mathcal{J}^q, \, \delta \mathcal{J} \subset \mathcal{J}^{q+1}\}, \qquad q \geq 0. \qquad (7)$$

The sequence (6) permits us to relate the cohomology of $\mathcal{T}$ to that of $\widetilde{\mathcal{T}}$. To calculate the cohomology of the latter sheaf, we use the exact sequence

$$0 \to \mathcal{A}_{q+1} \xrightarrow{\alpha} \widetilde{\mathcal{T}}_q \xrightarrow{\beta} \mathcal{B}_q \to 0. \qquad (8)$$

Here

$$\mathcal{A}_q = \mathcal{E}_\varphi^* \otimes \bigwedge^q \mathcal{E}_\varphi = \mathcal{E}_{\Phi_q}$$

with

(9) $$\Phi_q = \varphi^* \otimes \bigwedge^q \varphi,$$

and

$$\mathcal{B}_q = \Theta \otimes \bigwedge^q \mathcal{E}_\varphi = \mathcal{E}_{T_q}$$

with

(10) $$T_q = \tau \otimes \bigwedge^q \varphi.$$

The mapping β is the restriction of a derivation of degree q onto the structure sheaf $\mathcal{F}$ of M, and α identifies any sheaf homomorphism $\mathcal{E}_\varphi \to \bigwedge^{p+1} \mathcal{E}_\varphi$ with its extension which is a derivation of degree q and is zero on $\mathcal{F}$. In particular,

$$\mathcal{T}_{(-1)} \simeq \mathcal{A}_0 = \mathcal{E}_\varphi^* = \mathcal{E}_{\varphi^*}.$$

Now we make some remarks concerning the action of the group G on the sheaves under consideraion. Clearly, the action of G on the structure sheaf $\mathcal{O}$ induces an action of G on $\mathcal{T}$, preserving the parities. It follows that G preserves the filtrations (4) and (7), inducing an action on the sheaf $\widetilde{\mathcal{T}}$. Thus, $\widetilde{\mathcal{T}}_q$ for any q is a locally free analytic sheaf on M homogeneous with respect to G. One easily sees that the homomorphisms in the exact sequences (6) and (8) are G-equivariant.

To conclude these preliminaries, we shall write explicitly certain fundamental vector fields on $(M, \mathcal{O})$ associated with the action of G, using the local coordinates from (2). Let us denote by $X \rightsquigarrow X^*$ the Lie superalgebra homomorphism $\mathfrak{osp}_{2r|2s}(\mathbb{C}) \to H^0(M, \mathcal{T})$ induced by the action of $\mathrm{SOSp}_{2r|2s}(\mathbb{C})$ on $(M, \mathcal{O})$.

Let

$$H = \mathrm{diag}(\lambda_1, \ldots, \lambda_r, -\lambda_1, \ldots, -\lambda_r, \mu_1, \ldots, \mu_s, -\mu_1, \ldots, -\mu_s)$$

be the general diagonal matrix lying in $\mathfrak{g}$. Using (5), we get

(11) $$H^* = \sum_{\alpha < \beta}(\lambda_\alpha + \lambda_\beta) x_{\alpha\beta} \frac{\partial}{\partial x_{\alpha\beta}} + \sum_{i \leqslant j}(\mu_i + \mu_j) y_{ij} \frac{\partial}{\partial y_{ij}} + \sum_{\alpha,i}(\lambda_\alpha + \mu_i) \xi_{\alpha i} \frac{\partial}{\partial \xi_{\alpha i}}.$$

Now, for the elements $u, v \in \mathfrak{n}$ given by (1), we get, using (5) again,

$$u^* = \sum_{\alpha,\beta}(XUX)_{\alpha\beta} \frac{\partial}{\partial x_{\alpha\beta}} - \sum_{i,j}(\Xi^t U \Xi)_{ij} \frac{\partial}{\partial y_{ij}} + \sum_{\alpha,k}(XU\Xi)_{\alpha k} \frac{\partial}{\partial \xi_{\alpha k}},$$

$$v^* = -\sum_{\alpha,\beta}(\Xi V \Xi^t)_{\alpha\beta} \frac{\partial}{\partial x_{\alpha\beta}} + \sum_{i,j}(YVY)_{ij} \frac{\partial}{\partial y_{ij}} + \sum_{\alpha,k}(\Xi VY)_{\alpha k} \frac{\partial}{\partial \xi_{\alpha k}}.$$

Let us choose the following basis $X_{\alpha\beta}$ ($\alpha < \beta$), Y_{ij} ($i \leqslant j$) of $\mathfrak{n}$:

(12) $$\begin{aligned} X_{\alpha\beta} &= (E_{\alpha\beta} - E_{\beta\alpha})/2, \\ Y_{ij} &= (F_{ij} + F_{ji})/2 \quad (i \neq j), \\ Y_{ii} &= F_{ii}, \end{aligned}$$

where $E_{\alpha\beta}$ and F_{ij} are the natural bases of the vector spaces of matrices $M_r(\mathbb{C})$ and $M_s(\mathbb{C})$ respectively. Then, in particular, we have

$$X^*_{\alpha\beta} = \sum_{\gamma,\delta} x_{\gamma\alpha} x_{\beta\delta} \frac{\partial}{\partial x_{\gamma\delta}} - \sum_{i,j} \xi_{\alpha i} \xi_{\beta j} \frac{\partial}{\partial y_{ij}}$$
$$+ \frac{1}{2} \sum_{\gamma,k} (x_{\gamma\alpha} \xi_{\beta k} - x_{\gamma\beta} \xi_{\alpha k}) \frac{\partial}{\partial \xi_{\gamma k}},$$

(13)
$$Y^*_{ij} = -\sum_{\alpha,\beta} \xi_{\alpha i} \xi_{\beta j} \frac{\partial}{\partial x_{\alpha\beta}} + \sum_{k,l} y_{ki} y_{jl} \frac{\partial}{\partial y_{kl}}$$
$$+ \frac{1}{2} \sum_{\gamma,k} (y_{jk} \xi_{\gamma i} + y_{ik} \xi_{\gamma j}) \frac{\partial}{\partial \xi_{\gamma k}} \quad (i \neq j),$$

$$Y^*_{ii} = -\sum_{\alpha,\beta} \xi_{\alpha i} \xi_{\beta i} \frac{\partial}{\partial x_{\alpha\beta}} + \sum_{k,l} y_{ki} y_{il} \frac{\partial}{\partial y_{kl}} + \sum_{\gamma,k} y_{ik} \xi_{\gamma i} \frac{\partial}{\partial \xi_{\gamma k}}.$$

Now let $\mathfrak{n}^-$ be the nilpotent subalgebra of $\mathfrak{g}$ complementary to $\mathfrak{p}$; it has the form $\mathfrak{n}^- = \mathfrak{n}_0^- + \mathfrak{n}_1^-$, where $\mathfrak{n}_0^- \subset \mathfrak{g}_0$, $\mathfrak{n}_1^- \subset \mathfrak{g}_1$ consist of the matrices

$$u = \begin{pmatrix} 0 & U \\ 0 & 0 \end{pmatrix}, \quad v = \begin{pmatrix} 0 & V \\ 0 & 0 \end{pmatrix},$$

U and V being a skew-symmetric $r \times r$- and a symmetric $s \times s$-matrix respectively (cf. (1)). Consider the basis of $\mathfrak{n}^-$ formed by the elements $U_{\alpha\beta}$ $(\alpha < \beta)$, V_{ij} $(i < j)$, V_{ii} corresponding to the matrices $U = E_{\alpha\beta} - E_{\beta\alpha}$, $V = E_{ij} + E_{ji}$ $(i < j)$; E_{ii} respectively. One easily sees that

(14)
$$U^*_{\alpha\beta} = \frac{\partial}{\partial x_{\alpha\beta}}, \quad V^*_{ij} = \frac{\partial}{\partial y_{ij}}.$$

§2. The cohomology of $\mathcal{A}_q$ and $\mathcal{B}_q$

In this section we shall calculate the 0- and 1-cohomology of the sheaves $\mathcal{A}_q$ and $\mathcal{B}_q$. As in [6, 7], we use the theorem of Bott (see [1, Theorem IV']), which permits us to calculate the cohomology of the homogeneous sheaf $\mathcal{E}_\psi$ on M defined by a completely reducible representation ψ of P. More precisely, this theorem gives an algorithm for determining the highest weights of the G-modules $H^p(M, \mathcal{E}_\psi)$ in terms of the highest weights of ψ. To apply it, we must introduce some notation related to weights and roots of G.

We choose the Cartan subalgebra $\mathfrak{t} = \mathfrak{t}_0 \oplus \mathfrak{t}_1$ in the tangent Lie algebra $\mathfrak{g} = \mathfrak{g}_0 \oplus \mathfrak{g}_1$ of G such that $\mathfrak{t}_0$ and $\mathfrak{t}_1$ are the Cartan subalgebras of $\mathfrak{g}_0$ and $\mathfrak{g}_1$, respectively, formed by all diagonal matrices

$$H_0 = \text{diag}(\lambda_1, \ldots, \lambda_r, -\lambda_1, \ldots, -\lambda_r),$$
$$H_1 = \text{diag}(\mu_1, \ldots, \mu_s, -\mu_1, \ldots, -\mu_s).$$

We consider the system of positive roots $\Delta^+ = \Delta_0^+ \cup \Delta_1^+$, where

$$\Delta_0^+ = \{\lambda_i - \lambda_j, \lambda_i + \lambda_j \ (i < j)\},$$
$$\Delta_1^+ = \{\mu_p - \mu_q \ (p < q), \mu_p + \mu_q \ (p \leqslant q)\}.$$

Half of the sum of all positive roots of $\mathfrak{g}_0$, $\mathfrak{g}_1$, $\mathfrak{g}$ will be denoted by γ_0, γ_1, γ respectively; we have $\gamma = \gamma_0 + \gamma_1$. The corresponding system of simple roots of $\mathfrak{g}$ is $\Pi = \Pi_0 \cup \Pi_1$, where $\Pi_0 = \{\alpha_1, \ldots, \alpha_r\}$, $\Pi_1 = \{\beta_1, \ldots, \beta_s\}$ are the systems of simple roots of $\mathfrak{g}_0$, $\mathfrak{g}_1$ respectively; here we denote

$$\alpha_1 = \lambda_1 - \lambda_2, \ldots, \alpha_{r-1} = \lambda_{r-1} - \lambda_r, \alpha_r = \lambda_{r-1} + \lambda_r,$$
$$\beta_1 = \mu_1 - \mu_2, \ldots, \beta_{s-1} = \mu_{s-1} - \mu_s, \beta_s = 2\mu_s.$$

We denote by $\mathfrak{t}^*(\mathbb{R})$ the real subspace of $\mathfrak{t}^*$ spanned by all λ_i, μ_p, and define the scalar product on $\mathfrak{t}^*(\mathbb{R})$ such that λ_i, μ_p form its orthonormal basis. As usual, $\lambda \in \mathfrak{t}^*(\mathbb{R})$ is called *dominant* if $(\lambda, \alpha) \geq 0$ for all $\alpha \in \Delta^+$ or, equivalently, for all $\alpha \in \Pi$. Following Bott [1], we say that λ has *index* 1 if $(\lambda, \alpha) > 0$ for all $\alpha \in \Delta^+$ except for one root $\beta \in \Delta^+$, for which $(\lambda, \beta) < 0$. Now, λ is called *singular* if $(\lambda, \alpha) = 0$ for a certain $\alpha \in \Delta$. These definitions will be used with respect to $\mathfrak{g}_0$, $\mathfrak{g}_1$ as well.

Clearly, the subgroup $P = G_o$ defined above is a parabolic subgroup of G containing the Borel subgroup B^- corresponding to $-\Delta^+$. The system of simple roots of its reductive part R is $\Sigma = \Pi - \{\alpha_r, \beta_s\}$. An element $\lambda \in \mathfrak{t}^*(\mathbb{R})$ is called *R-dominant* if $(\lambda, \alpha) \geq 0$ for all $\alpha \in \Sigma$.

It is convenient to characterize an element $\lambda \in \mathfrak{t}^*(\mathbb{R})$ by the numbers $\lambda_\alpha = 2(\lambda, \alpha)/(\alpha, \alpha)$, $\alpha \in \Pi$, which are actually the coordinates of λ in the basis of the so-called fundamental weights. We have $\gamma_\alpha = 1$ for all $\alpha \in \Pi$. An element λ is dominant if and only if $\lambda_\alpha \geq 0$ for all $\alpha \in \Pi$.

The following proposition is well known and very easy to verify.

PROPOSITION 2. *An element*

$$\lambda = \sum_{i=1}^r k_i \lambda_i, \qquad k_i \in \mathbb{R},$$

is dominant if and only if $k_1 \geq k_2 \geq \ldots \geq |k_r|$. *It is R-dominant if and only if* $k_1 \geq k_2 \geq \ldots \geq k_r$.

An element

$$\lambda = \sum_{j=1}^s l_j \mu_j, \qquad l_j \in \mathbb{R},$$

is dominant if and only if $l_1 \geq l_2 \geq \ldots \geq l_s \geq 0$. *It is R-dominant if and only if* $l_1 \geq l_2 \geq \ldots \geq l_s$.

We must study the highest weights of the representations Φ_q and T_q of P defined by (9) and (10), respectively. It follows from Proposition 1 that

$$\Phi_q | R = (\rho_0 \otimes \rho_1) \bigwedge^q (\rho_0^* \otimes \rho_1^*).$$

Denote by i, i_α indices running over $1, \ldots, r$, and by j, j_β those running over $1, \ldots, s$. The weights of Φ_q have the form

(15) $$\Lambda = \Lambda_0 + \Lambda_1,$$

where

(16) $$\Lambda_0 = \lambda_i - \lambda_{i_1} - \cdots - \lambda_{i_q}, \qquad \Lambda_1 = \mu_j - \mu_{j_1} - \cdots - \mu_{j_q}.$$

Similarly, (3) implies that $T_q = T'_q + T''_q$, where

$$T'_q|R = \left(\bigwedge^2 \rho_0\right) \bigwedge^q (\rho_0^* \otimes \rho_1^*), \qquad T''_q|R = (S^2 \rho_1) \bigwedge^q (\rho_0^* \otimes \rho_1^*).$$

The weights of T'_q, T''_q have the form

(17) $$\Lambda = \Lambda_0 + \Lambda_1,$$

where for T'_q we have

(18) $$\begin{aligned}\Lambda_0 &= \lambda_i + \lambda_k - \lambda_{i_1} - \cdots - \lambda_{i_q}, & i < k, \\ \Lambda_1 &= -\mu_{j_1} - \cdots - \mu_{j_q},\end{aligned}$$

and for T''_q

(19) $$\begin{aligned}\Lambda_0 &= -\lambda_{i_1} - \cdots - \lambda_{i_q}, \\ \Lambda_1 &= \mu_j + \mu_l - \mu_{j_1} - \cdots - \mu_{j_q}, & j \leqslant l.\end{aligned}$$

We denote by Id_0, Id_1 the standard representations and by Ad_0, Ad_1 the adjoint representations of G_0, G_1 respectively. Note that in the case $r = 1$ we have $G_0 = R_0 \simeq \mathrm{GL}_1(\mathbb{C})$, and $\mathrm{Id}_0 = \rho_0 + \rho_0^*$.

PROPOSITION 3. *Let $r \geqslant 2$, $s \geqslant 1$. Then the G-module $H^0(M, \mathcal{A}_0) \simeq \mathbb{C}^{2r} \otimes \mathbb{C}^{2s}$ is irreducible and the corresponding representation is $\mathrm{Id}_0 \otimes \mathrm{Id}_1$. For $r = 1$, $s \geqslant 1$, the G-module $H^0(M, \mathcal{A}_0) \simeq \mathbb{C}^{2s}$ is irreducible with the representation $\rho_0 \otimes \mathrm{Id}_1$.*
We have $H^p(M, \mathcal{A}_0) = 0$ for any $p \geqslant 1$.

PROOF. The highest weight of $\Phi_0 = \varphi^*$ is $\lambda_1 + \mu_1$. It is dominant and is the highest weight of the representation $\mathrm{Id}_0 \otimes \mathrm{Id}_1$ (for $r \geqslant 2$) or $\rho_0 \otimes \mathrm{Id}_1$ (for $r = 1$) of G. Our assertions follow from the theorem of Bott.

PROPOSITION 4. *Suppose that $r \geqslant 1$, $r \neq 2$, $s \geqslant 1$. Then $H^0(M, \mathcal{A}_1) \simeq \mathbb{C}$ (the trivial G-module). In the case $r = 2$, $s \geqslant 1$ we have $H^0(M, \mathcal{A}_1) \simeq \mathbb{C} \oplus \mathfrak{sl}_2(\mathbb{C})$, where the first summand is the trivial G-module and the second one is the irreducible G-module with highest weight $\lambda_1 - \lambda_2$. In both cases we have*

$$H^p(M, \mathcal{A}_1) = 0, \qquad p \geqslant 1.$$

PROOF. Clearly, for $r \geqslant 2$, $s \geqslant 2$ we have

$$\Phi_1|R = (\rho_0 \rho_0^*) \otimes (\rho_1 \rho_1^*) = (1 + \sigma_0) \otimes (1 + \sigma_1) = 1 + \sigma_0 + \sigma_1 + \sigma_0 \otimes \sigma_1.$$

The trivial component gives the 1-dimensional trivial G-module. The highest weights of the nontrivial components are

$$\Lambda_0 = \lambda_1 - \lambda_r, \quad \Lambda_1 = \mu_1 - \mu_s, \quad \Lambda_0 + \Lambda_1.$$

The weight $\Lambda_0 + \gamma$ is singular for $r \geqslant 3$, since

$$(\Lambda_0 + \gamma)_{\alpha_r} = (\Lambda_0 + \gamma_0)_{\alpha_r} = -1.$$

For $r = 2$ the weight $\Lambda_0 = \lambda_1 - \lambda_2$ is dominant and determines the restriction of Ad_0 onto one of the simple ideals of $\mathfrak{g}_0 \simeq \mathfrak{sl}_2(\mathbb{C}) \oplus \mathfrak{sl}_2(\mathbb{C})$ (which coincides actually with $[\mathfrak{r}_0, \mathfrak{r}_0]$). Now, $\Lambda_1 + \gamma$ is singular for $s \geqslant 2$, since

$$(\Lambda_1 + \gamma)_{\beta_s} = (\Lambda_1 + \gamma_1)_{\beta_s} = -1.$$

Therefore, $\Lambda_0 + \Lambda_1 + \gamma$ is singular, too.

Thus, the proposition follows from the Bott theorem. In the cases $r = 1$ or $s = 1$ the corresponding adjoint representation does not enter into the expression of Φ_1, and we get the same result.

PROPOSITION 5. *For any $r \geqslant 1$, $s \geqslant 1$ we have*

$$H^0(M, \mathcal{A}_q) = H^1(M, \mathcal{A}_q) = 0, \qquad q \geqslant 2.$$

PROOF. Let Λ be a highest weight of Φ_q. Using its expression given by (15) and (16), we easily see from Proposition 2 that Λ_0 and Λ_1 cannot be dominant. Therefore the situation when Λ is dominant or $\Lambda + \gamma$ has index 1 is impossible.

PROPOSITION 6. *For $r \geqslant 3$, $s \geqslant 1$, the G-module $H^0(M, \mathcal{B}_0) \simeq \mathfrak{so}_{2r}(\mathbb{C}) \oplus \mathfrak{sp}_{2s}(\mathbb{C})$ splits into the sum of two irreducible components with the representations Ad_0, Ad_1. In the case $r = 2$, $s \geqslant 1$ the G-module $H^0(M, \mathcal{B}_0) \simeq \mathfrak{sl}_2(\mathbb{C}) \oplus \mathfrak{sp}_{2s}(\mathbb{C})$ splits into the sum of two irreducible components the first of which has the highest weight $\lambda_1 + \lambda_2$ while the second one is Ad_1. In the case $r = 1$, $s \geqslant 1$ we have the irreducible G-module $H^0(M, \mathcal{B}_0) \simeq \mathfrak{sp}_{2s}(\mathbb{C})$ with the representation Ad_1. Finally we have*

$$H^p(M, \mathcal{B}_0) = 0$$

for any $p \geqslant 1$ and all $r \geqslant 1$, $s \geqslant 1$.

PROOF. By (3), the highest weights of $T_0 = \tau$ are $\lambda_1 + \lambda_2$ (for $r \geqslant 2$) and $2\mu_1$. These are the highest weights of Ad_0 (if $r \geqslant 3$) and Ad_1. If $r = 2$, then $\lambda_1 + \lambda_2$ is the highest weight of the restriction of Ad_0 onto a simple ideal of $\mathfrak{g}_0$ (the complement to the ideal considered in Proposition 4).

PROPOSITION 7. *If $r \geqslant 2$, $s \geqslant 1$, then we have $H^p(M, \mathcal{B}_1) = 0$ for any $p \geqslant 0$. If $r = 1$, $s \geqslant 1$, then $H^0(M, \mathcal{B}_1) \simeq \mathbb{C}^{2s}$ is the irreducible G-module with the representation $\rho_0^* \otimes \mathrm{Id}_1$ and $H^p(M, \mathcal{B}_1) = 0$ for any $p \geqslant 1$.*

PROOF. One see easily that for $r \geqslant 2$,

$$T_1|R = \left(\bigwedge^2 \rho_0 \rho_0^*\right) \otimes \rho_1^* + \rho_0^* \otimes (S^2 \rho_1)\rho_1^*.$$

Clearly, $\lambda_r + \gamma$ and $\mu_s + \gamma$ are singular, and hence $\Lambda + \gamma$ is singular for any weight of T_1. The Bott theorem implies our assertion.

In the case $r = 1$ we have
$$T_1 | R = \rho_0^* \otimes (S^2 \rho_1) \rho_1^*.$$
The highest weights of this representation are $-\lambda_1 + \mu_1$ and (for $s \geqslant 2$) $2\mu_1 - \mu_s$. The first weight is dominant and gives the representation $\rho_0^* \otimes \mathrm{Id}_1$, while the sum of the second one with γ is singular.

PROPOSITION 8. *Suppose that* $r \geqslant 2$, $s \geqslant 1$. *Then*
$$H^0(M, \mathcal{B}_2) = 0, \qquad H^1(M, \mathcal{B}_2) \simeq \mathbb{C}^2$$
(*the trivial G-module*). *If* $r = 1$, $s \geqslant 1$, *then*
$$H^p(M, \mathcal{B}_2) = 0, \qquad p = 0, 1.$$

PROOF. By (3) we have
$$T_2 | R = \left(\bigwedge^2 \rho_0 + S^2 \rho_1\right) \bigwedge^2 (\rho_0^* \otimes \rho_1^*)$$
$$= \left(\bigwedge^2 \rho_0 + S^2 \rho_1\right) \left(\bigwedge^2 \rho_0^* \otimes S^2 \rho_1^* + S^2 \rho_0^* \otimes \bigwedge^2 \rho_1^*\right)$$
$$= \left(\bigwedge^2 \rho_0\right) \left(\bigwedge^2 \rho_0^*\right) \otimes S^2 \rho_1^* + \left(\bigwedge^2 \rho_0\right) (S^2 \rho_0^*) \otimes \bigwedge^2 \rho_1^*$$
$$+ \left(\bigwedge^2 \rho_0\right) \otimes (S^2 \rho_1)(S^2 \rho_1^*) + (S^2 \rho_0^*) \otimes (S^2 \rho_1) \left(\bigwedge^2 \rho_1^*\right).$$

The first three of these four summands exist only when $r \geqslant 2$. For the first one, any highest weight has the form (see (17)–(19)) $\Lambda = \Lambda_0 + \Lambda_1$, where
$$\Lambda_0 = \lambda_i + \lambda_j - \lambda_k - \lambda_l, \qquad \Lambda_1 = -2\mu_s.$$
Clearly, $r_{\beta_s}(\Lambda_1 + \gamma_1) = r_{\beta_s}(-\beta_s + \gamma_1) = \beta_s + \gamma_1 - \beta_s = \gamma_1$. Hence, $\Lambda_1 + \gamma_1$ has index 1. Therefore, we are interested only in the case when Λ_0 is dominant. Using Proposition 2, one sees easily that this is possible only for $\Lambda_0 = 0$ (which is a highest weight indeed!). Then $\Lambda + \gamma$ has index 1. By the algorithm of Bott, to Λ there corresponds an irreducible component of the G-module $H^1(M, \mathcal{B}_2)$ with highest weight $r_{\beta_s}(\Lambda + \gamma) - \gamma = 0$. Quite similarly, the third summand gives (if $r \geqslant 2$) only the 1-dimensional trivial component of $H^1(M, \mathcal{B}_2)$.

Now let $\Lambda = \Lambda_0 + \Lambda_1$ be a highest weight of one of two remaining summands. One easily sees from Proposition 2 that neither Λ_0 nor Λ_1 is dominant ($\Lambda_0 = 0$ is not a highest weight in these cases!). Therefore Λ cannot be dominant, nor can $\Lambda + \gamma$ have index 1.

PROPOSITION 9. *Suppose that* $r \geqslant 1$, $s \geqslant 1$. *Then* $H^0(M, \mathcal{B}_q) = H^1(M, \mathcal{B}_q) = 0$ *for any* $q \geqslant 3$.

PROOF. Let Λ be a weight of T_q'. Using (18), we see, by Proposition 3, that Λ_0 cannot be dominant if $q \geqslant 3$ and that Λ_1 cannot be dominant if $q \geqslant 1$. Quite similarly, for any weight Λ of T_q'' we see, using (19), that Λ_0 cannot be dominant if $q \geqslant 1$ and that Λ_1 cannot be dominant if $q \geqslant 3$. Thus, Λ cannot be dominant, nor can $\Lambda + \gamma$ have index 1. The proposition follows now from the theorem of Bott.

§3. The cohomology of $\widetilde{\mathcal{T}}$

As in [6], we shall use here some further results from Bott's paper [1]. Let E be a holomorphic P-module. Then (see [1], Theorem I and Corollary 2 of Theorem W_2) we have an isomorphism $H^p(M, \mathcal{E})^G \simeq H^p(\mathfrak{n}, E)^\tau$ between the G-invariants and the τ-invariants of the corresponding cohomology groups. This isomorphism is compatible with the homomorphisms induced by homomorphisms of P-modules.

These considerations can be applied to calculate the cohomology of $\mathcal{A}_q$ and $\mathcal{B}_q$ by expressing explicitly the cocycles that represent the basic cohomology classes. We need such an expression for the group $H^1(M, \mathcal{B}_2)$.

We shall use the standard coordinate system on $\mathrm{IGr}_{2r|r, 2s|s}$ in a neighborhood of o given by (2). As in [6], we note that the adjoint action of $\mathfrak{p}$ on $\mathfrak{n}$ coincides with τ^*; hence $\mathfrak{n}$, as a $\mathfrak{p}$-module, is isomorphic to the cotangent space $T_o(M)^*$ of M. By this isomorphism, the basis $dx_{\alpha\beta}$ ($\alpha < \beta$), dy_{ij} ($i \leqslant j$) of $T_o(M)^*$ corresponds to the basis (12) of $\mathfrak{n}$.

The result of Bott mentioned above gives the identification

$$H^1(M, \mathcal{B}_2) = H^1\left(\mathfrak{n}, T_o(M) \otimes \bigwedge^2 E_\phi\right)^\tau.$$

Since τ and ϕ are completely reducible, $\mathfrak{n}$ acts on the coefficients trivially, and hence the coboundary δ of the cochain complex $C(\mathfrak{n}, T_o(M) \otimes \bigwedge^2 E_\phi)$ is zero. It follows that

(20)
$$H^1\left(\mathfrak{n}, T_o(M) \otimes \bigwedge^2 E_\phi\right)^\tau = C^1\left(\mathfrak{n}, T_o(M) \otimes \bigwedge^2 E_\phi\right)^\tau$$
$$\simeq \left(T_o(M) \otimes T_o(M) \otimes \bigwedge^2 E_\phi\right)^\tau.$$

We are going to describe this vector space explicitly in terms of 1-cochains.

PROPOSITION 10. *The following two cochains c_0, c_1 form a basis of the space* $C^1(\mathfrak{n}, T_o(M) \otimes \bigwedge^2 E_\phi)^\tau$:

$$c_0(X_{\alpha\beta}) = \sum_{i,j} \frac{\partial}{\partial y_{ij}} \otimes \xi_{\alpha i}\xi_{\beta j} + \sum_i \frac{\partial}{\partial y_{ii}} \otimes \xi_{\alpha i}\xi_{\beta i}, \qquad c_0(Y_{ij}) = 0,$$

$$c_1(Y_{ij}) = \sum_{\alpha,\beta} \frac{\partial}{\partial x_{\alpha\beta}} \otimes \xi_{\alpha i}\xi_{\beta j}, \qquad c_1(X_{\alpha\beta}) = 0.$$

PROOF. By Proposition 1, the P-module E_ϕ is identified with $(\mathbb{C}^r)^* \otimes (\mathbb{C}^s)^*$ in such a way that $\xi_{\alpha i} = x_\alpha \otimes y_i$, where x_α, y_i are the standard coordinates. Then $\bigwedge^2 E_\phi = \bigwedge^2((\mathbb{C}^r)^* \otimes (\mathbb{C}^s)^*)$ contains an irreducible P-submodule isomorphic to $\bigwedge^2(\mathbb{C}^r)^* \otimes S^2(\mathbb{C}^s)^*$ which is spanned by the elements

$$(x_\alpha \otimes x_\beta - x_\beta \otimes x_\alpha) \otimes (y_i \otimes y_j + y_j \otimes y_i)$$
$$= \xi_{\alpha i} \otimes \xi_{\beta j} - \xi_{\beta i} \otimes \xi_{\alpha j} + \xi_{\alpha j} \otimes \xi_{\beta i} - \xi_{\beta i} \otimes \xi_{\alpha j}$$
$$= 2(\xi_{\alpha i}\xi_{\beta j} - \xi_{\beta j}\xi_{\alpha i}).$$

Then, by (20), $H^1(\mathfrak{n}, T_o(M) \otimes \bigwedge^2 E_\phi)^\tau$ contains the invariants of the submodule

$T_o(M) \otimes T_o(M) \otimes \bigwedge^2(\mathbb{C}^r)^* \otimes S^2(\mathbb{C}^s)^*$. Using (3), we see that precisely two linearly independent invariants lie there, while the complementary submodule does not contain any nonzero invariant. Since the basis $\partial/\partial x_{\alpha\beta}$ ($\alpha < \beta$), $\partial/\partial y_{ij}$ ($i \leq j$) is dual to (12), we get the basic invariants c_0, c_1 given by:

$$c_0(X_{\alpha\beta}) = \sum_{i<j} \frac{\partial}{\partial y_{ij}} \otimes (\xi_{\alpha i}\xi_{\beta j} + \xi_{\alpha j}\xi_{\beta i}) + 2\sum_i \frac{\partial}{\partial y_{ii}} \otimes \xi_{\alpha i}\xi_{\beta i}$$

$$= \sum_{i,j} \frac{\partial}{\partial y_{ij}} \otimes \xi_{\alpha i}\xi_{\beta j} + \sum_i \frac{\partial}{\partial y_{ii}} \otimes \xi_{\alpha i}\xi_{\beta i},$$

$$c_0(Y_{ij}) = 0,$$

$$c_1(Y_{ij}) = \sum_{\alpha<\beta} \frac{\partial}{\partial x_{\alpha\beta}} \otimes (\xi_{\alpha i}\xi_{\beta j} + \xi_{\alpha j}\xi_{\beta i}) = \sum_{\alpha,\beta} \frac{\partial}{\partial x_{\alpha\beta}} \otimes \xi_{\alpha i}\xi_{\beta j},$$

$$c_1(Y_{ii}) = 2\sum_{\alpha<\beta} \frac{\partial}{\partial x_{\alpha\beta}} \otimes \xi_{\alpha i}\xi_{\beta i} = \sum_{\alpha,\beta} \frac{\partial}{\partial x_{\alpha\beta}} \otimes \xi_{\alpha i}\xi_{\beta i},$$

$$c_1(X_{\alpha\beta}) = 0.$$

Now we can calculate $H^p(M, \widetilde{\mathcal{T}})$, $p = 0, 1$.

THEOREM 1. *Suppose that $r \geq 2$, $s \geq 2$ or $r \geq 3$, $s \geq 1$. Then the G-modules $H^p(M, \widetilde{\mathcal{T}}_q)$, $p = 0, 1$, $q \geq -1$ are indicated in the following table:*

$q =$	-1	0	1	2	≥ 3
$p = 0$	$\mathfrak{osp}_{2r\vert 2s}(\mathbb{C})_{\bar{1}}$	$\mathfrak{osp}_{2r\vert 2s}(\mathbb{C})_{\bar{0}} \oplus \mathbb{C}$	0	0	0
$p = 1$	0	0	0	$\mathbb{C}$	0

Here $\mathfrak{osp}_{2r\vert 2s}(\mathbb{C})_{\bar{0}}$ and $\mathfrak{osp}_{2r\vert 2s}(\mathbb{C})_{\bar{1}}$ are endowed with the adjoint representation of G, and $\mathbb{C}$ is the trivial G-module.

If $r = 2$, $s = 1$, then the table has the form

$q =$	-1	0	1	2	≥ 3
$p = 0$	$\mathfrak{osp}_{4\vert 2}(\mathbb{C})_{\bar{1}}$	$\mathfrak{osp}_{4\vert 2}(\mathbb{C})_{\bar{0}} \oplus \mathbb{C}$	0	0	0
$p = 1$	0	0	0	$\mathbb{C}^2$	0

Here $\mathbb{C}^2$ is the trivial G-module.

If $r = 1$, $s \geq 1$, then the corresponding table is as follows:

$q =$	-1	0	1	2	≥ 3
$p = 0$	$\mathbb{C}^{2s}$	$\mathfrak{sp}_{2s}(\mathbb{C}) \oplus \mathbb{C}$	$\mathbb{C}^{2s}$	0	0
$p = 1$	0	0	0	0	0

Here $\mathfrak{sp}_{2s}(\mathbb{C})$ is endowed with the adjoint representation of G, $\mathbb{C}$ is the trivial G-module and $\mathbb{C}^{2s}$ for $q = -1$, 1 is endowed with the representation $\rho_0 \otimes \mathrm{Id}_1$ or $\rho_0^* \otimes \mathrm{Id}_1$ respectively.

PROOF. We use the cohomology exact sequences associated with (8). In almost all cases the mappings in these sequences are determined uniquely. The only difficulty occurs when we try to calculate $H^1(M, \widetilde{\mathcal{T}}_2)$ with the help of the exact sequence

$$0 \to \mathcal{A}_3 \xrightarrow{\alpha} \widetilde{\mathcal{T}}_2 \xrightarrow{\beta} \mathcal{B}_2 \to 0.$$

By Proposition 5, we have the exact sequence

$$0 \to H^1(M, \widetilde{\mathcal{T}}_2) \xrightarrow{\beta^*} H^1(M, \mathcal{B}_2).$$

If $r = 1$ then, by Proposition 8, we have $H^1(M, \mathcal{B}_2) = 0$. Hence, $H^1(M, \widetilde{\mathcal{T}}_2) = 0$ in this case. In what follows we assume that $r \geq 2$.

By Proposition 8, $H^1(M, \mathcal{B}_2) \simeq \mathbb{C}^2$ (the trivial G-module). The sheaves $\widetilde{\mathcal{T}}_2$ and $\mathcal{B}_2$ are the sheaves of holomorphic sections of homogeneous vector bundles $\widetilde{\mathbf{T}}_2$ and $\mathbf{B}_2 = T(M) \otimes \bigwedge^2 \mathbf{E}_\phi$, and β is induced by a homomorphism of these bundles. As we have seen at the beginning of this section, β^* is interpreted as the homomorphism of the invariant 1-cohomology of the Lie algebra $\mathfrak{n}$:

$$H^1(\mathfrak{n}, (\widetilde{\mathbf{T}}_2)_o)^\tau \to H^1\left(\mathfrak{n}, T_o(M) \otimes \bigwedge^2 E_\phi\right)^\tau,$$

where $(\widetilde{\mathbf{T}}_2)_o$ is the fiber of $\widetilde{\mathbf{T}}_2$ at the point o endowed with a natural structure of the p-module. The group $H^1(\mathfrak{n}, (\widetilde{\mathbf{T}}_2)_o)^\tau$ coincides with the 1-cohomology of the complex $C(\mathfrak{n}, (\widetilde{\mathbf{T}}_2)_o)^\tau$ of τ-invariant cochains. Since $H^1(M, \mathcal{A}_3) = 0$ by Proposition 5, the vector space $C^1(\mathfrak{n}, (\widetilde{\mathbf{T}}_2)_o)^\tau$ is mapped isomorphically onto $C^1(\mathfrak{n}, T_o(M) \otimes \bigwedge^2 E_\phi)^\tau$. It follows from Proposition 10 that the cochains $c \in C^1(\mathfrak{n}, (\widetilde{\mathbf{T}}_2)_o)^\tau$ have the form

$$c(X_{\alpha\beta}) = a\left(\sum_{i,j} \xi_{\alpha i}\xi_{\beta j}\frac{\partial}{\partial y_{ij}} + \sum_i \frac{\partial}{\partial y_{ii}}\right), \quad c(Y_{ij}) = b\sum_{\alpha,\beta} \xi_{\alpha i}\xi_{\beta j}\frac{\partial}{\partial x_{\alpha\beta}},$$

where $a, b \in \mathbb{C}$. Clearly,

$$H^1(\mathfrak{n}, (\widetilde{\mathbf{T}}_2)_o)^\tau \simeq \{c \in C^1(\mathfrak{n}, (\widetilde{\mathbf{T}}_2)_o)^\tau \mid \delta c = 0\}.$$

By the definition of δ we have

$$(\delta c)(x, y) = xc(y) - yc(x), \quad x, y \in \mathfrak{n}.$$

The action of $\mathfrak{n}$ on $(\widetilde{\mathbf{T}}_2)_o$ is induced by commuting the fundamental vector fields of the action of G on $\mathrm{IGr}_{2r|r, 2s|s}$ with the elements of $\widetilde{\mathcal{T}}_2$, and then evaluating the commutator at $X = 0$, $Y = 0$. It follows from (13) that

$$(\delta c)(X_{\alpha\beta}, X_{\gamma\delta}) = (\delta c)(Y_{ij}, Y_{kl}) = 0,$$
$$(\delta c)(X_{\alpha\beta}, Y_{ij}) = (b-a)\sum_{\gamma,k}(\xi_{\alpha j}\xi_{\beta k}\xi_{\gamma i} + \xi_{\alpha k}\xi_{\beta j}\xi_{\gamma i}$$
$$+ \xi_{\alpha i}\xi_{\beta k}\xi_{\gamma j} + \xi_{\alpha k}\xi_{\beta j}\xi_{\gamma i})\frac{\partial}{\partial \xi_{\gamma i}}.$$

One easily sees that if $r \geq 2$, $s \geq 2$, then $\delta c = 0$ is equivalent to $a = b$. The same is true if $s = 1$, $r \geq 3$. In the remaining case $r = 2$, $s = 1$, we have $\delta c = 0$ for any invariant cochain c. Thus,

$$H^1(M, \widetilde{\mathcal{T}}_2) \simeq H^1(\mathfrak{n}, (\widetilde{\mathbf{T}}_2)_o)^\tau \simeq \begin{cases} \mathbb{C} & \text{if } r \geq 2, s \geq 2 \text{ or } r \geq 3, s = 1, \\ \mathbb{C}^2 & \text{if } r = 2, s = 1. \end{cases}$$

§4. The cohomology of $\mathcal{T}$

In this section, we prove our main theorem about 0- and 1-cohomology of the isotropic super-Grassmannian with values in the tangent sheaf. The proof repeats that of Theorem 2 of [6]. First we state a proposition that will play the main role in it.

It is clear that on the split supermanifold $(M, \text{gr}\,\mathcal{O})$ there exists a vector field $\varepsilon \in H^0(M, \widetilde{\mathcal{T}}_0)$ such that $\varepsilon(f) = qf$ for any $f \in \text{gr}_q \mathcal{O}$. This vector field commutes with any X^*, $X \in \mathfrak{g}$, and hence is a basis element of the trivial G-submodule $\mathbb{C} \subset H^0(M, \widetilde{\mathcal{T}}_0)$ (see Theorem 1).

PROPOSITION 11. *If $r \geq 2$, then ε does not lie in the image of the canonical mapping $H^0(M, \mathcal{T}_{(0)}) \to H^0(M, \widetilde{\mathcal{T}}_0)$.*

PROOF. As odd coordinates in a neighborhood of o in $(M, \text{gr}\,\mathcal{O})$ we take the elements $\tilde{\xi}_{\alpha i} = \xi_{\alpha i} + \mathcal{J}^2$. Then, clearly, ε is expressed in this neighborhood as

$$\varepsilon = \sum_{\alpha,i} \tilde{\xi}_{\alpha i} \frac{\partial}{\partial \tilde{\xi}_{\alpha i}}.$$

Suppose that there exists $\hat{\varepsilon} \in H^0(M, \mathcal{T}_{(0)})$ inducing the vector field ε. One may suppose that $\hat{\varepsilon} \in (H^0(M, \mathcal{T}_{(0)})_{\bar{0}})^G$. Then $[\hat{\varepsilon}, X^*] = 0$ for any $X \in \mathfrak{g}$. Consider the action of the derivation $\hat{\varepsilon}$ in $\mathcal{O}_o$. The mapping $X \to X^*$ is a linear representation of the Cartan subalgebra $\mathfrak{t}$ of $\mathfrak{g}$, commuting with $\hat{\varepsilon}$. We see from (11) that $x_{\alpha\beta}$, y_{ij}, $\xi_{\alpha i}$ lie in the weight subspaces of this representation, corresponding to the weights $\lambda_\alpha + \lambda_\beta$, $\mu_i + \mu_j$, $\lambda_\alpha + \mu_i$ respectively. It is clear that all these weight subspaces have dimension 1. Since $\hat{\varepsilon}$ maps any weight subspace into itself, we have

$$\hat{\varepsilon} = \sum_{\alpha,i} \xi_{\alpha i} \frac{\partial}{\partial \xi_{\alpha i}} + \sum_{\alpha<\beta} a_{\alpha\beta} x_{\alpha\beta} \frac{\partial}{\partial x_{\alpha\beta}} + \sum_{i \leq j} b_{ij} y_{ij} \frac{\partial}{\partial y_{ij}},$$

where $a_{\alpha\beta}$, $b_{ij} \in \mathbb{C}$. Now we have $[\hat{\varepsilon}, U^*_{\alpha\beta}] = [\hat{\varepsilon}, V^*_{ij}] = 0$ which, by (14), implies that $a_{\alpha\beta} = b_{ij} = 0$ for all $\alpha < \beta$, $i \leq j$. Thus,

$$\hat{\varepsilon} = \sum_{\alpha,i} \xi_{\alpha i} \frac{\partial}{\partial \xi_{\alpha i}}.$$

By (13) we see that

$$[\hat{\varepsilon}, X^*_{\alpha\beta}](y_{ij}) = 2\xi_{\alpha i}\xi_{\beta j}.$$

This cannot be 0 if $r \geq 2$, giving a contradiction.

As a corollary, we want to characterize the split isotropic super-Grassmannians.

COROLLARY. *The super-Grassmannian $\text{I}^\circ\text{Gr}_{2r|r, 2s|s}$ is split if and only if $r = 1$.*

PROOF. Proposition 11 shows that the super-Grassmannian is nonsplit if $r \geq 2$. For $r = 1$ we have $H^1(M, \mathcal{B}_q) = 0$ for all $q \geq 2$, by Propositions 8 and 9. Thus, all the obstructions to $\text{I}^\circ\text{Gr}_{2|1, 2s|s}$ being split are 0 (see [4], Chapter 4, §2), and hence it is split.

THEOREM 2. *We have, for any $r \geq 1$, $s \geq 1$,*

$$H^0(M, \mathcal{T}) \simeq \mathfrak{osp}_{2r|2s}(\mathbb{C})$$

as Lie superalgebras, the isomorphism being defined by the standard action of $\mathrm{OSp}_{2r|2s}(\mathbb{C})$. *Also*

$$H^1(M, \mathcal{T}) = \begin{cases} 0 & \text{if } (r, s) \neq (2, 1), \\ \mathbb{C}^{1|0} & \text{if } r = 2, s = 1. \end{cases}$$

PROOF. Suppose first that $(r, s) \neq (1, s)$ and $\neq (2, 1)$. Then the proof goes exactly as in [6]. Using Theorem 1 and the cohomology exact sequence corresponding to (6), we see that $H^0(M, \mathcal{T}_{(q)}) = H^1(M, \mathcal{T}_{(q)}) = 0$ for $q \geq 3$. For $q = 2$ this exact sequence shows that $H^0(M, \mathcal{T}_{(2)}) = 0$ and that $H^1(M, \mathcal{T}_{(2)})$ is mapped injectively into $H^1(M, \widetilde{\mathcal{T}}_2) \simeq \mathbb{C}^{1|0}$. Thus, $H^1(M, \mathcal{T}_{(2)}) \simeq \mathbb{C}^{k|0}$, $k \leq 1$. For $q = 1$ the exact sequence shows that $H^0(M, \mathcal{T}_{(1)}) = 0$ and that $H^1(M, \mathcal{T}_{(1)}) \simeq \mathbb{C}^{k|0}$. For $q = 0$ we get the exact sequence

(21)
$$0 \to H^0(M, \mathcal{T}_{(1)}) \to H^0(M, \mathcal{T}_{(0)}) \to H^0(M, \widetilde{\mathcal{T}}_0)$$
$$\to H^1(M, \mathcal{T}_{(1)}) \to H^1(M, \mathcal{T}_{(0)}) \to H^1(M, \widetilde{\mathcal{T}}_0).$$

This implies that $H^0(M, \mathcal{T}_{(0)})$ is mapped injectively into $H^0(M, \widetilde{\mathcal{T}}_0)$. By Proposition 11, the trivial submodule $\mathbb{C}$ does not lie in the image. Therefore $H^1(M, \mathcal{T}_{(1)}) \neq 0$, and hence $H^1(M, \mathcal{T}_{(1)}) \simeq \mathbb{C}^{1|0}$, $H^1(M, \mathcal{T}_{(0)}) = 0$. Also, $H^0(M, \mathcal{T}_{(0)}) \simeq \mathfrak{osp}_{2r|2s}(\mathbb{C})_{\bar{0}}$. Now, for $q = -1$ we get the exact sequence

$$0 \to H^0(M, \mathcal{T}_{(0)}) \to H^0(M, \mathcal{T}) \to H^0(M, \widetilde{\mathcal{T}}_{-1})$$
$$\to H^1(M, \mathcal{T}_{(0)}) \to H^1(M, \mathcal{T}) \to H^1(M, \widetilde{\mathcal{T}}_{-1}).$$

It implies that

$$H^0(M, \mathcal{T}) \simeq H^0(M, \mathcal{T}_{(0)}) \oplus H^0(M, \widetilde{\mathcal{T}}_{-1}) \simeq \mathfrak{osp}_{2r|2s}(\mathbb{C}), \qquad H^1(M, \mathcal{T}) = 0.$$

For the 0-cohomology, here we mean an isomorphism of G-modules. Since $\mathfrak{osp}_{2r|2s}(\mathbb{C})$ is simple [3], the homomorphism $X \rightsquigarrow X^*$ of this superalgebra into $H^0(M, \mathcal{T})$ is injective. Therefore it is an isomorphism of Lie superalgebras.

Suppose that $r = 2$, $s = 1$. Then the super-Grassmannian has dimension $2|2$. Using Theorem 1, we see that $H^1(M, \mathcal{T}_{(1)}) \simeq H^1(M, \mathcal{T}_{(2)}) \simeq \mathbb{C}^{2|0}$. Then the exact sequence (21) and Proposition 11 give that $H^1(M, \mathcal{T}_{(0)}) \simeq \mathbb{C}^{1|0}$. It follows that $H^1(M, \mathcal{T}) \simeq \mathbb{C}^{1|0}$.

The case $r = 1$ being the simplest one, we omit the proof.

It follows from Theorem 2 that the supermanifold $\mathrm{I}^\circ\mathrm{Gr}_{2r|r, 2s|s}$ is rigid if $(r, s) \neq (2, 1)$ (see [8]). The remaining case $r = 2$, $s = 1$ was actually studied before. It is easy to see that $\mathrm{I}^\circ\mathrm{Gr}_{4|2, 2|1}$ is precisely the supermanifold $\mathbb{G}(1, 1)$ from the family $\mathbb{G}(t_1, t_2)$ constructed in [2], where the corresponding part of Theorem 2 was proved. By Theorem 4 of [2], this family is a versal deformation of $\mathrm{I}^\circ\mathrm{Gr}_{4|2, 2|1}$. Thus, we get the following result.

COROLLARY. *The super-Grassmannian* $\mathrm{I}^\circ\mathrm{Gr}_{2r|r, 2s|s}$ *is a rigid supermanifold if and only if* $(r, s) \neq (2, 1)$.

References

1. R. Bott, *Homogeneous vector bundles*, Ann. Math. **66** (1957), 203–248.
2. V. A. Bunegina and A. L. Onishchik, *Two families of flag supermanifolds*, Diff. Geom. Appl. (to appear).
3. V. G. Kac, *Lie superalgebras*, Adv. Math. **26** (1977), 8–96.
4. Yu. I. Manin, *Gauge field theory and complex geometry*, Springer-Verlag, Berlin, 1988.
5. A. L. Onishchik, *Transitive Lie superalgebras of vector fields*, Reports Dep. Math. Univ. Stockholm **26** (1987), 1–21.
6. A. L. Onishchik, *On the rigidity of super-Grassmannians*, Ann. Global Anal. Geom. **11** (1993), 361–372.
7. A. L. Onishchik and A. A. Serov, *Holomorphic vector fields on super-Grassmannians*, Lie Groups, Their Discrete Subgroups, and Invariant Theory, Adv. in Soviet Math., vol. 8, Amer. Math. Soc., Providence, RI, 1992, pp. 113–129.
8. A. Yu. Vaintrob, *Deformations of complex superspaces and coherent sheaves on them*, J. Soviet Math. **51** (1990), 2069–2083.

Yaroslavl' University, Sovetskaya 14, 150 000 Yaroslavl', Russia

Tver' University, Zhelyabova 33, 170 000 Tver', Russia

A_∞ Algebras and the Cohomology of Moduli Spaces

MICHAEL PENKAVA AND ALBERT SCHWARZ

§1. Introduction

Let us consider an A_∞ algebra with an invariant inner product. The main goal of this paper is to classify the infinitesimal deformations of this A_∞ algebra preserving the inner product and to apply this result to the construction of homology classes on the moduli spaces of algebraic curves. With this aim, we define cyclic cohomology of an A_∞ algebra and show that it classifies the deformations we are interested in. To make the reading of our paper more independent of other works, we include a short review of Hochschild and cyclic cohomology of associative algebras, and explain the definition of A_∞ algebras.

Our constructions are based on ideas and results of Maxim Kontsevich; moreover, he has informed us that he also has given a definition of the cyclic cohomology of A_∞ algebras in a different manner than we do and has proved the results mentioned above as well. Another definition of cyclic cohomology of A_∞ algebras was given by Ezra Getzler and John D.S. Jones [4]. We did not study its relation to our definition.

In this paper we make the notational convention when dealing with $\mathbb{Z}_2$ grading, that if a is a homogeneous element with parity $|a|$, in superscripts we will use a in place of $|a|$, so that $(-1)^a$ stands for $(-1)^{|a|}$, and similarly, $(-1)^{ab} = (-1)^{|a||b|}$, not $(-1)^{|ab|}$, which is of course given by $(-1)^{a+b}$.

§2. Cohomology of associative algebras

In this section, we recall the definition of Hochschild cohomology of associative algebras. We relate this notion to the theory of deformations of the associative algebra structure. Then we discuss the theory of deformations of an associative algebra preserving an invariant inner product, and relate this notion to cyclic cohomology. The purpose of this is to motivate the definition of A_∞ algebras, and to set the stage for the more general discussion of cohomology of A_∞ algebras.

In this section, suppose that A is an algebra over a field k. For simplicity, we suppose that A is finite-dimensional over k. Let $C^n(A) = \mathrm{Hom}(A^n, A)$ be the space of n-multilinear functions on A with values in A; we call $C^n(A)$ the module

---

1991 *Mathematics Subject Classification.* Primary 14J15, 19D55, 58D27.

The work of the first author was partially supported by NSF Grant DMS-9404111.

The work of the second author was partially supported by NSF Grant DMS-9201366. Research at MSRI is supported by NSF Grant DMS-9022140.

©1995, American Mathematical Society

of cochains of degree n on A with values in A. We define a coboundary operator $b : C^n(A) \to C^{n+1}(A)$ by

$$\begin{aligned} bf(a_1, \ldots, a_{n+1}) &= a_1 f(a_2, \ldots, a_{n+1}) \\ &\quad + \sum_{i=1}^{n} (-1)^i f(a_1, \ldots, a_i a_{i+1}, \ldots, a_{n+1}) \\ &\quad + (-1)^{n+1} f(a_1, \ldots, a_{n+1}). \end{aligned} \tag{1}$$

Then

$$HH^n(A) = \ker(b : C^n(A) \to C^{n+1}(A))/\operatorname{Im}(b : C^{n-1}(A) \to C^n(A))$$

is the Hochschild cohomology of A with coefficients in A. In this paper we do not consider cohomology with other coefficients. The connection between Hochschild cohomology and (infinitesimal) deformations of A is given by $HH^2(A)$. If we denote the product in A by m, an infinitesimally deformed product by m_t, and express $m_t = m + t\phi$, where $t^2 = 0$, then the map $\varphi : A^2 \to A$ is a Hochschild cocycle and the trivial deformations are given by Hochschild coboundaries. To show the first assertion note that

$$\begin{aligned} m_t(m_t(a_1, a_2), a_3) &= a_1 a_2 a_3 + t(\varphi(a_1, a_2)a_3 + \varphi(a_1 a_2, a_3)), \\ m_t(a_1, m_t(a_2, a_3)) &= a_1 a_2 a_3 + t(a_1 \varphi(a_2, a_3) + \varphi(a_1, a_2 a_3)). \end{aligned}$$

Associativity of m_t is equivalent to the condition

$$a_1 \varphi(a_2, a_3) - \varphi(a_1 a_2, a_3) + \varphi(a_1, a_2 a_3) - \varphi(a_1, a_2) a_3 = 0. \tag{2}$$

But this last condition is simply the condition $b\varphi = 0$.

On the other hand, the notion of a trivial deformation is given by the condition that A with the new multiplication is isomorphic to the original algebra structure. This means that there is a linear bijection $\rho_t : A \to A$ such that $m_t(\rho_t(a_1), \rho_t(a_2)) = \rho_t(a_1 a_2)$. We can express $\rho_t = I + t\lambda$, where $\lambda : A \to A$ is a linear map. Then

$$\begin{aligned} m_t(a_1, a_2) &= \rho_t(\rho_t^{-1}(a_1), \rho_t^{-1}(a_2)) \\ &= a_1 a_2 - t(a_1 \lambda(a_2) - \lambda(a_1 a_2) + \lambda(a_1) a_2) = a_1 a_2 - t(b\lambda)(a_1, a_2). \end{aligned}$$

Thus coboundaries give rise to trivial deformations.

Next, consider an invariant non-degenerate inner product on A, by which we mean an inner product $\langle \cdot, \cdot \rangle$ that satisfies $\langle ab, c \rangle = \langle a, bc \rangle$. Note that for an invariant inner product, we also have that $\langle ab, c \rangle = \langle ca, b \rangle$, so that it is invariant under the cyclic permutations of a, b, c. We consider deformations of A preserving this inner product, and these are governed by cyclic cohomology. To see this connection we first note that in the presence of an inner product, there is a natural isomorphism $\operatorname{Hom}(A^n, A) \xrightarrow{\sim} \operatorname{Hom}(A^{n+1}, k)$, given as follows. If we denote the image of $f \in \operatorname{Hom}(A^n, A)$ in $\operatorname{Hom}(A^{n+1}, k)$ by $\widetilde{f}$, then

$$\widetilde{f}(a_1, \ldots, a_{n+1}) = \langle f(a_1, \ldots, a_n), a_{n+1} \rangle.$$

If we define an element $\widetilde{f}$ in $\operatorname{Hom}(A^n, k)$ to be cyclic whenever
$$\widetilde{f}(a_1, \ldots, a_n) = (-1)^{n+1}\widetilde{f}(a_n, a_1, \ldots, a_{n-1}),$$
then we see that a cyclic element $\widetilde{\varphi}$ in $\operatorname{Hom}(A^3, k)$ corresponds to a deformation φ in $C^2(A)$ preserving the inner product, because

(3) $\qquad \langle \varphi(a, b), c \rangle = \widetilde{\varphi}(a, b, c) = \widetilde{\varphi}(b, c, a) = \langle a, \varphi(b, c) \rangle.$

A trivial deformation preserving the inner product is determined by a linear map $\rho_t = I + t\lambda$ as before, but in addition we assume that

(4) $\qquad \langle \rho_t(a_1), \rho_t(a_2) \rangle = \langle a_1, a_2 \rangle.$

This is equivalent to the condition $\langle \lambda(a_1), a_2 \rangle = -\langle a_1, \lambda(a_2) \rangle$. This latter condition is precisely the condition that $\widetilde{\lambda}(a_1, a_2) = -\widetilde{\lambda}(a_2, a_1)$. In other words, λ is cyclic.

The Hochschild coboundary operator b induces a coboundary operator $\widetilde{b}: \operatorname{Hom}(A^{n+1}, k) \to \operatorname{Hom}(A^{n+2}, k)$ by $\widetilde{b}\widetilde{f} = \widetilde{bf}$. As we shall show later, the coboundary operator takes cyclic elements to cyclic elements. If we denote the submodule of $\operatorname{Hom}(A^{n+1}, k)$ consisting of cyclic elements by $CC^n(A)$, then the cyclic cohomology of A is defined by

$$HC^n(A) = \ker(\widetilde{b}: CC^n(A) \to CC^{n+1}(A))/\operatorname{Im}(\widetilde{b}: CC^{n-1}(A) \to CC^n(A)).$$

This definition depends on the choice of the inner product. However, one can express the coboundary operator $\widetilde{b}$ without reference to the inner product. If $f \in \operatorname{Hom}(A^{n+1}, k)$, then we see that

(5)
$$\widetilde{b}\widetilde{f}(a_1, \ldots, a_{n+2}) = \widetilde{bf}(a_1, \ldots, a_{n+2}) = \langle bf(a_1, \ldots, a_{n+1}), a_{n+2} \rangle$$
$$= \langle a_1 f(a_2, \ldots, a_{n+1}), a_{n+2} \rangle + \sum_{i=1}^{n}(-1)^i \langle f(a_1, \ldots, a_i a_{i+1}, \ldots, a_{n+1}), a_{n+2} \rangle$$
$$+ (-1)^{n+1} \langle f(a_1, \ldots, a_n) a_{n+1}, a_{n+2} \rangle$$
$$= \langle f(a_2, \ldots, a_{n+1}), a_{n+2} a_1 \rangle + \sum_{i=1}^{n}(-1)^i \langle f(a_1, \ldots, a_i a_{i+1}, \ldots, a_{n+1}), a_{n+2} \rangle$$
$$+ (-1)^{n+1} \langle f(a_1, \ldots, a_n), a_{n+1} a_{n+2} \rangle$$
$$= \widetilde{f}(a_2, \ldots, a_{n+1}, a_{n+2} a_1) + \sum_{i=1}^{n+1}(-1)^i \widetilde{f}(a_1, \ldots, a_i a_{i+1}, \ldots, a_{n+2}).$$

If $\widetilde{f}$ is cyclic, then we see that
(6)
$$\widetilde{b}\widetilde{f}(a_1, \ldots, a_{n+2})$$
$$= (-1)^n \widetilde{f}(a_{n+2}a_1, a_2, \ldots, a_n) + \sum_{i=1}^{n+1}(-1)^{ni+n+1} \widetilde{f}(a_i a_{i+1}, \ldots, a_{i-1})$$
$$= \sum_{i=0}^{n+1}(-1)^{n[i]+n+1} \widetilde{f}(a_{[i]}a_{[i+1]}, \ldots, a_{[i-1]}),$$

where $[i] = i \pmod{n+2}$, and we make the convention that $a_0 = a_{n+2}$. The fact that $\tilde{b}f$ is cyclic follows easily from this formula for the differential.

Thus the cyclic cohomology characterizes the deformations of A that preserve an invariant inner product, independently of the particular inner product involved. We note that cyclic homology is usually based on applying the operator b to the complex $C^n(A, A^*) = \text{Hom}(A^n, A^*)$. However, the description we give here is equivalent to this one, because the inner product induces a natural isomorphism between $\text{Hom}(A^n, A^*)$ and $\text{Hom}(A^n, A)$. Of course, the description in terms of $C^n(A, A^*)$, being independent of any inner product, makes completely transparent the fact that the cyclic homology does not depend on the inner product. However, it does not elucidate the connection between $HC^2(A)$ and deformations preserving an inner product. As was pointed out to us by Getzler, this relation was first shown by Connes-Flato-Sternheimer.

§3. Cohomology of $\mathbb{Z}_2$-graded algebras

Let us consider a $\mathbb{Z}_2$-graded algebra A over a ring k. The notion of an invariant inner product needs to be modified in this case. The definition of a graded symmetric inner product is given by the formula

$$\langle a, b \rangle = (-1)^{ab} \langle b, a \rangle$$

for homogeneous elements a, b in A. Let k be a commutative ring. Consider elements of k as having degree 0, making k a $\mathbb{Z}_2$-graded ring in a trivial sense. If we require the inner product to be an even map when considered as a map from $A \otimes A$ to k, then $\langle a, b \rangle = 0$ unless a and b have the same parity. On elements of even parity the inner product is symmetric, and on elements of odd parity the inner product is a skew symmetric form. This definition is for an even inner product. One can also consider the case when k is a supercommutative ring. Odd inner products can also be defined, but we will not consider them in this paper. The notion of an invariant inner product is given by the same relation $\langle ab, c \rangle = \langle a, bc \rangle$, but now we have

(7)
$$\langle ab, c \rangle = \langle a, bc \rangle = (-1)^{(b+c)a} \langle bc, a \rangle = (-1)^{(b+c)a} \langle b, ca \rangle$$
$$= (-1)^{(b+c)a + (c+a)b} \langle ca, b \rangle = (-1)^{c(a+b)} \langle ca, b \rangle.$$

We also assume that the multiplication is an even map, so that $|ab| = |a| + |b|$. We are only interested in deformations of A that preserve this property. If we denote the deformed multiplication by $m_t = m + t\varphi$, and consider t as an even parameter, then φ must be an even map, but if t is an odd parameter, then φ must be odd. For parameters, we assume the property of graded commutativity, so that $ta = (-1)^{ta} at$. The associativity condition (2) is modified by this consideration, so that now we have the formula

(8) $$(-1)^{a_1 \varphi} a_1 \varphi(a_2, a_3) - \varphi(a_1 a_2, a_3) + \varphi(a_1, a_2 a_3) - \varphi(a_1, a_2) a_3 = 0$$

which is the deformation condition for any homogeneous bilinear map φ.

Now we consider a trivial deformation, which, as before, is given by a linear map $p_t = I + t\lambda : A \to A$, but the parameter is allowed to be odd, in which case λ is odd as well. In this case, a trivial deformation is given by

$$m_t(a_1, a_2) = a_1 a_2 - t(a_1\lambda(a_2) - \lambda(a_1 a_2) + (-1)^{a_1} a_1\lambda(a_2)).$$

These two results suggest that we should define the Hochschild coboundary operator in the $\mathbb{Z}_2$-graded case as the map $b : C^n(A) \to C^{n+1}(A)$ given by

$$\begin{aligned}(9) \quad bf(a_1, \ldots, a_{n+1}) &= (-1)^{a_1 f} a_1 f(a_2, \ldots, a_{n+1}) \\ &+ \sum_{i=1}^{n} (-1)^i f(a_1, \ldots, a_i a_{i+1}, \ldots, a_{n+1}) \\ &+ (-1)^{n+1} f(a_1, \ldots, a_n) a_{n+1}\end{aligned}$$

for homogeneous f. It is easily checked that $b^2 = 0$. Then deformations of A correspond to cocycles, while trivial deformations correspond to coboundaries of this new Hochschild coboundary operator.

The definition of a cyclic element of $CC^n(A)$ is adjusted to be consistent with the grading, which makes the notion compatible with the definition of the invariance of the inner product. This is accomplished by defining an element $\widetilde{f} \in CC^n(A)$ to be cyclic if and only if

$$(10) \quad \widetilde{f}(a_1, \ldots, a_{n+1}) = (-1)^{n + a_{n+1}(a_1 + \cdots + a_n)} \widetilde{f}(a_{n+1}, a_1, \ldots, a_n).$$

Note that if $\widetilde{m} \in CC^2(A)$ is given by $\widetilde{m}(a, b, c) = \langle ab, c\rangle$, then $\widetilde{m}$ is cyclic precisely when the inner product is invariant with respect to the multiplication. Also, we see that $\widetilde{f}$ is cyclic precisely when

$$(11) \quad \langle f(a_1, \ldots, a_n), a_{n+1}\rangle = (-1)^{n + a_1 f} \langle a_1, f(a_2, \ldots, a_{n+1})\rangle$$

for all homogeneous elements.

We want to express the Hochschild differential of cyclic homology in terms of this new notion of cyclicity. Let us restrict ourselves to the case when k is a field. Then, as in the nongraded case, there is a natural isomorphism between $\mathrm{Hom}(A^n, A)$ and $\mathrm{Hom}(A^{n+1}, k)$ defined in the same manner as before, inducing a coboundary operator $\widetilde{b} : \mathrm{Hom}(A^n, k) \to \mathrm{Hom}(A^{n+1}, A)$ given by $\widetilde{b}\widetilde{f} = \widetilde{bf}$. The proof of the formula below is straightforward, and cyclicity is easily seen from this formula.

LEMMA. *The Hochschild differential $\widetilde{b}$ takes cyclic elements to cyclic elements. For a cyclic element $\widetilde{f}$, the Hochschild differential can be expressed in the form*

$$\begin{aligned}(12) \quad &\widetilde{b}\widetilde{f}(a_1, \ldots, a_{n+1}) \\ &= \sum_{i=0}^{n} -1^{ni + n + 1 + (a_1 + \cdots + a_{[i-1]})(a_{[i]} + \cdots + a_{n+1})} \widetilde{f}(a_{[i]} a_{[i+1]}, a_{[i+2]}, \ldots, a_{[i-1]}),\end{aligned}$$

where $[i] = i \pmod{n+1}$ *and by convention,* $a_0 = a_{n+1}$.

The sign in the expression above, as in the formulas below, follow from the exchange rule which is given by the principle that when two elements are exchanged in an expression, the corresponding sign is -1 to the power equal to the product of the parities of the two elements.

As in the nongraded case, one sees that the cyclic cohomology $HC^2(A)$ classifies the deformations of A which preserve a graded symmetric inner product. Cyclic cohomology of $\mathbb{Z}_2$-graded algebras was considered by D. Kastler in [7].

§4. Definition of A_∞ algebras

Suppose that W is a $\mathbb{Z}_2$-graded space over k, and denote the tensor algebra over W by $T(W) = \bigoplus_{n=1}^{\infty} W^n$, where $W^1 = W$ and $W^{n+1} = W \otimes W^n$. Then $T(W)$ is an associative algebra under the tensor product. As usual, the parity of a product $w = a_1 \otimes \cdots \otimes a_n$ is given by $|w| = |a_1| + \cdots + |a_n|$.

In this picture, we let $\hat{d}$ be a (super) derivation on $T(W)$; by this we mean that $|\hat{d}\omega| = |\omega| + |d|$ for homogeneous $\omega \in T(W)$, where $|d|$ is the parity of the map d, and also $\hat{d}(\omega \otimes \eta) = \hat{d}(\omega) \otimes \eta + (-1)^{\omega d} \omega \otimes \hat{d}\eta$, which is the derivation law for (super) derivations. If we denote the restriction of $\hat{d}$ to W by d, then the derivation law shows that $\hat{d}$ is uniquely determined by d. Denote $d = d_1 + d_2 + \cdots$, where $d_k : W \to W^k$ is the induced map, and similarly denote $\hat{d} = \hat{d}_1 + \hat{d}_2 + \cdots$, where $\hat{d}_l$ is the component of $\hat{d}$ of degree $l - 1$ and where for each k, the restriction $d_{l,k}$ of $\hat{d}_l$ to W^k is the map $d_{l,k} : W^k \to W^{k+l-1}$ induced by $\hat{d}$. We can express $d_{l,k}$ in terms of d_l by

$$d_{l,k} = \sum_{i+j=k-1} I_i \otimes d_l \otimes I_j$$

where $I_i : W^i \to W^i$ and $I_j : W^j \to W^j$ are the identity maps, with the obvious convention when either i or j is zero. It should be noted that if d is an odd (even) map from W to $T(W)$, the maps d_k, $d_{l,k}$, and $\hat{d}_k$ are all odd (even) as well. If $d : W \to T(W)$ is any map, then it extends uniquely to a derivation of $T(W)$, so that there is a one-to-one correspondence between derivations on $T(W)$ and maps from W to $T(W)$.

We say that $\hat{d}$ is a differential if $\hat{d}^2 = 0$. This condition is also determined by the mapping d alone. Since

(13) $\qquad \hat{d}^2(a \otimes b) = \hat{d}(da \otimes b + (-1)^{ad} a \otimes db) = \hat{d}d(a) \otimes b + a \otimes \hat{d}db$,

it is clear that the necessary condition $\hat{d}d = 0$ is sufficient for $\hat{d}^2 = 0$ as well. Hence,

(14) $$\sum_{k+l=n} d_{k,l} d_l = 0$$

for all $n > 1$. This yields an infinite set of relations which are necessary for $\hat{d}^2 = 0$, and these are evidently sufficient as well. These relations are simpler than the more complete set of relations $\sum_{k+l=n} \widehat{d}_k \widehat{d}_l = 0$ for all $n > 1$. Since $d_{1,1} = d_1$, the first relation yields $d_1^2 = 0$, so that d_1 determines a differential on W. Clearly, $\widehat{d}_1^2 = 0$, which shows that $\widehat{d}_1$ is itself a differential.

Now the actual object we want to study is the dual of the object that we have been considering. More precisely, let $V = \Pi(W^*)$, where Π denotes the parity reversion of W. (The parity reversion ΠW of a superspace W is given by taking the same underlying space, but assigning opposite parity to each element.) Then the map $\pi : W \to \Pi W$ given by the identity mapping is an odd map. A derivation $\widehat{d}$ induces a dual map $\widehat{m} : T(V) \to T(V)$, which is completely determined by the dual $m : T(V) \to V$ of d. We omit the details of this construction, but note that the signs which occur in the formulas below are obtained from this dualization of the map, using the exchange rule. Similarly, d_k and $d_{l,k}$ induce dual maps $m_k : V^k \to V$ and $m_{l,k} : V^{k+l-1} \to V^k$. If $\widehat{d}$ is a differential, these maps satisfy the condition

$$\sum_{k+l=n} m_k m_{l,k} = 0. \tag{15}$$

We can express $m_{l,k}$ in terms of m_l as follows:

$$m_{l,k} = \sum_{i+j=k-1} (-1)^{i(l+1)+m(k-1)} I_i \otimes m_l \otimes I_j,$$

where $m = |\widehat{d}|$ is the parity of m. Notice that for an odd derivation $\widehat{d}$, the associated maps m_l and $m_{l,k}$ are odd for odd l and even for even l, while for $\widehat{d}$ is even the situation is reversed. When $\widehat{d}$ is an odd differential, the condition $\sum_{k+l=n+1} m_k m_{l,k} = 0$ yields the relations

$$\sum_{\substack{k+l=n+1 \\ i+j=k-1}} (-1)^{s_{i,l}} m_k(v_1, \ldots, v_i, m_l(v_{i+1}, \ldots, v_{i+l}), v_{i+l+1}, \ldots, v_n) = 0, \tag{16}$$

where $s_{i,l} = l(v_1 + \cdots + v_i) + i(l+1) + n - l$. This set of relations defines the structure of a strongly homotopy associative algebra, also called an A_∞ algebra, on V. (This structure was introduced by Stasheff [14, 15].) The consideration above shows that in the case when V is finite-dimensional, an A_∞ algebra structure on V can be defined in terms of operators acting on the tensor algebra $T(W)$ of $W = (\Pi V)^*$. Namely, the A_∞ structure on V can be specified by means of an odd derivation $\widehat{d}$ of this algebra satisfying $\widehat{d}^2 = 0$.[1]

For $n = 1$, the relation (16) becomes $m_1^2(v) = 0$, which means that m_1 is a differential on V. For $n = 2$ the relation becomes

$$m_1(m_2(a, b)) - m_2(m_1(a), b) - (-1)^a m_2(a, m_1(b)) = 0 \tag{17}$$

---

[1] When V is not finite-dimensional, one can give a similar definition of an A_∞ algebra by replacing $T(W)$ with its completion, but it is more convenient to dualize the definition by passing from algebras to coalgebras. Namely, one should introduce a coalgebra structure on $\bigoplus_{k=1}^\infty (\Pi V)^k$, and define an A_∞ algebra by means of a codifferential on this coalgebra, i.e., a coderivation with square zero.

which says that m_1 is a derivation on the product on V determined by m_2. For $n = 3$, the relation yields

(18)
$$m_2(m_2(a, b), c) - m_2(a, m_2(b, c)) = m_1(m_3(a, b, c)) + m_3(m_1(a), b, c)$$
$$+ (-1)^a m_3(a, m_1(b), c) + (-1)^{a+b} m_3(a, b, m_3(c))$$

which says that m_2 is an associative product up to homotopy, which explains the name given to such a structure.

§5. Deformations of A_∞ algebras

Now we consider the case of an infinitesimal deformation of the operator $\widehat{d}$. Thus we consider a derivation of the form $\widehat{d} + t\widehat{\delta}$, where t is an infinitesimal parameter whose parity should be opposite to the parity of $\widehat{\delta}$, so that the resulting operator remains odd. This operator is a differential when $(\widehat{d} + t\widehat{\delta})^2 = 0$, which is precisely the condition $\widehat{d}\widehat{\delta} - (-1)^{\widehat{\delta}}\widehat{\delta}\widehat{d} = 0$, or, in other words, $[\widehat{d}, \widehat{\delta}] = 0$, where $[\cdot, \cdot]$ is the superbracket on the superderivation algebra of $T(W)$. We use the fact that the (super) derivations on $T(W)$ are in one-to-one correspondence with $\text{Hom}(W, T(W))$ to introduce a differential on $\text{Hom}(W, T(W))$ by $\widehat{D(\delta)} = [\widehat{d}, \widehat{\delta}]$. It is easy to check that $D^2 = 0$. If we denote $\widehat{\rho} = D(\widehat{\delta})$, then we calculate that $\rho_n = \sum_{k+l=n+1} d_{k,l}\delta_l - (-1)^{\widehat{\delta}}\delta_{l,k}d_l$.

Now if $\widehat{\delta}$ is any derivation on $T(W)$, then by the same construction as we used to associate m to $\widehat{d}$, we associate an element μ of $\text{Hom}(T(V), V)$ to $\widehat{\delta}$. We define the parity of the associated map to be the same as the derivation it is associated to, but as a map we note that m_l has parity l, and more generally, the parity of μ_l is $|\delta| + l + 1$. If $\widehat{\nu}$ is the map associated to $\widehat{\rho}$, then we compute that

(19)
$$\nu_n = \sum_{k+l=n+1} m_l \mu_{k,l} - (-1)^{\widehat{\mu}} \mu_l m_{k,l},$$

and this process defines a differential $D(\mu) = \nu$ on $\text{Hom}(T(V), V)$. We use the same notation for the differential on $\text{Hom}(T(V), V)$ as on $\text{Hom}(W, T(W))$. Thus we can consider the homology groups determined by these differentials. These homology groups coincide in our finite-dimensional case. We say that the homology obtained in this manner is the Hochshild cohomology of the A_∞ algebra V, and denote it by $HH(V)$.

For convenience in the formulas to follow, we make the following sign convention:

(20) $\quad s_{i,l,\mu,n} = (l + \mu + 1)(v_1 + \cdots + v_i) + (l + 1)i + \mu(n - l), \quad (i \geq 1).$

If $\nu = D(\mu)$, then we have

$\nu_n(v_1, \ldots, v_n)$
$$= \sum_{\substack{k+l=n+1 \\ 0 \leq i \leq k-1}} (-1)^{s_{i,l,\mu,n}} m_k(v_1, \ldots, v_i, \mu_l(v_{i+1}, \ldots, v_{i+l}), v_{i+l+1}, \ldots, v_n)$$
$$- \sum_{\substack{k+l=n+1 \\ 0 \leq i \leq l-1}} (-1)^{s_{i,k,m,n}+\mu} \mu_l(v_1, \ldots, v_i, m_k(v_{i+1}, \ldots, v_{i+k}), v_{i+k+1}, \ldots, v_n).$$

The kernel of this differential is the space of all infinitesimal deformations of the A_∞ algebra. If m is the collection of maps $m_k : V^k \to V$, which we call the multiplications in V, and $\mu \in \mathrm{Hom}(T(V), V)$, then μ determines an infinitesimal deformation $m + t\mu$ of m precisely when $D(\mu) = 0$. This interpretation is the same as in the case of an associative algebra. However, when we consider what a trivial deformation is, we note that it no longer is determined by a map $\rho_t : V \to V$, as in the case of associative algebras. Namely, we define a trivial deformation of an A_∞ algebra by means of infinitesimal automorphisms of the tensor algebra $T(W)$, or, equivalently, infinitesimal automorphisms of the cotensor algebra $\bigoplus_{k=1}^\infty (\Pi V)^k$. Such an automorphism has the form $\widehat{\rho}_t = I + t\widehat{\lambda}$, where $\widehat{\lambda}$ is a derivation of $T(W)$, and it is easy to check that the corresponding change of the differential is given by the formula $\widehat{d} \to \widehat{d} + tD(\widehat{\lambda})$. This means that the infinitesimal deformations are classified by the homology $HH(V)$.

§6. Cyclic cohomology of A_∞ algebras

We give a definition of the cyclic cohomology of an A_∞ algebra and prove that the infinitesimal deformations of an A_∞ algebra preserving an invariant inner product are classified by the cyclic cohomology. Suppose that $\widetilde{\mu} \in \mathrm{Hom}(T(V), k)$. Then we define $\widehat{D}(\widetilde{\mu})$ by

$$\widehat{D}(\widetilde{\mu})_{n+1}(v_1, \ldots, v_{n+1}) = \sum_{\substack{k+l=n+1 \\ 0 \le i \le n}} (-1)^{(v_1+\cdots+v_{[i+l]})(v_{[i+l+1]}+\cdots+v_{n+1})+l(n+1)+ni+\mu} \tag{21}$$
$$\times \widetilde{\mu}_l(m_k(v_{[i+l+1]}, \ldots, v_i), v_{[i+1]}, \ldots, v_{[i+l]}).$$

An element $\widetilde{m}$ of $\mathrm{Hom}(T(V), k)$ is said to be cyclic if

$$\widetilde{\mu}_n(v_1, \ldots, v_{n+1}) = (-1)^{n+v_{n+1}(v_1+\cdots+v_n)} \widetilde{\mu}_n(v_{n+1}, v_1, \ldots, v_n). \tag{22}$$

One can check the following fact: *If $\widetilde{\mu}$ is cyclic, then $\widetilde{v} = \widehat{D}(\widetilde{\mu})$ is also cyclic.*

The fact that $\widehat{D}$ is a differential on the cyclic elements of $\mathrm{Hom}(T(V), k)$ will follow from the considerations below. We define the cyclic cohomology $HC(V)$ to be the homology determined by this differential.

Suppose that an A_∞ algebra V is equipped with an inner product $\langle \cdot, \cdot \rangle$, and that $\mu \in \mathrm{Hom}(T(V), k)$ is induced by the maps $\mu_k : V^k \to V$. Then $\langle \cdot, \cdot \rangle$ is said to be invariant with respect to μ if the maps $\widetilde{\mu}_k : V^{k+1} \to k$, given by

$$\widetilde{\mu}_k(v_1 \otimes \cdots \otimes v_{k+1}) = \langle \mu_k(v_1 \otimes \cdots \otimes v_k), v_{k+1} \rangle \tag{23}$$

are cyclic.

The inner product induces a map $\mathrm{Hom}(T(V), V) \to \mathrm{Hom}(T(V), k)$, by associating the map $\widetilde{\mu}$ to μ, where $\widetilde{\mu}_k$ is given by formula (23). When this is an isomorphism, we can use it to define a differential $\widetilde{D}$ on $\mathrm{Hom}(T(V), k)$, by the rule $\widetilde{D}(\widetilde{\mu}) = \widetilde{D(\mu)}$.

It follows that $\widetilde{v} = \widetilde{D}(\widetilde{\mu})$ is given by the formula

$$\widetilde{v}(v_1, \ldots, v_{n+1})$$
$$= \sum_{\substack{k+l=n+1 \\ i \leq k-1}} (-1)^{s_{i,l,\mu,n}} \widetilde{m}_k(v_1, \ldots, v_i, \mu_l(v_{i+1}, \ldots, v_{i+l}), v_{i+k+1}, \ldots, v_{n+1})$$
$$+ \sum_{\substack{k+l=n+1 \\ i \leq l-1}} (-1)^{s_{i,k,m,n}+\mu+1} \widetilde{\mu}_l(v_1, \ldots, v_i, m_k(v_{i+1}, \ldots, v_{i+k}), v_{i+k+1}, \ldots, v_{n+1}).$$

If we assume temporarily that both m and μ are cyclic with respect to the inner product, then we can express the differential as follows:

$$\widetilde{D}(\widetilde{\mu})(v_1, \ldots, v_{n+1})$$
$$= \sum_{\substack{k+l=n+1 \\ 0 \leq i \leq k-1}} (-1)^{(v_1+\cdots+v_{i+l})(v_{i+l+1}+\cdots+v_{n+1})+l(n+1)+ni+\mu}$$
(24)
$$\times \widetilde{\mu}_l(m_k(v_{i+l+1}, \ldots, v_{n+1}, v_1, \ldots, v_i), v_{i+1}, \ldots, v_{i+l})$$
$$+ \sum_{\substack{k+l=n+1 \\ 0 \leq i \leq l-1}} (-1)^{(v_1+\cdots+v_i)(v_{i+1}+\cdots+v_{n+1})+ni+l+\mu}$$
$$\times \widetilde{\mu}_l(m_k(v_{i+1}, \ldots, v_{i+k}), v_{i+k+1}, \ldots, v_{n+1}, v_1, \ldots, v_i).$$

Now we drop the assumption that m is cyclic and define a new operator $\widehat{D}$ on cyclic elements of $\mathrm{Hom}(TV, k)$ by the formula above. This map coincides with the map defined in the beginning of this section. It is straightforward to see that $\widehat{D}$ is a differential.

From the foregoing, we see that deformations of an A_∞ algebra that preserve an invariant inner product are classified by cyclic homology. Trivial deformations are given by infinitesimal automorphisms of the coalgebra associated to V preserving the inner product.

§7. A_∞ deformations of associative algebras

Suppose that V is actually an associative algebra, so that m_2 is the associative product and all other multiplications vanish. Then we can consider the deformations of V into an A_∞ algebra. These deformations are given by the coboundary operator of A_∞ cohomology. Now suppose that μ has only one term μ_k. Comparing the coboundary operator with the Hochschild coboundary operator in the $\mathbb{Z}_2$-graded associative algebra case, one sees that the Hochschild coboundary coincides with the A_∞ coboundary. Therefore, we see that A_∞ deformations of an associative algebra are actually classified by the Hochshild cohomology. In other words, an A_∞ deformation is determined by an element in $\prod_{k=1}^{\infty} HH^k(V)$. Similarly, one sees that deformations of an associative algebra with an (invariant) inner product to an A_∞ algebra with an inner product are given by the cyclic cohomology $\prod_{k=1}^{\infty} HC^k(V)$.

§8. Second order deformations

We show that the cohomology ring $HH(V)$ possesses a natural structure of a Lie (super) algebra. To see this, consider the dual picture again, with $\widehat{d}$ being a derivation of the tensor algebra $T(W)$. We saw that elements of $\operatorname{Hom}(W, T(W))$ correspond to derivations of $T(W)$. Thus we have a bracket on $\operatorname{Hom}(W, T(W))$ given by the bracket of derivations. Recall that the differential is given in terms of this bracket by $D(\delta) = [d, \delta\}$. It is easy to check that the bracket descends to a bracket on the cohomology. In our finite-dimensional picture, this induces a bracket on the cohomology $HH(V)$. In the general situation, this is still true, but one uses the the bracket of coderivations to define the bracket structure. Actually, more is true. It is known (see [11, 10]) that the cohomology group of a differential Lie algebra has the natural structure of an L_∞ algebra (strongly homotopy Lie algebra). Applying this statement to the situation above we can provide $HH(V)$ with the structure of an L_∞ algebra. Similar considerations give a Lie algebra structure (and moreover an L_∞ structure) on the cyclic cohomology $HC(V)$ of an A_∞ algebra V with an invariant inner product. It is important to stress that this structure depends on the choice of the inner product.

Now we consider a second order deformation of $\widehat{d}$. It is given by $\widehat{d}_t = d + t\delta + t^2 \varepsilon$, where we set $t^3 = 0$. We assume here that the parameter t is even, although there exists a more general definition. Then the condition $d_t^2 = 0$ is equivalent to $D(\delta) = 0$ and $[\delta, \delta] = 2D(\varepsilon)$. Thus δ is a cocycle. Denote its image in homology, by $\bar{\delta}$. Then the second condition means that $[\bar{\delta}, \bar{\delta}] = 0$, which is a necessary and sufficient condition for extending the first-order deformation $d + t\delta$ to a second-order one. In other words, $[\bar{\delta}, \bar{\delta}]$ is the first obstruction to extending the infinitesimal deformation to a formal deformation of the algebra.

§9. Ribbon graphs

Let us consider a ribbon graph with each vertex having at least three edges. By definition, a ribbon graph (fatgraph) is a graph together with a fixed cyclic order of the edges at each vertex. Since all graphs we consider will be ribbon graphs, we will omit the word ribbon from now on. We say that a graph is equipped with a metric if a positive number is assigned to each edge. The set σ_Γ of all metrics on the ribbon graph can be identified with $\mathbb{R}_+^k$, where k is the number of edges in Γ. (Here $\mathbb{R}_+^k$ denotes the subset of $\mathbb{R}^k$ consisting of points with positive coordinates.) Therefore, σ_Γ is topologically a cell. There is a standard construction of a closed surface corresponding to a ribbon graph. If the graph is equipped with a metric then the surface can be provided with a complex structure. Let $\mathcal{R}_{g,n}^{\mathrm{met}}$ be the union of the cells σ_Γ, where Γ corresponds to a surface of genus g with n punctures. This set has a natural topology. The limit when the length of one of the edges tends to zero corresponds to the contraction of this edge. It is well known [5, 13] that $\mathcal{R}_{g,n}^{\mathrm{met}}$ is topologically equivalent to $\mathcal{M}_{g,n} \times \mathbb{R}_+^n$, where $\mathcal{M}_{g,n}$ is the moduli space of compact complex curves of genus g with n marked points. The decomposition of $\mathcal{R}_{g,n}^{\mathrm{met}}$ into the cells σ_Γ is not a cell complex; however, one can define the boundary of a cell σ_Γ in the usual manner and thus define the corresponding homology. This homology is closely related to the homology of $\mathcal{M}_{g,n}$.

To define the homology, we need to examine the complex more carefully. First, we notice that there is an obvious method of deciding when two graphs are equivalent, determining the same cell. We also need a notion of an orientation of the cell corresponding to a graph, which is given by choosing a labeling of the edges and an ordering of the holes in the graph. Then two oriented cells are equivalent if there is a graph equivalence between them such that the orientation induced by the mapping agrees with the chosen orientation of the cell.

It is known that $\mathcal{M}_{g,n}$ has a natural complex structure, so that given a choice of the ordering of the holes in the graph, which determines an orientation of $\mathbb{R}_+^n$, one obtains a canonical orientation on the highest dimensional cells in $\mathcal{R}_{g,n}^{met}$. Note that the highest dimensional cells are those corresponding to trivalent graphs.

In [8], there is a general construction that uses an arbitrary A_∞ algebra to associate a function to oriented graphs. Before considering this general construction, we examine a simplified version, which is related to the construction given in [2].

Suppose that V is an associative (super) algebra with a nondegenerate (super) symmetric bilinear form h. Suppose that $\{e_i\}_{i \in I}$ is a basis of V, and the structure constants m_{ij}^k are given by $e_i e_j = m_{ij}^k e_k$. We use the matrix $h_{ij} = h(e_i, e_j)$, and its inverse h^{ij} to raise and lower the indices. The lower structure constants $m_{ijk} = h(e_i e_j, e_k) = m_{ij}^l h_{kl}$ are cyclically (graded) symmetric, so that $m_{ijk} = (-1)^{e_k(e_i+e_j)} m_{kij}$. Similarly, we note that $h^{ij} = (-1)^{e_i e_j} h^{ji}$. We assign a number to a trivalent graph as follows. To each edge in the graph we associate two indices, one for each vertex, and the tensor h^{ij}, where i, j are the indices associated to the edge. To each vertex we assign the tensor m_{ijk}, where i, j, k are the indices associated to the incident edges, and the order is chosen to be consistent with the cyclic order of the edges at the vertex. We multiply all of these symbols together and sums over repeating indices to obtain a number $Z(\Gamma)$.[2]

The resulting function $Z(\Gamma)$ depends only on the genus g and the number n of holes in the graph. This is easy to see, since every trivalent graph with the same number of holes and the same genus can be obtained from one such graph by performing a series of simple transformations, called the fusion move which is illustrated below.

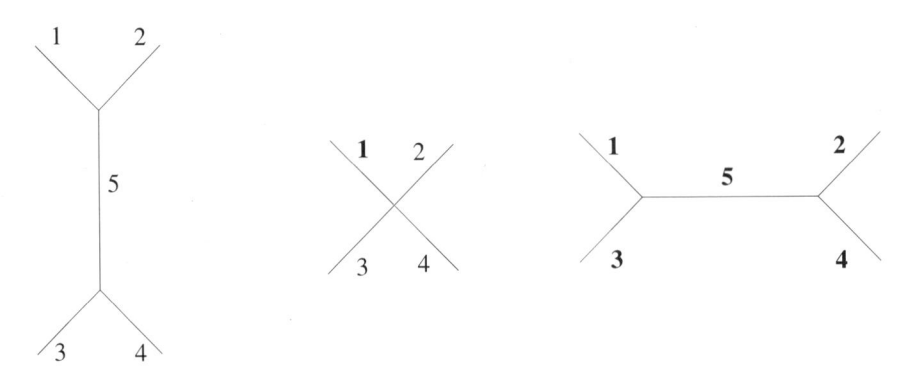

FIGURE 1. Fusion diagram.

---

[2]In the graded case, one also picks up a sign in each term depending on the parity of the basis elements, since what we are really doing here is performing a series of graded contractions of a tensor according to the prescription which is dictated by the graph.

It is easy to calculate how the functions of two graphs differing by a fusion move are related, and the associativity of the algebra guarantees that the functions will have the same value. Thus if V is equipped with the structure of an associative algebra, we can use the structure constants $m_{ijk} = \widetilde{m}_2(e_i, e_j, e_k)$ to obtain a function that depends only on the genus and number of holes in the graph. We note that in writing down the function $Z(\Gamma)$, an order of the vertices and an order of the edges must be chosen, but the resulting function is independent of these orders because the tensors m_{ijk} and h^{ij} are even, so that the corresponding contractions are the same. We also see that the form of the function $Z(\Gamma)$ depends on the starting edge at each vertex, which determines the order in which the indices for m_{ijk} are listed, but the cyclic symmetry of this tensor again ensures that the function is independent of this choice. Finally, the same arguments show that the function is independent of the order in which the tensor h^{ij} is presented, due to the symmetry of the inner product.

We would like to construct homology classes of the complex $\mathcal{R}_{g,n}^{\mathrm{met}}$. With this goal in mind, we construct chains in this complex, that is, linear combinations of oriented cells. Since a change of the orientation corresponds to a change of sign in the complex, we want to define a function on oriented cells that changes sign when the orientation of the cell is reversed. For simplicity, we identify cells with graphs, and therefore replace oriented cells with oriented graphs. Note that this is not the usual notion of an oriented graph. In the discussion above, the function $Z(\Gamma)$ was defined on the set of trivalent graphs. What we really want is to define a function on the set of oriented graphs Γ_{or}, which we shall denote $Z(\Gamma_{\mathrm{or}})$. For an oriented trivalent graph Γ_{or}, we define $Z(\Gamma_{\mathrm{or}}) = Z(\Gamma)$ if the orientation is the canonical one, and $Z(\Gamma_{\mathrm{or}}) = -Z(\Gamma)$ otherwise. It is easy to see that the element

$$(25) \qquad Z_{\max} = \sum Z(\Gamma_{\mathrm{or}}) \Gamma_{\mathrm{or}},$$

where the sum is taken over all oriented trivalent graphs, is a cycle in the homology of $\mathcal{R}_{g,n}^{\mathrm{met}}$ discussed above, since $Z(\Gamma_{\mathrm{or}})$ is constant on all graphs with the canonical orientation. This result is just the assertion that the cell complex is orientable. Of course, the orientation in this simple case is somewhat superfluous, but in the more general construction to follow, the orientation is quite relevant.

Now suppose that φ_k is a cyclic k-cocycle on the algebra V. We construct a cycle of dimension $k-2$ less than the maximal dimension on $\mathcal{R}_{g,n}^{\mathrm{met}}$. Consider graphs in which all vertices except one are trivalent, and the exceptional vertex has $k+1$ incident edges. As before, we assign to each trivalent vertex the tensor m_{ijk}, and to the exceptional edge we assign the tensor $\varphi_{i_1 \cdots i_{k+1}} = \widetilde{\varphi}(e_{i_1}, \ldots, e_{i_{k+1}})$. One constructs $Z(\Gamma)$ in the same manner as before, but now there is a problem in the definition if k is odd, because in this case the cyclicity of $\widetilde{\varphi}_k$ means that the formula for $Z(\Gamma)$ depends on which edge one starts with. Of course, the function is determined up to the total sign, since in the expression

$$(26) \qquad \varphi_{i_1 \cdots i_{k+1}} = (-1)^{k + e_{i_{k+1}}(e_{i_1} + \cdots + e_{i_k})} \varphi_{i_{k+1} i_1 \cdots i_k}$$

the sign $(-1)^{e_{i_{k+1}}(e_{i_1} + \cdots + e_{i_k})}$ cancels because this is a graded contraction. Thus only the $(-1)^k$ plays a role. Therefore, in order to assign a fixed value we must consider a starting edge for the vertex.

To explain this, we introduce the notion of a ciliated graph Γ_{cil}, which is a ribbon graph with a preferred edge chosen for each vertex. This terminology was suggested by Fock and Rosly, see [1]. Given a ciliated graph with at most one nontrivalent edge, one obtains a well-defined formula for $Z(\Gamma_{cil})$ by using the preferred edge to determine the order in which to write down the terms for the vertices. Of course, for the trivalent vertices, the choice of the cilia does not affect the outcome, but for the exceptional vertex the choice is relevant if k is odd. Now we also note that in this case the order of the vertices is not important, because there is at most one vertex corresponding to an odd tensor, so this case is still independent of the order.

We want to define a partition function on oriented graphs that takes opposite signs on graphs of opposite parity. To do this we also use the canonical orientations of trivalent graphs as follows. For the exceptional vertex, we can expand the vertex by inserting edges, using the ciliation as a starting point, to obtain a trivalent graph, as illustrated in the figure below.

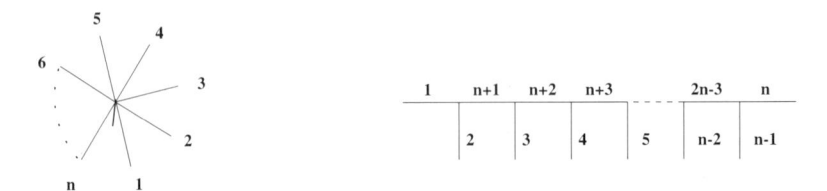

FIGURE 2. Expanding a ciliated vertex.

An orientation of the original graph induces an orientation on the trivalent graph, which we can compare to the canonical orientation. If the number of incident edges to the vertex is even, the comparison will depend upon where the cilia is placed, and will alternate as one moves the cilia one edge at a time. On the other hand, if the number of edges is odd, then the orientation of the expanded graph will not depend on the placement of the cilia. Thus we can define the function $Z(\Gamma_{or}) = \pm Z(\Gamma_{cil})$, where the sign is plus if the expanded graph associated to Γ_{or} with the given ciliation has the canonical orientation. Note that result does not depend on the ciliation. One can prove that the element

$$(27) \qquad Z = \sum Z(\Gamma_{or})\Gamma_{or},$$

where the sum runs over all distinct[3] oriented graphs with exactly one nontrivalent vertex with $k+1$ incident edges, is a cycle. We omit a direct proof of this statement. Instead we will derive this result from the fact that a cyclic cocycle determines an infinitesimal deformation of an associative algebra with an invariant inner product into an A_∞ algebra with an invariant inner product. Applying a general result of Kontsevich which we shall explore below, one sees immediately that Z is a cycle.

Now let us consider the case when V is an A_∞ algebra, and graphs with only the restriction that each vertex has at least three edges. We can repeat the construction of the function $Z(\Gamma)$ as before, associating to each vertex with $k+1$ edges the tensor

---

[3]Note that in this construction a graph with opposite orientation is the negative of the oriented graph, so is not considered as distinct in our consideration.

$\widetilde{m}_{i_1\cdots i_{k+1}}$. The result will depend on the order in which the tensors corresponding to the vertices are listed, more precisely, the order in which the vertices with an even number of incident edges are listed. However, we can resolve this in a similar way by considering graphs with a fixed ciliation and a fixed order of vertices, for which the partition function is well defined, and then multiplying by a sign that depends on the order, ciliation, and orientation, in such a manner that the resulting partition function depends on the orientation alone. To do this, we expand those vertices that are not trivalent, as we did before, but now we must also keep track of the order of the vertices when adding the new labels, so that we not only get an orientation of the graph, but an order of the new holes as well. The canonical orientation is determined by the canonical ordering of the holes as well, so that interchanging the order of the expansion for two vertices will actually produce a reversal of the sign precisely when both vertices are even. But this reversal of sign is mirrored in the contraction as well, so the effect cancels. The ciliation effect is treated as before, so that again we are able to define a function $\mathbb{Z}(\Gamma_{\text{or}})$ depending only on the orientation of the graph.

A result formulated by Kontsevich in this case is that the chain

$$(28) \qquad Z = \sum Z(\Gamma_{\text{or}})\sigma_{\Gamma_{\text{or}}}$$

is a cycle. To see why this is true, note that it is sufficient to restrict the sum to graphs with a fixed number of edges, and to show that this partial sum is a cycle. Taking a graph with one edge less than this fixed number, we consider the ways to expand this graph by adding one edge. We can consider the graphs obtained by expanding each nontrivalent vertex separately. Restricting attention to those obtained by expanding a single vertex by inserting an edge in one of the possible manners, we obtain expressions for the function that differ only in the contribution from the vertex. If the vertex has n edges, then we will obtain expressions of the form $m_{i_1\ldots i_l} m_{i_{l+1}\ldots i_{n+2}}$, multiplied by a sign that is determined by the expansion into a trivalent graph. We will show that the sign coincides with the signs given in the expressions for the relations satisfied by the A_∞ algebra.

The defining relations for an A_∞ algebra with an invariant inner product can be stated in the form $\widehat{D}(\widetilde{m}) = 0$, from which, using equation (21), we obtain

$$(29) \qquad \sum_{\substack{k+l=n+1 \\ 0 \le i \le n}} (-1)^{(v_1+\cdots+v_{[i+1]})(v_{[i+1]}+\cdots+v_{n+1})+l(n+1)+ni+\mu}$$

$$\times \widetilde{m}_l(m_k(v_{[i+l+1]},\ldots,v_i), v_{[i+1]},\ldots,v_{[i+l]}) = 0.$$

These relations yield the expression

$$(30) \qquad \sum_{\substack{k+l=n+1 \\ 0 \le i \le n}} -1^s m_{a, j_{[i+1]},\ldots, j_{[i+l]}} m_{j_{[i+l+1]},\ldots, j_i, b} = 0$$

where

$$s = (e_1 + \cdots + e_{[i+l]})(e_{[i+l+1]} + \cdots + e_{n+1}) + ni + l + 1.$$

We use these relations to explain the Kontsevich result. We shall concern ourselves here with the factor $ni + l + 1$ which appears in the exponent, as the other part of this exponent is cancelled because this is a graded contraction.

Let us consider a fixed graph, and consider all possible ways to expand this graph into a graph with one additional edge inserted at a fixed vertex. Suppose that this fixed vertex has $n + 1$ edges, where of course, $n \geq 3$. The insertion of an edge will create a graph with two new vertices, having $l + 1$ and $k + 1$ edges, where $k + l = n + 1$. Let us suppose that the original vertex had incident edges $1, \ldots, n + 1$ with associated labels $j_1, \ldots, j_{n+1}$. We insert an edge in such a manner that the new graph will will have a vertex with labels $a, j_{[i+1]}, \ldots, j_{[i+l]}$, and one with labels $j_{[i+l+1]}, \ldots, j_i, b$, where a and b are the indices attached to the new edge, which has label $n + 2$. The figure below illustrates the labeling of the expanded graph.

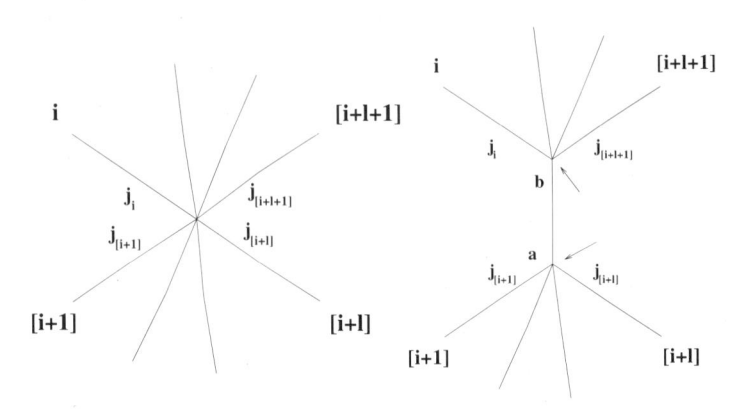

FIGURE 3. Splitting a vertex.

Inserting cilia between b and $j_{[i+l+1]}$, and $j_{[i+l]}$ and a respectively, the contribution of the two vertices to the function $Z(\Gamma_{\text{cil}})$ is $m_{a, j_{[i+1]}, \ldots, j_{[i+l]}} m_{j_{[i+l+1]}, \ldots, j_{[i]}, b}$. When expanding the graph according to this ciliation, one sees that the expanded graph is identical to the one which would be obtained from the original graph, with a cilia placed between $j_{[i+l]}$ and $j_{[i+l+1]}$. However, the orientation of the expanded graph is affected by the fact that the inserted edge has been labeled first, instead of in the order in which it would have been labeled by the usual expansion. In case the original vertex had an odd number of edges, this is the only factor affecting the orientation, because the placement of the cilia in the original vertex is irrelevant. Since n is even, the factor ni does not contribute in the sign $ni + l + 1$. Thus one picks up a factor which depends only on l. On the other hand, when the original vertex had an even number of edges, the factor ni does contribute, which reflects the fact that the placement of the cilia is relevant to the sign of the function $Z(\Gamma_{\text{cil}})$. When we sum over all graphs which result from the insertion of an edge at the fixed vertex, these considerations and equation (30) show that the sum is zero, and this is precisely the incidence number of the graph in the boundary of the chain Z. This shows that this chain is a cycle.

We have tacitly assumed that the algebra in question does not have any product of degree one, as the terms involving such a product are not accounted for. However, there are some cases when one can include such a multiplication. We shall not go into this matter here.

The construction of a cycle in $\mathcal{R}_{g,n}^{\mathrm{met}}$ by a cyclic cocycle of an associative algebra can be considered as a limiting case of Kontsevich's construction. We can consider this as an infinitesimal deformation of the algebra into an A_∞ algebra. On the other hand, the proof can be carried out directly because the signs in equation (12) coincide with those in equation (30). We can also apply second-order deformations in the context above to construct other interesting examples of cocycles.

We wish to thank Dmitry Fuchs and Jim Stasheff, who read this manuscript and made helpful suggestions. We also would like to express our gratitude to Ezra Getzler and Maxim Kontsevich for useful conversations. A. Schwarz is grateful to the MSRI and the I. Newton Institute for Mathematical Sciences for their hospitality.

References

1. V. Fock and A. Rosly, *Poisson structure on moduli of flat connections on riemann surfaces*, Preprint ITEP-92-72, ITEP, Moscow, 1992.
2. M. Fukuma, S. Hosono, and H. Kawai, *Lattice topological field theory in two dimensions*, Preprint, 1992.
3. M. Gerstenhaber, *The cohomology structure of an associative ring*, Ann. of Math. **78** (1963), 267–288.
4. E. Getzler and J. D. S. Jones, *A_∞-algebras and the cyclic bar complex*, Illinois J. Math. **34** (1990), 256–283.
5. J. Harer, *The cohomology of the moduli space of curves*, Lecture Notes in Math., vol. 1337, Springer-Verlag, Berlin, Heidelberg, and New York, 1980, pp. 138-221.
6. G. Hochschild, *On the cohomology groups of an associative algebra*, Ann. of Math. **46** (1945), 58–67.
7. D. Kastler, *Cyclic cohomology within the differential envelope*, Hermann, Paris, 1988.
8. M. Kontsevich, *Feynman diagrams and low dimensional topology*, Preprint, 1991.
9. _____, *Intersection theory on the moduli space of curves and the matrix airy function*, Comm. Math. Phys. **147** (1992), 1–23.
10. T. Lada and M. Markl, *Strongly homotopy lie algebras*, Preprint hep-th 9406095, 1994.
11. T. Lada and J. Stasheff, *Introduction to sh lie algebras for physicists*, Preprint hep-th 9209099, 1990.
12. J. Loday, *Cyclic homology*, Springer-Verlag, Berlin, Heidelberg, and New York, 1992.
13. R. Penner, *The decorated Teichmüller space of punctured surfaces*, Comm. Math. Phys. **113** (1987), 299–339.
14. P. Seibt, *Cyclic homology of algebras*, World Scientific, Singapore, 1987.
15. J. D. Stasheff, *On the homotopy associativity of H-spaces.* I, Trans. Amer. Math. Soc. **108** (1963), 275–292.
16. _____, *On the homotopy associativity of H-spaces.* II, Trans. Amer. Math. Soc. **108** (1963), 293–312.

DEPARTMENT OF MATHEMATICS, UNIVERSITY OF CALIFORNIA, DAVIS, CA 95616
E-mail address: michae@math.ucdavis.edu, schwarz@math.ucdavis.edu

Amer. Math. Soc. Transl.
(2) Vol. **169**, 1995

On Hamburger's Theorem

I. PIATETSKI-SHAPIRO AND RAVI RAGHUNATHAN

To Eugene Dynkin: On the occasion of his seventieth birthday

1. Introduction. In 1919 Hamburger formulated and proved the following celebrated theorem:

Let $f(s) = \sum_{n=1}^{\infty} \frac{a_n}{n^s}$ *be absolutely convergent for* $\mathrm{Re}(s) = \sigma > 1$ *and suppose* $f(s) = G(s)/P(s)$, *where* $G(s)$ *is an entire function of finite growth and* $P(s)$ *is a polynomial. Suppose further that*

$$\text{(i)} \qquad \Gamma\left(\frac{s}{2}\right)\pi^{-s/2}f(s) = \Gamma\left(\frac{1-s}{2}\right)\pi^{-\frac{(1-s)}{2}}g(1-s)$$

where $g(1-s) = \sum_{n=1}^{\infty} \frac{b_n}{n^{1-s}}$, *the series being convergent for* $\sigma < -\alpha < 0$. *Then* $f(s) = c\zeta(s)$, *where* c *is a constant.*

The purpose of Theorem 1 of this paper is to give a simpler proof of the results of Hamburger which appeared in three papers in *Mathematische Zeitschrift* between 1919 and 1921 [H1, H2, H3]. We give a short outline of the proof of Theorem 1 in the simplest case – which is the result of Hamburger stated above.

Outline of the proof. We first observe that each side of the equation (i) can be expressed as a Mellin transform. Consider the functions $\Phi(x) = \sum_{n=1}^{\infty} a_n\varphi(nx)$ and $\tilde{\Phi}(x) = \sum_{n=1}^{\infty} b_n\hat{\varphi}(nx)$; here φ is any Schwartz function with the property $\hat{\varphi}(0) = 0$ and $\hat{\varphi}(x)$ is defined by

$$\hat{\varphi}(x) = \int_{\mathbb{R}} \varphi(y)e^{-2\pi ixy}\,dy.$$

If we denote by $M\Phi(s)$ (resp. $M\tilde{\Phi}(s)$) the Mellin transform of $\Phi(x)$ (resp. $\tilde{\Phi}(x)$), then a simple computation shows that

$$M\Phi(s) = f(s)L(\varphi, s) \text{ and } M\tilde{\Phi}(1-s) = g(1-s)L(\hat{\varphi}, 1-s)$$

1991 *Mathematics Subject Classification.* Primary 11M41; Secondary 42A38.

©1995, American Mathematical Society

where $L(\varphi, s) = \int_0^\infty \varphi(x)|x|^s d^*x$ and $L(\hat{\varphi}, 1-s) = \int_0^\infty \hat{\varphi}(x)|x|^{1-s} d^*x$. Using Tate's local functional equation, (i) gives us

(ii) $$M\Phi(s) = M\tilde{\Phi}(1-s).$$

The left-hand side of (ii) is a Mellin transform at s and the right-hand side a Mellin Transform at $1-s$. However, we can use the following analytic trick. We integrate both sides along the line $\text{Re}(s) = \sigma + 1$. Then by the Phragmen-Lindelof Principle we will be able to shift the line of integration of the right-hand side to $\text{Re}(s) = \sigma$. After that we can take the inverse Mellin transform to get $\Phi(x) = \tilde{\Phi}(x)$ or, putting $x = 1$,

(iii) $$\sum_{n=1}^\infty a_n \varphi(n) = \sum_{n=1}^\infty b_n \hat{\varphi}(n).$$

This equality holds for any $\varphi \in w$, where w is a finite codimension subspace of the Schwartz space. Then we prove a lemma showing that $a_n = b_n = c$ under the above circumstances. This is actually a statement of "uniqueness" of the Poisson summation formula. This will complete the proof since

$$a_n = b_n = c \Rightarrow f(s) = \sum_{n=1}^\infty \frac{c}{n^s} = c\zeta(s)$$

and

$$g(1-s) = \sum_{n=1}^\infty \frac{c}{n^{1-s}} = c\zeta(1-s).$$

The techniques described above are used to prove the more general Theorem 1 with a slight weakening of the assumptions.

To prove his theorem Hamburger had to assume the convergence of the Dirichlet series $L(s)$ for $\text{Re}(s) > 1$. In 1922 Siegel established that it is sufficient to assume the convergence of $L(s)$ for $\text{Re}(s) > 2 - \delta$ for some $\delta > 0$ [S]. We note that the proof given in this paper requires the convergence of $L(s)$ only in some right half plane (this is also true for Ehrenpreis' and Kawai's proof). Ehrenpreis and Kawai [E-K] treat the case of Hamburger's theorem for the Riemann zeta-function over $\mathbb{Q}$ and $\mathbb{Q}(\sqrt{-1})$.

Theorem 2 gives a complete characterization of Dirichlet series with multiplicative coefficients and satisfying a "Hamburger type" functional equation with one gamma factor. Theorem 2 has been proved by Conrey and Ghosh [C-G] under stronger assumptions; in particular, they assume a more restrictive kind of functional equation for the Dirichlet series $L(s)$. They also assume that $L(s)$ has just one pole of order m at $s = 1$ and that $L(s)$ is Ramanujan, i.e., $a_n \ll n^\varepsilon$ for all $\varepsilon > 0$. We are able to prove Theorem 2 assuming a more general form of the functional equation, an arbitrary location of a finite number of poles of any order, and convergence of $L(s)$ in some right half plane. The method of proof is also substantially different. After assuming multiplicativity of the coefficients of $L(s)$ we get a converse theorem for $GL_1(\mathbb{Q})$, that is, we are able to characterize all such series essentially as series arising out of primitive Dirichlet characters.

We note that the methods of proof in this paper are in keeping with the spirit of Tate's thesis and the modern theory of automorphic forms. The key ideas are the equivalence of the "uniqueness" of the Poisson summation formula and the functional equation, and the use of Tate's local functional equation to obtain the more general form of Tate's functional equation for any Schwartz-Bruhat function. In the context of the theory of automorphic forms this "uniqueness" corresponds to a multiplicity one theorem for the three-dimensional Heisenberg group. The "uniqueness" of the Poisson summation formula was first observed and used by Gurevic [Gu] and by Ehrenpreis and Kawai in their proof of Hamburger's theorem.

2. Let $L(s) = \sum_{n=1}^{\infty} \frac{a_n}{n^s}$ be a Dirichlet series converging uniformly and absolutely in some right half plane, i.e., for all $s \in \mathbb{C}$ such that $\text{Re}(s) > c_1$, $c_1 > 0$. Suppose further that

(i) $L(s)$ is holomorphic for $s \in \mathbb{C}$ except for a finite number of poles.

(ii) $L(s)$ satisfies one of the following functional equations

$$\pi^{-s/2}\Gamma\left(\frac{s}{2}\right)L(s) = \varepsilon(s)\pi^{-\left(\frac{1-s}{2}\right)}\Gamma\left(\frac{1-s}{2}\right)D(1-s), \tag{1}$$

$$\pi^{-s/2}\Gamma\left(\frac{s+1}{2}\right)L(s) = \varepsilon(s)\pi^{-\left(\frac{1-s}{2}\right)}\Gamma\left(\frac{2-s}{2}\right)D(1-s), \tag{2}$$

where $\varepsilon(s) = k^{-s}$, $k > 0$, $k \in \mathbb{R}$.

Here $\Gamma(s)$ is the usual gamma function and $D(s) = \sum_{n=1}^{\infty} \frac{b_n}{n^s}$ is an arbitrary Dirichlet series converging uniformly and absolutely for all s such that $\text{Re}(s) > c_2 > 0$.

(iii) $L(s)$ has finite growth on lines parallel to the imaginary axis in the complex plane for $|\text{Re}(s)| \leq c$, where $c = \max\{c_1, c_2\} + 1$, i.e., for $-c \leq \text{Re}(s) \leq c$ we have

$$|L(\sigma + it)| \leq Ke^{|t|^p}, \quad K > 0.$$

Under the above assumptions we have the following theorem.

THEOREM 1. (i) *If* $k \notin \mathbb{Z}^+$ *then* (1) *and* (2) *have no nontrivial solutions.*

(ii) *If* $k \in \mathbb{Z}^+$ *let* n^+ (*resp.* n^-) *be the dimension of the space of solutions of equation* (1) (*resp.* (2)). *Then*

$$n^+ + n^- = k.$$

(iii) *If* $L(s)$ *satisfies* (1), *then* $L(s)$ *has at most two poles at the points* $s = 0$ *or* $s = 1$. *If* $L(s)$ *satisfies* (2), *then it is an entire function.*

REMARKS. 1. Since $D(s)$ converges in some right half plane, $D(1-s)$ converges in some left half plane.

2. In Theorem 1 (ii), solutions of equations (1) and (2) will be shown to be precisely the Dirichlet Series "of period k", i.e., those for which $a_{n+k} = a_n$.

3. The number $k \in \mathbb{R}$ that appears in (1) (resp. (2)) is called the conductor of the functional equation (1) (resp. (2)).

PROOF OF THEOREM 1. Consider the functions

$$\Phi(x) = \sum_{n=1}^{\infty} a_n \varphi(nx),$$

$$\tilde{\Phi}(x) = \sum_{n=1}^{\infty} b_n \hat{\varphi}(nx),$$

where $\varphi \in \mathcal{S}(\mathbb{R})$ is an arbitrary Schwartz function on $\mathbb{R}$ and

$$\hat{\varphi}(x) = \int_{-\infty}^{\infty} \varphi(y) e^{-2\pi i x y} dy.$$

We denote by Mf the Mellin transform of a function f. We note that if φ is chosen in $\mathcal{S}_0 \subset \mathcal{S}(\mathbb{R})$, where $\mathcal{S}_0 = \{\varphi \in \mathcal{S}(\mathbb{R}) | \varphi(0) = 0 = \hat{\varphi}(0)\}$, then $\sum_{n=1}^{\infty} a_n \varphi(nx) |x|^s$ and $\sum_{n=1}^{\infty} b_n \hat{\varphi}(nx) |x|^{1-s}$ converge uniformly for all $x \in \mathbb{R}$ for $\operatorname{Re}(s) > \max\{c_1, c_2\}$. So for $\operatorname{Re}(s) \geq c_1$ we have

$$M\Phi(s) = \int_{-\infty}^{\infty} \Phi(x) |x|^s d^*x = \int_{-\infty}^{\infty} \sum_{n=1}^{\infty} a_n \varphi(nx) |x|^s d^*x$$

$$= \sum_{n=1}^{\infty} a_n \int_{-\infty}^{\infty} \varphi(nx) |x|^s d^*x \quad \text{for } \operatorname{Re}(s) > c_1.$$

Putting $y = nx$ we obtain for $\operatorname{Re}(s) > c_1$

(3) $$M\Phi(s) = \sum_{n=1}^{\infty} \frac{a_n}{n^s} \int_{-\infty}^{\infty} \varphi(x) |x|^s d^*x = L(s) L(\varphi, s),$$

where $L(\varphi, s) = \int_{-\infty}^{\infty} \varphi(x) |x|^s d^*x$.

Similarly, for $\operatorname{Re}(s) \geq c_2$, we have

(4) $$M\tilde{\Phi}(1-s) = D(1-s) L(\hat{\varphi}, 1-s).$$

We recall that from the local functional equation of Tate's thesis we know that there exist entire functions $E(\varphi, s)$ and $E'(\varphi, s)$ such that

(5) $L(\varphi, s) = E(\varphi, s) \pi^{-s/2} \Gamma\left(\frac{s}{2}\right)$, $L(\hat{\varphi}, 1-s) = E(\varphi, s) \pi^{-(\frac{1-s}{2})} \Gamma\left(\frac{1-s}{2}\right)$

and

(6) $$L(\varphi, s) = E'(\varphi, s) \pi^{-s/2} \Gamma\left(\frac{s+1}{2}\right),$$
$$L(\hat{\varphi}, 1-s) = E'(\varphi, s) \pi^{-(\frac{1-s}{2})} \Gamma\left(\frac{2-s}{2}\right).$$

Thus equations (3) and (4) yield

(7)
$$M\Phi(s) = \pi^{-s/2}E(\varphi, s)\Gamma\left(\frac{s}{2}\right)L(s),$$
$$M\tilde{\Phi}(1-s) = \pi^{-(\frac{1-s}{2})}E(\varphi, s)\Gamma\left(\frac{1-s}{2}\right)D(1-s).$$

when (5) is used, while (6) gives

(8)
$$M\Phi(s) = E'(\varphi, s)\Gamma\left(\frac{s+1}{2}\right)L(s)$$
$$M\tilde{\Phi}(1-s) = E'(\varphi, s)\Gamma\left(\frac{2-s}{2}\right)D(1-s)$$

In either of the above cases the functional equations (1) and (2) give

(9)
$$M\Phi(s) = \varepsilon(s)M\tilde{\Phi}(1-s).$$

CLAIM. $\varphi \in \mathcal{S}_0(\mathbb{R})$ can be chosen so that $M\Phi(s)$ is entire.

PROOF OF CLAIM. The proof follows from the lemma.

LEMMA. Let $n_1, \ldots, n_k \in \mathbb{N}$ and $a_1, \ldots, a_k \in \mathbb{C}$. If $W \subset \mathcal{S}(\mathbb{R})$ is an infinite dimensional subspace, we can find $\varphi \in W$ such that $M\varphi(s)$ has zeros of order at least n_i at a_i for all $1 \leq i \leq k$.

PROOF OF LEMMA. Consider the collection of linear functionals

$$T_{ij} : W \to \mathbb{C},$$
$$T_{ij}(\varphi) = \frac{d^i}{ds^i}M\varphi(a_j), \qquad 0 \leq i \leq n_j, \ 1 \leq j \leq k.$$

This is a finite collection of functional, each with a kernel of codimension 1 in W. Since dim $W = \infty$, $\bigcap_{j=1}^{k}\bigcap_{i=0}^{n_j} \ker(T_{ij}) \neq \varnothing$.

Hence there exists $\varphi \in W$ such that $T_{ij}(\varphi) = 0$ for all $0 \leq i \leq n_j, 1 \leq j \leq k$, i.e., there exists $\varphi \in W$ such that $M\varphi(s)$ has zeros of order at least n_j at a_j for all $1 \leq i \leq k$. This proves the lemma.

Now equation (3) gives us

$$M\Phi(s) = L(s)L(\varphi, s).$$

By assumption, $L(s)$ has only finitely many poles. Suppose $L(s)$ has poles of order n_i at the points $s = a_i$ for $1 \leq i \leq k$. Then, by the preceding lemma, we can find $\varphi \in \mathcal{S}_0(\mathbb{R})$ such that $L(\varphi, s)$ has zeros of order n_i at the points $s = a_i$, $1 \leq i \leq k$. Thus the function $L(s)L(\varphi, s)$ is entire, i.e., $M\Phi(s)$ is entire. This proves the claim.

Now we rewrite equation (9) in the form

(10) $$k^s M\Phi(s) = M\tilde{\Phi}(1-s).$$

We assume from now on that $\varphi \in \mathcal{S}(\mathbb{R})$ has been chosen so that the left-hand side of (10) (and hence the right-hand side) is entire. The left-hand side of (10) can be written as

$$k^s \int_{-\infty}^{\infty} \Phi(x)|x|^s d^*x = \int_{-\infty}^{\infty} \Phi(x)|kx|^s d^*x = \int_{-\infty}^{\infty} \Phi\left(\frac{x}{k}\right) |x|^s d^*x$$

(after a change of variables) and the right-hand side becomes

$$\int_{-\infty}^{\infty} \tilde{\Phi}(x)|x|^{(1-s)} d^*x.$$

Now we integrate both sides of equation (10) along the line $\mathrm{Re}(s) = c$ to get

$$\int_{-\infty}^{\infty}\int_{-\infty}^{\infty} \Phi\left(\frac{x}{k}\right) |x|^{c+it} d^*x \, dt = \int_{-\infty}^{\infty}\int_{-\infty}^{\infty} \tilde{\Phi}(x)|x|^{1-c-it} d^*x \, dt.$$

Next, we have $k^s M\Phi(s) = k^s L(s) L(\varphi, s)$. All the factors k^s, $L(s)$ (by assumption) and $L(\varphi, s)$ (since $\varphi \in \mathcal{S}(\mathbb{R})$) have finite growth on the line $\mathrm{Re}(s) = c$. Hence we may apply the Phragmen-Lindelof principle to the integral on the right and shift the line of integration to the line $\mathrm{Re}(s) = 1 - c$. This gives

(11) $$\int_{-\infty}^{\infty}\int_{-\infty}^{\infty} \Phi\left(\frac{x}{k}\right) |x|^{c+it} d^*x \, dt = \int_{-\infty}^{\infty}\int_{-\infty}^{\infty} \tilde{\Phi}(x)|x|^{c-it} d^*x \, dt.$$

Taking inverse Mellin transforms of both sides gives

$$\sum_{n=1}^{\infty} a_n \varphi\left(\frac{nx}{k}\right) = \sum_{n=1}^{\infty} b_n \hat{\varphi}(nx^{-1})$$

and putting $x = 1$ gives

(12) $$\sum_{n=1}^{\infty} a_n \varphi\left(\frac{n}{k}\right) = \sum_{n=1}^{\infty} b_n \hat{\varphi}(n).$$

We note that the right-hand side of equation (12) is invariant under translation by $\mathbb{Z}$, since

$$\sum_{n=1}^{\infty} b_n \hat{\varphi}_\ell(n) = \sum_{n=1}^{\infty} b_n e^{2\pi i \ell n} \hat{\varphi}(n) = \sum b_n \hat{\varphi}(n), \text{ for all } \varphi \in \mathcal{S}(\mathbb{R}),$$

where $\varphi_\ell(x) = \varphi(x + \ell)$. Hence $\sum_{n=1}^{\infty} a_n \varphi\left(\frac{n}{k}\right)$ is also invariant under translation by $\mathbb{Z}$. Choose φ such that $\varphi\left(\frac{n}{k}\right) = 1$ and $\varphi\left(\frac{m}{k}\right) = 0$ for $m \neq n$. By the translational invariance, we have

$$a_n = \sum_{n=1}^{\infty} a_n \varphi\left(\frac{n}{k}\right) = \sum_{n=1}^{\infty} a_n \varphi\left(\frac{n}{k} - 1\right) = \sum_{n=1}^{\infty} a_n \varphi\left(\frac{n-k}{k}\right).$$

If $k \notin \mathbb{Z}^+$, we can choose φ to get $a_n = 0$. If $k \in \mathbb{Z}^+$, then

$$a_n = \sum_{n=1}^{\infty} a_n \varphi\left(\frac{n-k}{k}\right) = a_{n+k}.$$

Since the space of functions φ satisfying the above condition and belonging to $\mathcal{S}_0(\mathbb{R})$ is infinite dimensional, the lemma shows that we can choose φ such that $M\Phi(s)$ is an entire function. This proves part (i) of the theorem. Note that this also shows that the sum of the dimensions of the spaces of solutions to (1) and (2) does not exceed k, since $a_{n+k} = a_n$.

3. Now we prove the second part of Theorem 1.

PROPOSITION 1. *If $L(s)$ satisfies the functional equation* (1) *(resp.* (2)*) with two different conductors k_1 and k_2 where $(k_1, k_2) = 1$, then $L(s)$ satisfies* (1) *(resp.* (2)*) with conductor* 1.

COROLLARY. *If $(k_1, k_2) = d$, then $L(s)$ satisfies* (1) *(resp.* (2)*) with conductor d.*

PROOF OF PROPOSITION 1. Assume without loss of generality that $L(s)$ satisfies equation (1) with two different conductors k_1 and k_2, $(k_1, k_2) = 1$. So

(13) $$\pi^{-s/2} \Gamma\left(\frac{s}{2}\right) L(s) = k_1^{-s} \pi^{-\left(\frac{1-s}{2}\right)} \Gamma\left(\frac{1-s}{2}\right) \sum_{n=1}^{\infty} \frac{b_{n,1} k_2 k_2^{-s}}{(nk_2)^{1-s}}$$

for some series $\sum_{n=1}^{\infty} \frac{b_{n,1}}{n^{1-s}} = D_1(1-s)$. Setting $D_1'(1-s) = \sum_{n=1}^{\infty} \frac{b_{n,1} k_2}{(nk_2)^{1-s}}$, we see that $L(s)$ satisfies equation (1) for the conductor $k_1 k_2$ and we also note that the indices of all nonzero coefficients of $D_1'(1-s)$ are divisible by k_2. Similarly, from

$$\pi^{-s/2} \Gamma\left(\frac{s}{2}\right) L(s) = k_2^{-s} \pi^{-\left(\frac{1-s}{2}\right)} \Gamma\left(\frac{1-s}{2}\right) D_2(1-s)$$

we have for $D_2'(1-s) = \sum_{n=1}^{\infty} \frac{b_{n,2} k_1}{(nk_1)^{1-s}}$ (where $D_2(1-s) = \sum_{n=1}^{\infty} \frac{b_{n,2}}{n^{1-s}}$) the following formula:

(14) $$\pi^{-s/2} \Gamma\left(\frac{s}{2}\right) L(s) = k_1^{-s} k_2^{-s} \pi^{-\left(\frac{1-s}{2}\right)} \Gamma\left(\frac{1-s}{2}\right) D_2'(1-s)$$

where the indices of all nonzero coefficients of $D_2'(1-s)$ are divisible by k_1. However, comparing (13) and (14) we see that $D_1'(1-s) = D_2'(1-s)$. It follows from the lemma below that the coefficients of D_1' and D_2' are equal.

LEMMA. *If $\sum_{n=1}^{\infty} \frac{a_n}{n^s} = \sum_{n=1}^{\infty} \frac{b_n}{n^s}$ for all s such that $\operatorname{Re}(s) > \sigma > 0$ for some σ, then $a_n = b_n$ for all $n \in \mathbb{N}$.*

PROOF. This lemma follows from the formula

$$\lim_{T \to \infty} \frac{1}{T} \int_0^T D(\sigma + it) n^{it} dt = a_n n^{\sigma},$$

which shows that $a_n = b_n$ for all n. Hence, the indices of all nonzero coefficients of $D_1'(1-s)$ are divisible by both k_1 and k_2 and hence by $k_1 k_2$. So $D_1'(1-s)$ has the form

$$D_1'(1-s) = \sum_{n=1}^{\infty} \frac{c_n}{(k_1 k_2 n)^{1-s}} = \frac{1}{(k_1 k_2)^{-s}} \sum_{n=1}^{\infty} \frac{c_n'}{n^{1-s}} = \frac{1}{(k_1 k_2)^{-s}} D'(1-s).$$

Thus $L(s)$ satisfies equation (1)

$$\pi^{-s/2} \Gamma\left(\frac{s}{2}\right) L(s) = \pi^{-\left(\frac{1-s}{2}\right)} \Gamma\left(\frac{1-s}{2}\right) D'(1-s)$$

with conductor 1. This proves Proposition 1.

REMARK 1. We note that the proof of Proposition 1 shows that if $L(s)$ satisfies equation (1) (resp. (2)) with conductor k, then it satisfies equation (1) (resp. (2)) with conductor $k\ell$ for any $\ell \in \mathbb{Z}^+$.

REMARK 2. Proposition 1 and, in particular, its corollary allow us to define the conductor of the Dirichlet series $L(s)$. We define it to be the smallest $k \in \mathbb{Z}^+$ such that $L(s)$ satisfies equation (1) (resp. (2)) with conductor k. Proposition 1 ensures that k is uniquely defined.

Now we prove the second part of Theorem 1. Consider the Hurwitz Zeta function defined by the formula (see [Ti])

$$\zeta(s, a) = \sum_{n=0}^{\infty} \frac{1}{(n+a)^s}, \qquad 0 < a \leq 1, \quad \sigma > 1, \quad s = \sigma + it.$$

It is easy to show that

$$\zeta(s, a) = \sum_{n=0}^{\infty} \frac{1}{\Gamma(s)} \int_0^{\infty} x^{s-1} e^{-(n+a)x} dx = \frac{1}{\Gamma(s)} \int_0^{\infty} \frac{x^{s-1} e^{-ax}}{1 - e^{-x}} dx.$$

This can be further transformed by expressing $\int_C \frac{z^{s-1}}{e^z - 1} dz$ as a loop integral (for a suitable constant e) to give

$$\zeta(s, a) = \frac{e^{-i\pi s} \Gamma(1-s)}{2\pi i} \int_C \frac{z^{s-1} e^{-az}}{1 - e^{-z}} dz.$$

Expanding the loop to infinity and evaluating the integral we get

$$(15) \qquad \zeta(s, a) = \frac{2\Gamma(1-s)}{(2\pi)^{1-s}} \left\{ \sin\frac{\pi s}{2} \sum_{m=1}^{\infty} \frac{\cos 2m\pi a}{m^{1-s}} + \cos\frac{\pi s}{2} \sum_{m=1}^{\infty} \frac{\sin 2m\pi a}{m^{1-s}} \right\}.$$

If we now use the formulas

$$\zeta(s) = 2 \sin\frac{\pi s}{2} (2\pi)^{s-1} \Gamma(1-s) \zeta(1-s)$$

and
$$\zeta(s) = \frac{2^{s-1}\pi^s}{\cos\frac{\pi s}{2}} \frac{\zeta(1-s)}{\Gamma(s)},$$

$$\zeta(s)\Gamma\left(\frac{s}{2}\right) = \pi^s \Gamma\left(\frac{1-s}{2}\right)\zeta(1-s),$$

we get
$$\frac{2\Gamma(1-s)}{(2\pi)^{1-s}} \sin\frac{\pi s}{2} = \pi^s \Gamma\left(\frac{1-s}{2}\right)/\Gamma\left(\frac{s}{2}\right)$$

and
$$\frac{2\Gamma(1-s)}{(2\pi)^{1-s}} \cos\frac{\pi s}{2} = \pi^s \Gamma\left(\frac{2-s}{2}\right)/\Gamma\left(\frac{1+s}{2}\right).$$

Now let us set $a = r/k$, $1 \leq r \leq [k/2]$. Then we obtain

(16)
$$\Gamma\left(\frac{s}{2}\right)\left[\zeta\left(s,\frac{r}{k}\right) + \zeta\left(s,\frac{k-r}{k}\right)\right] = \pi^s \Gamma\left(\frac{1-s}{2}\right)\left\{\sum_{m=1}^{\infty} 2\cos\frac{2\pi m(r/k)}{m^{1-s}}\right\}$$

and

(17)
$$\Gamma\left(\frac{1-s}{2}\right)\left[\zeta\left(s,\frac{r}{k}\right) - \zeta\left(s,\frac{k-r}{k}\right)\right] = \pi^s \Gamma\left(\frac{2-s}{2}\right)\left\{\sum_{m=1}^{\infty} 2\sin\frac{2\pi m(r/k)}{m^{1-s}}\right\}.$$

Note that
$$\zeta(s,\frac{r}{k}) = \sum_{n=0}^{\infty} \frac{1}{(n+\frac{r}{k})^s} = \sum_{n=0}^{\infty} \frac{k^s}{(kn+r)^s} = k^s \sum_{n=0}^{\infty} \frac{1}{(kn+r)^s} = k^s L'(s),$$

where $L'(s) = \sum_{n=0}^{\infty} \frac{1}{(kn+r)^s}$. Thus equations (16) and (17) are precisely of the same form as equations (1) and (2).

If k is odd, equation (i) has $\frac{k+1}{2}$ solutions and equation (ii) has $\frac{k-1}{2}$ solutions.

If k is even, equation (i) has $\frac{k}{2} + 1$ solutions and equation (ii) has $\frac{k}{2} - 1$ solutions.

Therefore, in either case, the sums of the dimensions n^+ and n^- of the solution spaces for equations (1) and (2) respectively is k.

Part (iii) of Theorem 1 follows from the explicit construction of the solutions above. This completes the proof of Theorem 1.

4. Now we formulate our next result.

THEOREM 2. *Suppose now that $L(s)$ satisfies all the conditions of* (i), (ii) *and* (iii). *Furthermore, suppose that $L(s) = \sum_{n=1}^{\infty} \frac{a_n}{n^s}$ has completely multiplicative coefficients, i.e., $a_m a_n = a_{mn}$ for all $m, n \in \mathbb{N}$. Then there exists a primitive character χ of $(\mathbb{Z}/m)^*$ for some $m \mid k$ such that*

$$L(s) = \prod_{\substack{p \mid k \\ p \nmid m}} \left(1 - \frac{\chi(p)}{p^s}\right) L(s, \chi).$$

PROOF OF THEOREM 2. Let k be the conductor of $L(s)$. By Theorem 1 we know that $a_{n+k} = a_n$ for all $n \in \mathbb{N}$.

LEMMA 1. *For each prime p we have $|a_p| = 0$ or 1.*

PROOF. Suppose that $p^\alpha \mid k$ and $p^{\alpha+1} \nmid k$. We can write $k = p^\alpha k_1$, where $(k_1, p^\alpha) = 1$. Then we claim that $(a_p)^{\alpha+\varphi(k_1)} = (a_p)^\alpha$. Notice that

$$p^{\alpha+\varphi(k_1)} - p^\alpha = p^\alpha(p^{\varphi(k_1)} - 1) \equiv p^\alpha k_1 \pmod{k} \equiv 0 \pmod{k}.$$

Hence $(a_p)^{\alpha+\varphi(k_1)} = (a_p)^\alpha$, so that $|a_p| = 0$ or $|a_p| = 1$.

LEMMA 2. *If $p \mid k$, then $|a_p| = 0$.*

PROOF. Let $S = \{p \text{ prime}, p \mid m, |a_p| = 1\}$.

Since the coefficients of $L(s)$ are multiplicative, there is a character $\tilde{\chi} : (\mathbb{Z}/k)^* \to \mathbb{C}$ such that $\tilde{\chi}(n) = a_n$. Let χ be a primitive character on $(\mathbb{Z}/m)^*$, $m \mid k$, such that χ induces $\tilde{\chi}$. Then for this χ we can write

$$(18) \qquad L(s) = L(s, \chi) \prod_{p \in S} \left(\frac{1}{1 - a_p/p^s} \right) \prod_{\substack{p \mid k \\ p \nmid m}} \left(1 - \frac{\chi(p)}{p^s} \right),$$

where $\chi(q)$ is defined by

$$\chi(q) = a_q \quad \text{when } q \nmid m,$$
$$\chi(q) = 0 \quad \text{when } q \mid m.$$

Equation (18) can be rewritten as follows:

$$(19) \qquad \prod_{\substack{p \mid k \\ p \nmid m}} \left(1 - \frac{\chi(p)}{p^s} \right)^{-1} L(s) \prod_{p \in S} \left(1 - \frac{a_p}{p^s} \right) = L(s, \chi).$$

Now

$$\left(1 - \frac{a_p}{p^s} \right) = 0 \Leftrightarrow s = \frac{\log |a_p|}{\log p} + \frac{2\pi n i}{\log p} = \frac{2\pi n i}{\log p},$$

so that the left-hand side of (18) has infinitely many zeros on the line $\text{Re}(s) = 0$. But this contradicts a theorem (of Hadamard and de la Vallée Poussin) which states that $L(s, \chi)$ has no zeros on the imaginary axis. Hence we must have $|a_p| = 0$, i.e., $a_p = 0$.

Hence, (18) reduces to

$$L(s) = L(s, \chi) \prod_{\substack{p \mid k \\ p \nmid m}} \left(1 - \frac{\chi(p)}{p^s} \right).$$

This proves Theorem 2.

PROPOSITION 2. *If $b_1 \neq 0$, then $L(s) = L(s, \chi)$, where χ is a primitive character of $(\mathbb{Z}/k)^*$ and k is the conductor of $L(s)$.*

PROOF. Let k be the conductor of $L(s)$. Since the coefficients of $L(s)$ are multiplicative, we know by Lemma 1 that $\chi(p) = a_p$ defines a character on $(\mathbb{Z}/k)^*$.

If χ is not primitive, then χ is induced by some primitive character $\tilde{\chi}$ on $(\mathbb{Z}/m)^*$. But this means that the coefficients of $L(s)$ have period m and hence $L(s)$ satisfies equation (1) or (2) with conductor $m < k$. Assume without loss of generality that $L(s)$ satisfies (1). So we have

$$\pi^{-s/2}\Gamma(s/2)L(s) = m^{-s}\pi^{-\left(\frac{1-s}{2}\right)}\Gamma\left(\frac{1-s}{2}\right)\sum_{n=1}^{\infty}\frac{b_n'}{n^{1-s}}$$

for some $D'(s) = \sum_{n=1}^{\infty}\frac{b_n'}{n^s}$. Then

$$\pi^{-s/2}\Gamma(s/2)L(s) = m^{-s}k'^{-s}\pi^{-\left(\frac{1-s}{2}\right)}\Gamma\left(\frac{1-s}{2}\right)\sum_{n=1}^{\infty}\frac{b_n'k'}{(k'n)^{1-s}}.$$

But this means that $D(s) = \sum_{n=1}^{\infty}\frac{b_n'k'}{(k'n)^{1-s}}$, where $mk' = k$. Hence $b_1 = 0$, which is a contradiction. Thus, if $b_1 \neq 0$, then $L(s)$ must arise from a primitive character.

We close this discussion with an example of a Dirichlet series whose conductor is not a rational integer. Hecke proved that the group Γ generated by the elements $\begin{pmatrix} 0 & -1 \\ 1 & 0 \end{pmatrix}$ and $\begin{pmatrix} 1 & \lambda \\ 0 & 1 \end{pmatrix}$ is discrete if $\lambda = 2\cos\frac{\pi}{k}$, $k = 3, 4, \ldots$. Hence for sufficiently large k (k even) there exists a modular form which transforms suitably under the action of Γ. The Fourier series of this form looks like

$$f(\tau) = \sum_{n=0}^{\infty} a_n e^{2\pi i n\tau/\lambda}$$

and satisfies

$$f(\tau) = \left(\frac{\tau}{i}\right)^{-k} f(-1/\tau).$$

If $a_n = O(n^c)$, consider $L(s) = \sum_{n=1}^{\infty} a_n n^{-s}$. Then it is well known that $L(s)$ satisfies the functional equation

$$\left(\frac{2\pi}{\lambda}\right)^{-s}\Gamma(s)L(s) = c\left(\frac{2\pi}{\lambda}\right)^{-(1-s)}\Gamma(1-s)D(1-s)$$

which, it is easy to see, has conductor λ^2. So if $\lambda \neq \sqrt{n}$, $n \in \mathbb{Z}$, then $L(s)$ has a conductor which is not a rational integer. In particular, we may choose $\lambda = 2\cos\frac{\pi}{5} = \frac{1+\sqrt{5}}{2}$. Note that this is in sharp contrast to the case of the Dirichlet series satisfying a "Hamburger type" functional equation.

References

[H1] H. Hamburger, *Über die Funktionaleichung der ζ-Funktion*, Math. Z. **10** (1921), 240–258.

[H2] _____, *Über die Funktionalgleichung der ζ-Funktion*, Math. Z. **11** (1921), 224–245.

[H3] _____, *Über die Funktionalgleichung der ζ-Funktion*, Math. Z. **13** (1922), 283–311.

[S] C.L. Siegel, *Bemerkung Zri einem Satz von Hamburger über die Funktionalgleichung der Riemannsche Zetafunktion*, Math. Ann. **86** (1922), 276–279.

[E-K] Leon Ehrenpreis and Takahiro Kawai, *Poisson summation formula and Hamburger's theorem*, Publ. Res. Inst. Math. Sci. Kyoto Univ. **18** (1982), no. 2.

[C-G] J. B. Conrey and A. Ghosh, *On the Selberg class of Dirichlet series: small weights*, Duke Math. J. **72** (1993), 673–693.

[Gu] M. I. Gurevich, *Determining L-series from their functional equations*, Mathematics of the USSR, Sbornik **14** (1971), 537–553.

[T] J. Tate, *Fourier analysis in number fields and Hecke's zeta function*, Thesis, Princeton Univ., 1950.

[Ti] E. C. Titchmarch, *The theory of the Riemann zeta function*, Clarendon Press, Oxford, 1986.

DEPARTMENT OF MATHEMATICS YALE UNIVERSITY NEW HAVEN, CT 06520-8283

An Analogue of M. Artin's Conjecture on Invariants for Nonassociative Algebras

VLADIMIR L. POPOV
To E. B. Dynkin

The problem of describing invariants of a system of d vectors in a vector space V with respect to a group of linear transformations $H \subseteq \mathrm{GL}(V)$ has the classical origin. If V has a structure of a (not necessarily associative) algebra and H acts by its automorphisms, there is a construction of invariants by means of traces of products of operators of left and right multiplications. In this paper I discuss "how many" invariants, polynomial and rational, can be obtained in this way, and "how many" actions are covered by this method of constructing invariants. This discussion shows that the method of traces provides a great deal of information about the analogues of Classical Invariant Theory for nonclassical groups (for instance, for E_8), in which case almost nothing was known before (moreover there are also some nonreductive groups which are covered by this method). I formulate in Section 3 some general conjectures which make more precise the resources of this approach.

This paper is based on the notes of my talk [Po2] distributed among the participants.

In the sequel k denotes the ground field, algebraically closed and of characteristic zero.

Notation and terminology.

- k^*, the multiplicative group of k,
- $\mathrm{Frac}(R)$, the field of fractions of a commutative integral domain R,
- $\mathrm{Mat}_n(C)$, the ring of $n \times n$ matrices over a ring C,
- $E_n = \mathrm{diag}(1, \ldots, 1) \in \mathrm{Mat}_n(C)$,
- $K \cdot S$, the linear span of a subset S of a vector space over a field K,
- $P(L)$, the projective space associated to a vector space L,
- $k[X]$, the algebra of regular functions on an algebraic variety X,
- $k(X)$, the field of rational functions on an irreducible algebraic variety X,
- M^G, the fixed point set of an action of a group G on a set M,
- id_M, the identity transformation of M,

1991 *Mathematics Subject Classification.* Primary 15A72, 17A99; Secondary 20G20.
Research supported in part by Grant # MQZ000 from the International Science Foundation.

- G°, the connected component of an algebraic group G containing its unit element,
- $\mathrm{Lie}(G)$, the Lie algebra of G,
- $R(\lambda)$, the simple G-module with the highest weight λ of a connected semisimple algebraic group G,
- $\omega_1, \ldots, \omega_l$, the fundamental weights of G (in the Bourbaki numbering, [Bo]).

In the sequel "algebra" means not necessarily associative algebra and "ideal" means two-sided ideal. "Simple algebra" means an algebra with nonzero multiplication and without proper ideals.

If not specified, all vector spaces, linear subspaces, algebras and linear mappings are considered over k.

Acknowledgement. I wish to thank A. V. Iltyakov for his comments.

1. History of the problem

(1.1) The source of our considerations lies in the following old problem:

Describe the algebra and the field of invariants of a system of d linear operators in an n-dimensional vector space.

(1.2) Using the matrix language and standard terminology of Modern Invariant Theory this problem can be reformulated as follows:

Let $A := \mathrm{Mat}_n(k)$. Group $G := \mathrm{GL}_n(k)$ acts on A by conjugation. Consider diagonal action of G on $V_{d,n} := A^d := A \oplus \ldots \oplus A$ (d summands). Describe $k[V_{d,n}]^G$ and $k(V_{d,n})^G$.

(1.3) The algebra $k[V_{d,n}]^G$ is finitely generated by the Hilbert Theorem on Invariants. Since the action of G on $V_{d,n}$ factors through the action of $\mathrm{PGL}_n(k)$ and the latter group has no nontrivial homomorphisms to k^*, one has $k(V_{d,n})^G = \mathrm{Frac}(k[V_{d,n}]^G)$, cf. [PV, 3.2].

(1.4) One obtains a number of nonconstant invariants by means of the following construction.

Let $i_1, \ldots, i_t$ be a sequence of numbers, not necessarily all different, taken from the set $\{1, \ldots, d\}$. Since the traces of conjugated matrices are equal, the function

$$(1.4.1) \qquad V_{d,n} \longrightarrow k, \quad (a_1, \ldots, a_d) \mapsto \mathrm{tr}\, a_{i_1} \ldots a_{i_t}$$

is constant on G-orbits. It is a polynomial function, i.e. an element of $k[V_{d,n}]^G$. In the coordinates it can be presented as follows.

Let $x_{ij}^{(s)} \in k[V_{d,n}]$ be the (i,j)th standard coordinate function on the sth summand of $V_{d,n}$, i.e.

$$x_{ij}^{(s)}(v) = a_{ij}^{(s)}, \quad v = (a_1, \ldots, a_d) \in V_{d,n}, \quad a_l = (a_{pq}^{(l)}) \in A.$$

We have $k[V] = k[\ldots, x_{ij}^{(s)}, \ldots]$. Consider the "generic matrix" X_s of the s th summand of $V_{d,n}$, i.e.

$$X_s = (x_{ij}^{(s)}) \in \operatorname{Mat}_n(k[V_{d,n}]), \quad s = 1, \ldots, d.$$

Then the invariant (1.4.1) has the appearance

(1.4.2) $\qquad\qquad\qquad \operatorname{tr} X_{i_1} \ldots X_{i_t}.$

(1.5) In [Ar] M. Artin conjectured that

(1.4.3) $\qquad k[V_{d,n}]^G$ *is generated by invariants of the form* (1.4.2)[1].

Minimal system of generators of $k[V_{d,n}]^G$ of the type (1.4.2) was found for $n = 2$ by the classical researchers, cf. [GY], and for $n = 3$ in [Sib, Sm] (see also the related references in [Pr1]). According to the classical theorem, if $d = 1$ and n is arbitrary, the coefficients of characteristic polynomial of $X := X_1$ are the (algebraically independent) generators of $k[A]^G$, cf. [PV], and therefore $\operatorname{tr} X, \ldots, \operatorname{tr} X^n$ are the generators as well (because of the Newton formulas); see also [Gu].

In the following more precise form the proof of (1.4.3) was given in [Pr1]:

THEOREM 1. $k[V_{d,n}]^G$ *is generated by invariants of the form* (1.4.2) *with* $t \leqslant 2^n - 1$.

In [Ra2] this estimate is sharpened to $t \leqslant n^2$.

(1.6) In fact it was proven much more, namely that all relations between invariants (1.4.2) are obtained (in a certain precise sense) from the Hamilton–Cayley Theorem, cf. [Ko, Ra2, Pr1].

One can show that $k[V_{d,n}]^G$ is a free algebra if and only if either $d = 1$ or $(d, n) = (2, 2)$, [AG, S1]. In the general case its structure seems to be rather complicated, [LBT].

(1.7) However, this does not exclude a hope that $k(V_{d,n})^G = \operatorname{Frac}(k[V_{d,n}]^G)$ has a simple structure, more precisely, that this field is *rational* or at least *stably rational* (over k).

No counterexamples to the conjecture on (stable) rationality of $k(V_{d,n})^G$ is known at present.[2]

Moreover, it is proved that $k(V_{d,n})^G$ is rational, if $n = 2$ ([Pr2]), $n = 3$ ([Fo2]), and $n = 4$ ([Fo3]), and *stably rational*, if $n = 5$ and 7 ([BLB]). It is also proved, [Ka1], [Sc], that $k(V_{d,st})^G$ is stably rational provided $k(V_{d,s})^G$ and $k(V_{d,t})^G$ are stably rational and s and t are relatively prime. This implies that $k(V_{d,n})^G$ is stably rational whenever n divides 420.

It is shown in [Pr2] that $k(V_{d,n})^G$ for $d \geqslant 2$ is rational over $k(V_{2,n})^G$. Therefore, $d = 2$ is the key case of the problem of (stable) rationality of $k(V_{d,n})^G$.

---

[1] As a matter of fact, by that time (1.4.3) was already proven, see [Gu].

[2] One can find in [Ros] the arguments used to show that this field is *not* rational (as a matter of fact, even stably). However, these arguments are *erroneous* (nonsurjectivity of $R_{n,F}$ does not follow from existence of generic algebras that are not crossed products, and in fact conflicts with the Merkur'ev-Suslin Theorem).

(1.8) It is worth mentioning that this problem is an (important, see (2.1), (2.2)) particular case of the following general open problem:

Is it true that the field of invariants of any connected linear algebraic group is (stably) rational?

According to [Sa], the assumption that the group is connected is essential. This problem is closely related to the problem of (stable) rationality of homogeneous spaces, [Po1].

2. Applications

(2.1) One can show, [Pr3], that if $k(V_{2,n})^G$ is stably rational, then for any field $K \supseteq k$ and any central simple associative K-algebra of dimension n^2 over K its class in the Brauer group $\mathrm{Br}(K)$ of K is a product of the classes of cyclic algebras. Therefore, stable rationality of $k(V_{2,n})^G$ for all n implies the Merkur'ev-Suslin Theorem for any field K containing k (i.e. that $\mathrm{Br}(K)$ is generated by the classes of cyclic algebras).

This explains the old-standing interest of the ring-theorists in the problem of (stable) rationality of $k(V_{d,n})^G$.

(2.2) There is an interpretation of $k(V_{d,n})^G$ as the field of rational functions on a moduli variety of the appropriate geometrical objects; this explains the interest of algebraic geometers in this problem.

Namely, one can show, [LB1], that $k(V_{2,n})^G$ is the field of rational functions on the moduli variety of stable vector bundles of rank n over $\mathbf{P}^2$ with the Chern numbers $(0, n)$ (in [Ka2] a more general result concerning a connection between the moduli variety of stable vector bundles of rank s, $1 < s \leqslant n$, over $\mathbf{P}^2$ with the Chern numbers $(0, n)$ and the field $k(V_{2,(s,n)})^G$ is obtained).

There is also another interpretation, [LB1, VdB, Vin]: $k(V_{2,n})^G$ is the field of rational functions on the variety of pairs $(Y, \mathcal{L})$, where Y is a smooth plane projective curve of degree n and $\mathcal{L}$ is a divisor of degree $n(n-1)/2$ on Y.

See [LB2, LBT, LBS] about other applications, including those to PDE.

3. Generalization

(3.1) The construction of invariants described in (1.4) is based essentially on the fact that $\mathrm{Mat}_n(k)$ is a finite dimensional algebra and $\mathrm{GL}_n(k)$ acts on it by automorphisms. Therefore, it can be applied in a much more general setting.

(3.2) Namely, let A be any finite dimensional algebra and $G := \mathrm{Aut}\, A$.

Consider the diagonal action of G on $V := A^d := A \oplus \ldots \oplus A$ (d summands). Let $\pi_i : A^d \longrightarrow A$ be the projection onto the i-th summand.

Denote by L_a and R_a the operators of left and right multiplication of A by $a \in A$ respectively. These operators linearly depend on a and

(3.2.1) $\qquad T_{g \cdot a} = g\, T_a g^{-1}, \quad g \in G, \ a \in A, \ T \in \{L, R\}.$

For any nonassociative polynomial f in noncommutative variables $t_1, \ldots, t_d$ with the coefficients in k and without constant term, and for any element $v \in V$ the substitution $t_i = \pi_i(v)$ defines an element $f(v) \in A$. One has

(3.2.2) $\qquad f(g \cdot v) = g \cdot f(v), \quad g \in G, \ v \in V.$

Now let $s, \ldots, r$ be a finite set of such polynomials. Then it follows from (3.2.1) and (3.2.2) that any function of the form

$$(3.2.3) \qquad V \longrightarrow k, \quad v \mapsto \operatorname{tr} T^{(s)}_{s(v)} \ldots T^{(r)}_{r(v)}, \quad T^{(s)}, \ldots, T^{(r)} \in \{L, R\}$$

is constant on G-orbits. It is a polynomial function, i.e., an element of $k[V]^G$.

Denote by $\operatorname{Tr} A^d$ the subalgebra with unit of $k[V]^G$ generated by all functions of the form (3.2.3).

(3.3) EXAMPLE. Let $A = \operatorname{Mat}_n(k)$. It follows from associativity of A that any invariant (3.2.3) is either of the form

$$v \mapsto \operatorname{tr} T_{f(v)}, \quad T \in \{L, R\},$$

or of the form

$$v \mapsto \operatorname{tr} L_{f(v)} R_{h(v)}.$$

It is easily seen that

$$\operatorname{tr} T_a = n \operatorname{tr} a, \quad \operatorname{tr} L_a R_b = \operatorname{tr} a \operatorname{tr} b, \quad a, b \in A.$$

Therefore, $\operatorname{Tr} A^d$ is generated by invariants of the form (1.4.2), i.e., $\operatorname{Tr} A^d = k[A^d]^{\operatorname{Aut} A}$ by Theorem 1.

(3.4) The general nature of the construction in (3.2) together with example (3.3) naturally lead one to ask whether or not the analogue of M. Artin's conjecture is true for algebras different from $\operatorname{Mat}_n(k)$.[1] The example of an algebra with zero multiplication shows that when expecting the affirmative answer one has to impose certain constraints on A. Algebra $\operatorname{Mat}_n(k)$ is simple and we restrict ourselves to consideration of *simple* algebras. The results stated below give an evidence in favour of this restriction.

As a matter of fact, when formulating an analogue of M. Artin's conjecture, one has to distinguish two aspects of it, regular and rational ones (cf. (1.3)):

Let A be a finite dimensional simple algebra.

(A) *Is it true that* $k[A^d]^{\operatorname{Aut} A} = \operatorname{Tr} A^d$?
(F) *Is it true that* $k(A^d)^{\operatorname{Aut} A} = \operatorname{Frac}(\operatorname{Tr} A^d)$?

If $\operatorname{Aut} A$ has no nontrivial homomorphisms to k^*, then the affirmative answer to (A) implies the affirmative answer to (F), cf. (1.3).

---

[1] I do not know where this question was explicitly formulated for the first time. It is natural to guess that the people who considered the concrete algebras had it in mind. For instance, A. V. Iltyakov informed me that some time ago I.P.Shestakov considered this question (in a preprint) for the Albert algebra and that apparently he had in mind the general formulation as well.

(3.5) At the time this paper was written, affirmative answers to (A) and (F) were obtained for some other types of algebras besides $\text{Mat}_n(k)$. For a simple Jordan algebra of a nondegenerate symmetric bilinear form such answers follow from Classical Invariant Theory. In [S2] the affirmative answer to (A) was obtained for the Cayley-Dickson algebra $A = \mathbf{O}$ (since $\text{Aut}\,\mathbf{O}$ is a simple connected linear algebraic group (of the type G_2), this gives the affirmative answer to (F) as well). Moreover, it is shown in [PS] that $k(\mathbf{O}^d)^{\text{Aut}\,\mathbf{O}}$ is rational if $d \geqslant 3$. In [I2] the Albert algebra $A = \mathbf{A}$ was considered (i.e. the 27-dimensional exceptional simple Jordan algebra of Hermitian 3×3-matrices over $\mathbf{O}$ in which case $\text{Aut}\,A$ is a simple connected linear algebraic group of the type F_4) and (F) was answered in the affirmative. Moreover, it was proven in [I2] that $k[\mathbf{A}^d]^{\text{Aut}\,\mathbf{A}}$ is integral over $\text{Tr}\,\mathbf{A}^d$ and $\text{Tr}\,\mathbf{A}^d$ separates closed orbits in $\mathbf{A}^d$. In [Pol] an algebraically independent system of generators of $\text{Tr}\,\mathbf{A}^d$ was found for $d \leqslant 2$. Thus, these two results give the affirmative answer to (A) for $A = \mathbf{A}$ and $d \leqslant 2$. It is proved in [IS] that $k(\mathbf{A}^d)^{\text{Aut}\,\mathbf{A}}$ is rational for any d.

(3.6) In the next section I shall show that "in the general case" (A) is answered in the *negative*. It is remarkable that the situation with question (F) appears to be *different*. Namely, recently Iltyakov proved in [I1] the following

THEOREM 2. *Let A be a finite dimensional simple algebra generated by $\leqslant d$ elements. Then*

$$(3.6.1) \qquad k(A^d)^{\text{Aut}\,A} = \text{Frac}(\text{Tr}\,A^d).$$

In §6 a proof of this theorem is given, which simplifies and, I believe, clarifies the original proof of [I1] (but retains its main idea to use Rosenlicht's criterion, cf. Theorem 9). For instance, in contrast to [I1], I prove and use some general statements concerning any subfields of $k(V)$ and separation of orbits, and do not use central polynomials.[1]

(3.7) The condition on d in Theorem 2 is not actually very restrictive in the known examples. For instance, $\text{Mat}_n(k)$ (and, more generally, any finite dimensional associative semi-simple algebra, [Re]), is generated by two elements. Algebra $\mathbf{O}$ is generated by three elements. Any simple Lie algebra $\mathfrak{g}$ is generated by two elements (if $\mathfrak{g} = \mathfrak{t} \oplus (\oplus_{\alpha \in \Delta} \mathfrak{g}_\alpha)$ is the canonical decomposition, let $h \in \mathfrak{t}$ be an element such that all values $\alpha(h)$, $\alpha \in \Delta$, are pairwise different and nonzero and let $x = \sum_{\alpha \in \Delta} e_\alpha$ where $e_\alpha \in \mathfrak{g}_\alpha$, $e_\alpha \neq 0$; then h and x generate $\mathfrak{g}$). In the general case no nontrivial upper bound of the minimal number of generators of a simple algebra is known to me; it would be interesting to obtain it. At any rate, the statement of Theorem 2 is definitely true for any $d \geqslant \dim A$.

(3.8) Although $\text{Tr}\,A^d$ is not always equal to $k[A^d]^{\text{Aut}\,A}$, Theorem 2 shows that these algebras are sufficiently "close" to each other.

CONJECTURE 1. *Let A be a finite dimensional simple algebra. Then $k[A^d]^{\text{Aut}\,A}$ is integral over $\text{Tr}\,A^d$.*

Using the terminology of [PV] we call a point $v \in A^d$ *nilpotent* if $f(v) = 0$ for any $f \in k[A^d]^{\text{Aut}\,A}$ such that $f(0) = 0$. If $\text{Aut}\,A$ is *reductive* (for instance,

---

[1] A. V. Iltyakov informed me that I. P. Shestakov explained to him the proofs of Theorems 8, 11, 12, sketched in [PS], which also do not use central polynomials.

if Aut A acts on A irreducibly), then, using a result of Hilbert [Hil] one can reformulate this conjecture as follows (see (3.2)):

CONJECTURE 1*. *A point $v \in A^d$ is nilpotent if and only if* $\operatorname{tr} T^{(s)}_{s(v)} \ldots T^{(r)}_{r(v)} = 0$ *for all polynomials* $s, \ldots, r$.

CONJECTURE 2. $\operatorname{Tr} A^d$ *separates closed orbits of* Aut A *in* A.

(3.9) REMARK. As A. V. Iltyakov explained to me, simple arguments show that the algebra $\operatorname{Tr} A^d$ is *finitely generated*.

(3.10) REMARK. There is an analogy between Theorem 2 and Conjecture 1 of the present paper and Theorem 1 and Conjecture of [Vi], where the invariants of normalizer of a connected reductive subgroup G of $\operatorname{GL}_n(k)$ acting diagonally on $G \times \ldots \times G$ were considered.

4. Algebras in general position

(4.1) Let L be an n-dimensional vector space (over k).

Fixing a structure of algebra on L is equivalent to fixing a bilinear mapping $L \times L \longrightarrow L$, which in turn is equivalent to fixing a linear mapping $L \otimes L \longrightarrow L$. Therefore, n^3-dimensional vector space $\operatorname{Hom}(L \otimes L, L)$ has a natural interpretation as the variety of structures of algebras on L. We canonically identify this space with $L^* \otimes L^* \otimes L$. An element $\sum f \otimes h \otimes v \in L^* \otimes L^* \otimes L$ corresponds to a structure of algebra on L with the multiplication given by

$$ab = \sum f(a)h(b)v, \quad a, b \in L.$$

Denote by $\{L, m\}$ the algebra defined by a structure $m \in L^* \otimes L^* \otimes L$.

(4.2) Using standard terminology of Invariant Theory, [PV], we say that an (n-dimensional) *algebra in general position has some property* if for any structure m of a certain (depending on the property under consideration) open dense subset of $L^* \otimes L^* \otimes L$ the algebra $\{L, m\}$ has this property.

(4.3) Group $G = \operatorname{GL}(L)$ acts naturally on $L^* \otimes L^* \otimes L$. It is easy to see that the structures $m_1, m_2 \in L^* \otimes L^* \otimes L$ lie in the same G-orbit if and only if $\{L, m_1\}$ and $\{L, m_2\}$ are isomorphic. In particular, the stabilizer of a point $m \in L^* \otimes L^* \otimes L$ is $\operatorname{Aut}\{L, m\}$.

(4.4) To answer question (A) in (3.4) I need a result which seems to belong to folklore:

THEOREM 3. *The automorphism group of an n-dimensional algebra in the general position is trivial.*

PROOF. According to (4.3), we have to show that the generic stabilizer (cf. [PV, §7]) of the action of G on $L^* \otimes L^* \otimes L$ is trivial.

The subgroup $S := \{\alpha \operatorname{id}_L \mid \alpha \in k^*\}$ of G acts on $L^* \otimes L^* \otimes L$ by scalar multiplication. Therefore, if an element $g = st$, $s \in S$, $t \in \operatorname{SL}(L)$, lies in the generic stabilizer of the natural action of G on $L^* \otimes L^* \otimes L$, then t lies in the generic stabilizer of the natural action of $\operatorname{SL}(L)$ on $P(L^* \otimes L^* \otimes L)$.

The SL(L)-module $L^* \otimes L^* \otimes L$ decomposes into irreducible components as follows

(4.4.1) $L^* \otimes L^* \otimes L = R(\omega_1 + 2\omega_{n-1}) \oplus R(\omega_1 + \omega_{n-2}) \oplus R(\omega_{n-1}) \oplus R(\omega_{n-1})$.

(One can obtain (4.4.1), say, by means of the tables in [OV]; the summand $R(\omega_1 + \omega_{n-2})$ does not occur for $n = 2$).

In [Pop] a list was obtained of all connected simple algebraic groups $H \subset \mathrm{GL}(U)$ such that the generic stabilizer of natural action of H on $P(U)$ contains a nonscalar transformation. By inspection, the module (4.4.1) does not occur in this list.

Hence, $t \in S$ and therefore $g \in S$ as well. Since S acts on $L^* \otimes L^* \otimes L$ by means of the character χ^{-1} given by $\chi(\alpha \operatorname{id}_L) = \alpha$, and g lies in the stabilizer of some nonzero point of $L^* \otimes L^* \otimes L$, one has $g \in \ker \chi^{-1} = \{e\}$. □

(4.5) Let r be an integer, $0 \leqslant r \leqslant n$, and

$$\mathcal{I}_r = \{ m \in L^* \otimes L^* \otimes L \mid \{L, m\} \text{ has an } r\text{-dimensional ideal} \}.$$

THEOREM 3. *$\mathcal{I}_r$ is a closed subset of $L^* \otimes L^* \otimes L$ of dimension $\leqslant n^3 - r(n-r)(2n-r-1)$.*

PROOF. Let $e_1, \ldots, e_n$ be a basis of L and $e^1, \ldots, e^n$ the dual basis of L^*. We can assume $r \geqslant 1$. Let L_r be the linear span of $\{e_1, \ldots, e_r\}$.

It is easily seen that L_r is an ideal of $\{L, m\}$ for $m = \sum c_{ij}^s e^i \otimes e^j \otimes e_s$ if and only if $c_{ij}^s = 0$ for $s > r$ and either $i \leqslant r$ or $j \leqslant r$. It follows from here that

$$\mathcal{L}_r = \{ m \in L^* \otimes L^* \otimes L \mid L_r \text{ is an ideal of } \{L, m\} \}$$

is a linear subspace of $L^* \otimes L^* \otimes L$ of dimension $n^3 - r^3 + 3nr^2 - 2n^2 r$.

Further, if I is an ideal of $\{L, m\}$ and $g \in G$, then gI is an ideal of $\{L, gm\}$. Since G acts transitively on the set of all r-dimensional linear subspaces of L, it follows that

(4.5.1) $\mathcal{I}_r = G \cdot \mathcal{L}_r$.

Consider now the subgroup $P_r := \{ g \in G \mid g \cdot L_r = L_r \}$ of G. We have

(4.5.2) $P_r \cdot \mathcal{L}_r = \mathcal{L}_r$.

Since P_r is a parabolic subgroup, it follows from (4.5.1) and (4.5.2) that $\mathcal{I}_r$ is closed and $\dim \mathcal{I}_r \leqslant \dim G - \dim P_r + \dim \mathcal{L}_r$. The desired estimate on $\dim \mathcal{I}_r$ now follows from the equalities $\dim G = n^2$ and $\dim P_r = r^2 - nr + n^2$. □

(4.6) It follows from Theorem 3 that an n-dimensional algebra in general position is simple. More precisely, one has the following

THEOREM 4. *The set $\{ m \in L^* \otimes L^* \otimes L \mid \{L, m\}$ is a simple algebra $\}$ is open and dense in $L^* \otimes L^* \otimes L$.*

PROOF. The statement follows from Theorem 3 due to the fact that $r(n-r) \cdot (2n-r-1) > 0$ for $1 \leqslant r \leqslant n-1$. □

(4.7) According to Theorems 3 and 4, for each n there exists an n-dimensional algebra A such that
 (a) A is simple;
 (b) $\operatorname{Aut} A = \{e\}$
(and moreover, a "typical" n-dimensional algebra has these properties).

It follows from (b) that for such A one has

(4.7.1) $$k[A^d]^{\operatorname{Aut} A} = k[A^d] \text{ for any } d.$$

The definition of $\operatorname{Tr} A^d$ implies that all homogeneous elements of degree 1 in $\operatorname{Tr} A^d \cap \pi_i^*(k[A])$ are linear combinations of two invariants

$$v \mapsto \operatorname{tr} L_{\pi_i(v)} \text{ and } v \mapsto \operatorname{tr} R_{\pi_i(v)}.$$

Therefore, for $n \geqslant 3$ one has

(4.7.2) $$k[A^d] \neq \operatorname{Tr} A^d$$

for an algebra A in general position (on the contrary, if $n \leqslant 2$, then $k[A^d] = \operatorname{Tr} A^d$ for such A).

Therefore, returning back to (3.4), we obtain from (4.7.1) and (4.7.2):

If $\dim A \geqslant 3$, then for any d the answer to question (A) in (3.4) is, "as a rule", negative.

5. When is Theorem 2 applicable?

(5.1) Let L be a finite dimensional vector space (over k) and H a subgroup of $\operatorname{GL}(L)$. The problem of describing polynomial H-invariants of the systems of vectors of L (i.e., H-invariants of the diagonal actions of H on $L \oplus \ldots \oplus L$) has the classical origin (its solution is called "the (First) Fundamental Theorem" of the theory of vector H-invariants, cf. [Gu]). As a matter of fact, the key case is to describe H-invariants of $n = \dim L$ vectors of L, see [PV, 9.2].

(5.2) Classical Invariant Theory provides a solution to this problem for classical linear groups H, cf. [PV, §9]. Except for these cases, its complete solution is known only in a few other cases. In [S2] a solution is given for the simplest 7-dimensional irreducible module of G_2 (in [Sim] the right answer was given, but the proof was incomplete), in [S2] for the simplest 8-dimensional irreducible module of Spin_7, and in [S3] for the SL_2-module of binary cubic forms.

(5.3) If there is a structure of algebra $m \in L^* \otimes L^* \otimes L$ on L such that $H \subseteq \operatorname{Aut} A$ for $A = \{L, m\}$, one can obtain invariants of the systems of vectors of L by the construction described in (3.2). If, moreover, $H = \operatorname{Aut} A$, then these invariants provide an essential approximation to the solution of the aforementioned problem because of Theorem 2 (see also Conjectures 1, 1^*, 2 in (3.8)).

(5.4) It is convenient to use the following terminology. Let $A = \{L, m\}$, F be a group, and $\phi : F \longrightarrow \mathrm{GL}(L)$ a homomorphism. If $\phi(F) \subseteq \mathrm{Aut}\, A$, we say that m is a *structure of F-invariant algebra* on L or that A is an *F-invariant algebra*. If L is a simple F-module (with respect to the module structure defined by ϕ), we say that A is an *irreducible F-invariant algebra*.

Thus our discussion leads naturally to the following questions:

(a) *When is there a nonzero structure of H-invariant algebra A on L?*
(b) *When is A simple?*
(c) *When is $H = \mathrm{Aut}\, A$?*

These questions were considered, directly or indirectly, in a series of papers. For instance, in [Di] the structures of SL_2-invariant algebras on certain SL_2-modules were investigated by means of transvectants and simplicity of these algebras was proved (the Cayley-Dickson algebra **O** and the exceptional Jordan algebra were constructed in this way). An extensive information on question (a) was obtained in [E1], cf. (5.10).

(5.5) First consider question (b).

THEOREM 5. *Let A be a finite dimensional algebra with a nonzero multiplication. If $\mathrm{Aut}\, A$ contains a connected algebraic subgroup H that acts irreducibly on A, then*

(i) *A is a simple algebra;*
(ii) *any algebraic subgroup S of $\mathrm{Aut}\, A$ containing H is semisimple and has a trivial center.*

PROOF. (i) Assume that A is not simple. Let I be a minimal proper ideal of A.

Since H acts irreducibly on A, the sum of all ideals $h \cdot I$, where h runs over H, coincides with A. Since each of the ideals $h \cdot L$ is minimal and intersection of any proper ideal with a minimal ideal which is not contained in it equals 0, it follows that there are $e = h_1, \ldots, h_l \in H$ such that

$$(5.5.1) \qquad A = h_1 \cdot I \oplus \ldots \oplus h_l \cdot I.$$

Since the product of different minimal ideals is 0, (5.5.1) is a direct sum of algebras.

Let J be another minimal proper ideal of A. Assume that $J \neq h_i \cdot I$ for each i. Then it follows from (5.5.1) that $ab = ba = 0$ for any $a \in A$, $b \in J$. In particular, J has zero multiplication. In the same way as for I, we can obtain for J a decomposition analogous to (5.5.1). Since $h \cdot J$ has zero multiplication for any $h \in H$ as well, this shows that A has zero multiplication, which contradicts the assumption.

Therefore, each minimal proper ideal of A coincides with one of $h_1 \cdot I, \ldots, h_l \cdot I$. Hence, H acts by permutations of $h_1 \cdot I, \ldots, h_l \cdot I$. This action is trivial because H is connected. Hence, each $h_i \cdot I$ is an H-invariant linear subspace of A, contradicting the assumption that H acts irreducibly on A.

(ii) Since $H \subseteq S$, the action of S on A is irreducible as well. Therefore, S is reductive and its center acts on A by scalar multiplications. Let $s = \alpha \, \mathrm{id}_A$, $\alpha \in k^*$, be an element of the center of S. Take $a, b \in A$ such that $ab \neq 0$. Then $s \cdot ab = \alpha ab = (s \cdot a)(s \cdot b) = (\alpha a)(\alpha b) = \alpha^2 ab$. Hence, $\alpha = 1$ and $s = \mathrm{id}_A$. □

(5.6) As it will become clear from the subsequent discussion of questions (a) and (c) from (5.4), Theorem 5 shows that one can use the technique of traces and Theorem 2 to describe H-invariants of the systems of vectors in L rather frequently, including the cases of many modules L of nonclassical algebraic groups H (for instance, exceptional simple groups) for which almost nothing was known about invariants before.

Having in mind Theorem 5, we shall assume further on that H *is reductive* (but we shall *not* assume that H acts irreducibly on L; there is a construction of simple H-invariant algebras generalizing construction of [Di], in which H acts reducibly on L, cf. [E1]. It has to be noted that it is *not true* in general that if A is a simple algebra, then Aut A is reductive, see an example in (5.18).

(5.7) Discuss now question (a) from (5.4).

Existence of a nonzero structure of H-invariant algebra on L is equivalent to existence of a nonzero H-equivariant bilinear mapping $L \times L \longrightarrow L$ or, which is the same, of a nonzero morphism of H-modules $L \otimes L \longrightarrow L$. Since H is reductive, this latter condition is equivalent to the condition that $L \otimes L$ and L have a common (up to an isomorphism) nonzero simple submodule. In particular, if H acts irreducibly on L, a nonzero structure of H-invariant algebra on L exists if and only if L occurs in the decomposition of $L \otimes L$ into irreducibles.

(5.8) Assume now that H is *connected and semisimple*. Let Λ be the monoid of its dominant weights written additively. According to (5.7) and (5.5)

$$\Lambda_{\text{alg}} = \{ \lambda \in \Lambda \mid R(\lambda) \text{ has a nonzero structure of } H\text{-invariant algebra} \}$$
$$= \{ \lambda \in \Lambda \mid R(\lambda) \text{ is a submodule of } R(\lambda) \otimes R(\lambda) \}.$$

It follows from Theorem 5 (ii) that

THEOREM 6. Λ_{alg} *is a finitely generated monoid.*

Closedness of Λ_{alg} with respect to addition was proven in [Kr1, Kr2]. In [E2] another proof was given and it was conjectured that Λ_{alg} is finitely generated. This conjecture was proven by M. Brion and F. Knop; see their proof in [E1] (simultaneously their arguments give a new proof of closedness).

(5.9) By Theorem 6, the set of all irreducible H-modules admitting a nonzero structure of a simple H-invariant algebra is described by finite data, namely, by a system of generators of Λ_{alg} (notice that there is only one minimal system, namely, $\Lambda_{\text{alg}} \setminus 2\Lambda_{\text{alg}}$).

Finding such a system is reduced to the cases of *simple* groups H, [E2].

The problem of explicit description of Λ_{alg} and of its generators for simple groups H was considered in [E1, E2], where one can find a lot of additional information (many of the results in [E1, E2] are just announced, apparently due to the fact that the proofs are based on the explicit bulky calculations). I shall adduce the formulations of some of these results as an illustration of the character of information which is available at this writing.

(5.10) A complete description of Λ_{alg} is known for the groups of types B_l, E_7, and E_8.

If $H = B_l$, then

$$\omega_i \text{ for } 1 \leqslant i \leqslant l-1 \text{ and } i \text{ even}, \text{ if } i \leqslant (2l+1)/3;$$
$$\omega_i + \omega_{i+2s}, \text{ for } i \text{ odd and } i + 2s < (2l+1)/3;$$
$$\omega_i + \omega_j, \text{ for } i < (2l+1)/3 \leqslant j \leqslant l-1;$$
$$\omega_i + 2\omega_l, \text{ for } 1 \leqslant i \leqslant l-1;$$
$$2\omega_l$$

is the minimal system of generators of Λ_{alg}.

If $H = E_7$, then $\Lambda_{\text{alg}} = \Lambda_{\text{rad}}$.

If $H = E_8$, then $\Lambda_{\text{alg}} = \Lambda$.

(5.11) For other groups only partial results are obtained, for instance:

If $H = C_l$, D_{2l}, G_2 or F_4, then $2\Lambda \subset \Lambda_{\text{alg}}$.

If $H = D_{2l+1}$ or E_6, then all self-contragredient weights from Λ_{rad} lie in Λ_{alg}.

(5.12) Case $H = A_l$ appears to be the most complicated one, see [E1, E2], where the minimal systems of generators of Λ_{alg} for certain small values of l are found.

(5.13) Concluding this section, I shall make several remarks concerning question (c) from (5.4) under the assumption that $H \subset \text{GL}(L)$ is a *connected semisimple* group acting *irreducibly* on L.

In the first place, it is *not true* in general that $H = \text{Aut}\,A$.

EXAMPLE. Let L be an irreducible D_4-module with the highest weight $2\omega_3$. There is a structure of D_4-invariant algebra on L. It is known that the subgroup B_3 of D_4 acts irreducibly on L (see, for instance, tables in [MP]). Therefore, if one takes H to be this subgroup, $H = B_3 \subsetneq D_4 \subseteq \text{Aut}\,L$.

(5.14) If, using the notation of the proof of Theorem 3, we take $A = \{L, m\}$ and

$$(5.14.1) \qquad m = \sum c_{ij}^s e^i \otimes e^j \otimes e_s,$$

then $\text{Lie}(\text{Aut}\,A)$ consists of all linear operators $X \in \text{End}(L)$ satisfying

$$(5.14.2) \quad \sum c_{ij}^s (X^* \cdot e^i \otimes e^j \otimes e_s + e^i \otimes X^* \cdot e^j \otimes e_s + e^i \otimes e^j \otimes X \cdot e_s) = 0,$$

where $X^* \in \text{End}(L^*)$ is the dual of X.

In the coordinates, condition (5.14.2) is equivalent to a system of linear equations in the entries of X. In principle, this enables one to find $\text{Lie}(\text{Aut}\,A)$ in each special case and hence to decide whether or not H coincides with $(\text{Aut}\,A)^\circ$ (in practice this may not be easy to perform; however, other considerations may be helpful, see Remark (5.17)).

(5.15) As for the group $\text{Aut}\,A$ itself, it is connected under some conditions:

THEOREM 7. *Let A be a finite dimensional algebra with a nonzero multiplication. Assume that*

(i) $\operatorname{Lie}(\operatorname{Aut} A)$ *acts irreducibly on* A,
(ii) $\operatorname{Lie}(\operatorname{Aut} A)$ *has no outer automorphisms.*

Then $\operatorname{Aut} A$ *is a connected semisimple group.*

PROOF. Semisimplicity of $\operatorname{Aut} A$ follows from (i) and Theorem 6.

Let $g \in \operatorname{Aut} A$. It follows from (ii) that conjugation by g is an inner automorphism of $(\operatorname{Aut} A)°$. Hence, there is $g_0 \in (\operatorname{Aut} A)°$ such that gg_0 lies in the center of $(\operatorname{Aut} A)°$. Since this center is trivial by Theorem 6, one has $g \in (\operatorname{Aut} A)°$. □

(5.16) REMARK. The same arguments show that without assumption (ii) natural homomorphism $G/G° \longrightarrow \operatorname{Aut} G°/\operatorname{Int} G°$, $G = \operatorname{Aut} A$, is injective.

(5.17) REMARK. Since H acts irreducibly on L, it follows from [Dy] that "as a rule" H is a maximal connected subgroup of $\operatorname{SL}(L)$ (all exceptions are found in [Dy]), and hence $H = (\operatorname{Aut} A)°$ by Theorem 5.

(5.18) Finally, I shall give an example of a family of simple algebras A such that $\operatorname{Aut} A$ is *not reductive*.

EXAMPLE. We maintain the notation of (4.1), (4.5), and (5.14) and set $n = 3$.

For any $\alpha = (\alpha_1, \ldots, \alpha_7) \in k^7$ denote by $m(\alpha)$ the algebra structure on L with the constants c_{ij}^s in (5.14.1) is determined from the following multiplication table:

$$e_1 e_1 = 0,$$
$$e_1 e_2 = \alpha_1 e_1,$$
$$e_1 e_3 = \alpha_2 e_1 + \alpha_1 e_2,$$
$$e_2 e_1 = -\alpha_1 e_1,$$
$$e_2 e_2 = \alpha_3 e_1,$$
$$e_2 e_3 = \alpha_4 e_1 + (\alpha_2 + \alpha_3) e_2 + \alpha_1 e_3,$$
$$e_3 e_1 = \alpha_5 e_1 - \alpha_1 e_2,$$
$$e_3 e_2 = \alpha_6 e_1 + (\alpha_3 + \alpha_5) e_2 - \alpha_1 e_3,$$
$$e_3 e_3 = \alpha_7 e_1 + (\alpha_4 + \alpha_6) e_2 + (\alpha_2 + 2\alpha_3 + \alpha_5) e_3.$$

Denote $\{L, m(\alpha)\}$ by $A(\alpha)$.

Let X be the linear operator on L defined by the conditions: $X \cdot e_1 = 0$, $X \cdot e_2 = e_1$, $X \cdot e_3 = e_2$. One can easily check that (5.14.2) is fulfilled for $m = m(\alpha)$. Therefore,

(5.14.3) $$X \in \operatorname{Lie}(\operatorname{Aut} A(\alpha)).$$

Let $\mathbf{1} = (1, 1, 1, 1, 1, 1, 1)$ and assume that there is a proper ideal I in $A(\mathbf{1})$.

If $\dim I = 1$ and $e = a_1 e_1 + a_2 e_2 + a_3 e_3 \in I$, $a_i \in k$, $e \neq 0$, then $ee_1 = (a_3 - a_2)e_1 - a_3 e_2 \in I$. Hence, $e = a_1 e_1 + a_2 e_2$. Therefore, $e_2 e = -a_1 e_1 + a_2 e_2$, whence either $e_1 \in I$ or $e_2 \in I$. Since $e_1e_3 = e_1 + e_2$ and $e_1e_2 = e_1$, both cases are impossible.

Thus, $\dim I = 2$. Since $\dim L = 3$ and there are no one-dimensional ideals, I is invariant with respect to $\operatorname{Aut} A(\mathbf{1})$ and hence with respect to X. Since X is nilpotent and $k \cdot e_1$ is its kernel, I has a basis e_1, $a_2e_2 + a_3e_3$, $a_i \in k$. Therefore, $(a_2e_2 + a_3e_3)e_1 = (a_3 - a_2)e_1 - a_3e_2 \in I$ whence $e_2 \in I$. Hence, $e_3e_2 = e_1 + 2e_2 - e_3 \in I$ which contradicts the condition $\dim I = 2$.

Therefore, $A(\mathbf{1})$ is a simple algebra.

It follows now from Theorem 4 that $\Omega := \{m(\alpha) \mid A(\alpha) \text{ is a simple algebra}\}$ is an open and dense subset of the linear subspace $\{m(\alpha) \mid \alpha \in k^7\}$ of $L^* \otimes L^* \otimes L$.

Let $R := \{m(\alpha) \mid \operatorname{Aut} A(\alpha) \text{ is reductive}\}$. If $\alpha \in R$, it follows from (5.14.3) and the Jacobson-Morozov Theorem that there is a homomorphism $\phi : \operatorname{SL}_2 \longrightarrow \operatorname{Aut} A(\alpha)$ such that $X \in d\phi(\operatorname{Lie}(\operatorname{SL}_2))$. Therefore, $A(\alpha)$ is an SL_2-invariant algebra. Since $\operatorname{rk} X = 2$, it is irreducible.

Since for any integer $m \geqslant 0$ there is a unique (up to an isomorphism) simple SL_2-module S_m of dimension $m + 1$ and $S_m \otimes S_m = S_{2m} \oplus S_{2m-2} \oplus S_{2m-4} \oplus \cdots$ (the Clebsch-Gordan formula), there is a unique (up to an isomorphism) irreducible SL_2-invariant algebra in each odd dimension and there are no such even dimensional algebras.

It follows that each $A(\alpha)$, $\alpha \in R$, is isomorphic to $\operatorname{Lie}(\operatorname{SL}_2)$ and therefore, there is a $\operatorname{GL}(L)$-orbit $\mathcal{O}$ in $L^* \otimes L^* \otimes L$ such that $R \subseteq \mathcal{O}$. Since $\operatorname{Aut} \operatorname{Lie}(\operatorname{SL}_2) = \operatorname{PSL}_2$, one has $\dim \mathcal{O} = \dim \operatorname{GL}(L) - \dim \operatorname{PSL}_2 = 6$. Since $\dim\{m(\alpha) \mid \alpha \in k^7\} = 7$, we obtain $\Omega \setminus R \neq \varnothing$.

(5.19) REMARK. This example shows that the method of traces provides a lot of information about the invariants of some *nonreductive* groups as well. It would be interesting to clarify how wide the class of nonreductive groups is which appear in this way (i.e., as the full automorphism groups of simple algebras).

More generally, our discussion leads naturally to the following

QUESTION. *What is a group theoretical characterization of the algebraic groups which may be realised as the full automorphism groups of the finite dimensional simple algebras over an algebraically closed field of characteristic zero ?*

It is known that some interesting *finite* groups may be realised in this way, for instance, the Monster, [Gr].

QUESTION. *Is it true that any finite group may be realised as the full automorphism group of a finite dimensional simple algebra A over an algebraically closed field k of characteristic zero ?*

It is known that the answer to the latter question is affirmative if either the assumption $\dim_k A < \infty$ or the assumption of algebraic closedness of k is dropped: one can even take A to be a field in this case, [DG, Ge].

6. The proof of Theorem 2

(6.1) We maintain the notation of (3.2).
Until explicitly specified, we *do not* assume that A is simple.

(6.2) Let
$$n = \dim A,$$
$e_1, \ldots, e_n$ be a basis of A and $x_i^{(s)}$ the i th coordinate function on the s th summand of $V := A^d$ with respect to this basis.

We identify $x_i^{(s)}$ with $\pi_s^*(x_i^{(s)}) \in k[V]$. Then
$$k[V] = k[\ldots, x_i^{(s)}, \ldots], \quad k(V) = k(\ldots, x_i^{(s)}, \ldots).$$

(6.3) Consider the $k(V)$-algebra
$$A_{k(V)} = k(V) \otimes_k A$$
and identify A with $1 \otimes A$. We also identify $A_{k(V)}$ with the $k(V)$-algebra of all rational mappings $V \longrightarrow A$ considering an element $\sum f_j e_j \in A_{k(V)}$ as the mapping $v \mapsto \sum f_j(v) e_j$. Then

(6.3.1) $$y_s = \sum_{j=1}^n x_j^{(s)} e_j$$

is the projection $\pi_s : V \longrightarrow A$.

(6.4) Denote by B the k-subalgebra of $A_{k(V)}$ generated over k by the elements (6.3.1). One can obtain the following interpretation of $\operatorname{Tr} A^d$ in terms of B.

Denote by $M(A_{k(V)})$ the k-algebra generated over k by all operators T_p, $T \in \{L, R\}$, of left and right multiplication of $A_{k(V)}$ by elements $p \in A_{k(V)}$. We identify these operators with their matrices in the basis $e_1, \ldots, e_n$ of $A_{k(V)}$ over $k(V)$, thus assuming that
$$M(A_{k(V)}) \subset \operatorname{Mat}_n(k(V)).$$

As a matter of fact, $M(A_{k(V)})$ is the $k(V)$-subalgebra of $k(V)$-algebra $\operatorname{Mat}_n(k(V))$ since the dependence of T_p on p is $k(V)$-linear.

Denote by $M(B)$ the k-subalgebra of $M(A_{k(V)})$ generated over k by all operators T_p, $p \in B$. It follows from the definition of B that

(6.4.1) $$M(B) \subset \operatorname{Mat}_n(k[V]).$$

Any element of B is obtained from an appropriate nonassociative polynomial $f = f(t_1, \ldots, t_d)$ in noncommutative variables $t_1, \ldots, t_d$ with the coefficients in k by means of the subtitution $t_i = y_i$. Considered as a mapping $V \longrightarrow A$, this element $f(y_1, \ldots, y_d)$ is of the form $v \mapsto f(v)$ (see (3.2) and (6.3)). Therefore, polynomial (3.2.3) is the trace of polynomial matrix

$$T^{(s)}_{s(y_1, \ldots, y_d)} \cdots T^{(r)}_{r(y_1, \ldots, y_d)}.$$

Therefore, $\operatorname{Tr} A^d$ is the subalgebra with the unit in $k[V]^G$ generated by traces of all of the matrices from $M(B)$.

(6.5) In the next three subsections it is shown that B plays an important role in the separation of orbits by invariants.

THEOREM 8. *Assume that algebra A is generated by $\leqslant d$ elements. Then*

(6.5.1) $$k(V) \cdot B = A_{k(V)}.$$

PROOF. Let $a_1, \ldots, a_d$ be a system of generators of algebra A. Then

(6.5.2) $$e_j = h_j(a_1, \ldots, a_d)$$

for some nonassociative polynomials $h_j(t_1, \ldots, t_d)$ in noncommutative variables $t_1, \ldots, t_d$ with the coefficients in k.

Let us show that n elements $h_j(y_1, \ldots, y_d) \in B$ are linearly independent over $k(V)$. Since $k(V)$-algebra $A_{k(V)}$ is n-dimensional, this will complete the proof.

Since $\{e_j\}$ is a basis,

(6.5.3) $$a_s = \sum_j \alpha_j^{(s)} e_j, \quad \alpha_j^{(s)} \in k.$$

It follows from (6.3.1) that

(6.5.4) $$h_i(y_1, \ldots, y_d) = \sum_j f_{ij} e_j, \quad f_{ij} \in k[V].$$

Formulas (6.5.3) and (6.5.2) show that sustituting $\alpha_j^{(s)}$ instead of $x_j^{(s)}$ in the right-hand side of (6.5.4) is equivalent to substituting a_s instead of y_s in the left-hand side, which gives e_i in view of (6.5.2). Hence, the polynomial matrix (f_{ij}) is specialized to E_n under this specialization of variables $x_j^{(s)}$. Therefore, (f_{ij}) is nondegenerate, and the proof is completed by virtue of (6.5.4). □

COROLLARY 1. *Assume that the algebra A is generated by $\leqslant d$ elements. Then*

$$\dim_K K \cdot B \geqslant n$$

for any subfield K of $k(V)$.

COROLLARY 2. *Assume that the algebra A is generated by $\leqslant d$ elements and K is a subfield of $k(V)$ such that*

$$\dim_K K \cdot B = n.$$

Let $b_1, \ldots, b_n$ be a basis of $K \cdot B$ over K. If $b_i = \sum_j t_{ij} e_j$, $t_{ij} \in k(V)$, one has $t := \det(t_{ij}) \neq 0$.

PROOF. It follows from the assumption, Theorem 8, and Corollary 1 that $b_1, \ldots, b_n$ is a basis of $A_{k(V)}$ over $k(V)$. The claim now follows since $e_1, \ldots, e_n$ is such a basis as well. □

COROLLARY 3. *Assume that the algebra A is generated by $\leq d$ elements. Then*

$$k(V) \cdot M(B) = M(A_{k(V)}).$$

(6.6) Let H be an algebraic group acting morphically on an irreducible algebraic variety X.

We say that a subfield K of $k(X)$ *separates orbits in general position in X*, if for points in general position in X each level variety of K is contained in an H-orbit. In other words, there exists a dense open subset of X such that any two points of it which are not separated by K (i.e. the values of each function from K at these points are either both defined and equal or both not defined) lie in the same H-orbit.

We *do not* assume in this definition that $K \subset k(X)^H$.

The following statement is a general result of Invariant Theory proved by Rosenlicht, [Rs] (see also [PV, 2.3]).

THEOREM 9. *Let $k \subseteq K \subseteq k(X)^H$. Then K separates G-orbits in general position in X if and only if $K = k(X)^H$.*

(6.7) In our case there is the following numerical criterion:

THEOREM 10. *Assume that the algebra A is generated by $\leq d$ elements and let K be a subfield of $k(V)$. If*

$$\dim_K K \cdot B = n,$$

then K separates G-orbits in general position in X.

PROOF. Let $b_1, \ldots, b_n \in B$ be a basis of $K \cdot B$ over K. Since B is a k-algebra, $K \cdot B$ is a K-algebra. Therefore,

(6.7.1) $$b_i b_j = \sum_s f_{ij}^{(s)} b_s, \quad f_{ij}^{(s)} \in K.$$

Since $y_j \in B$, one has

(6.7.2) $$y_j = \sum_s h_j^{(s)} b_s, \quad h_j^{(s)} \in K.$$

According to Corollary 2 of Theorem 8, a notation of which we shall use, there exists an open dense subset U of V such that for any point $v \in U$ all functions $f_{ij}^{(s)}$ and $h_j^{(s)}$ are regular at v, and $t(v) \neq 0$.

If $v \in U$, it follows from $b_i(v) = \sum_j t_{ij}(v) e_j$ and $t(v) = \det(t_{ij}(v)) \neq 0$ that $b_1(v), \ldots, b_n(v)$ is a basis of A over k. It follows from (6.7.1) that $b_i(v) b_j(v) = \sum_s f_{ij}^{(s)}(v) b_s(v)$. Therefore, the elements $f_{ij}^{(s)}(v) \in k$ are the structural constants of A in this basis.

Assume now that the points v and $u \in U$ are not separated by K. Then $f_{ij}^{(s)}(v) = f_{ij}^{(s)}(u)$, i.e., the structural constants of A in the bases $\{b_i(v)\}$ and $\{b_i(u)\}$ are equal. Therefore, the k-linear mapping

(6.7.3) $$g : A \longrightarrow A, \quad g \cdot b_i(v) = b_i(u)$$

is an automorphism of algebra A, i.e., $g \in G$.

Now we show that $g \cdot v = u$. We have $h_j^{(s)}(v) = h_j^{(s)}(u)$ because of the condition on u and v. Therefore, taking into account the interpretation of π_j as y_j, see (6.3), and formulas (6.7.2) and (6.7.3), we get

$$g \cdot \pi_j(v) = g \cdot y_j(v) = g \cdot \sum_s h_j^{(s)}(v) b_s(v)$$
$$= \sum_s h_j^{(s)}(v) \, g \cdot b_s(v) = \sum_s h_j^{(s)}(u) b_s(u) = y_j(u) = \pi_j(u),$$

and the proof is completed. □

QUESTION. *Is the converse to Theorem 10 true?*

(6.8) Assume that we can find a subfield K of $k(V)$ such that
(a) $K \subseteq \operatorname{Frac}(\operatorname{Tr} A^d)$,
(b) $\dim_K K \cdot B = n$.

In view of (b) and Theorem 10, K separates orbits in general position in V. Therefore, the same holds for the field $\tilde{K}$ generated by k and K. Since $k \subseteq \tilde{K} \subseteq \operatorname{Frac}(\operatorname{Tr} A^d) \subseteq k(V)^G$, it follows from Theorem 9 that $\tilde{K} = \operatorname{Frac}(\operatorname{Tr} A^d) = k(V)^G$.

Therefore, to prove Theorem 2 it suffices to find a field K having properties (a) and (b). This is done in the subsections (6.9)–(6.11). The methods used for that purpose are related entirely to Noncommutative Ring Theory which indicates, from my point of view, that its role in Invariant Theory is not undestood properly as yet.

(6.9) Using the notation of (6.4), consider in $M(A_{k(V)})$ the k-subalgebra $M(A)$ generated over k by all operators T_p, where $T \in \{L, R\}$, $p \in A$. Since $k(V) \cdot A = A_{k(V)}$, one has $k(V) \cdot M(A) = M(A_{k(V)})$.

THEOREM 11. (1) *If the algebra A is simple, then*

(6.9.1) $$M(A) = \operatorname{Mat}_n(k),$$
(6.9.2) $$M(A_{k(V)}) = \operatorname{Mat}_n(k(V)).$$

(2) *If the algebra A is simple and generated by $\leqslant d$ elements, then*

(6.9.3) $$k(V) \cdot M(B) = \operatorname{Mat}_n(k(V)),$$

and $K \cdot B$ and $K \cdot M(B)$ are prime K-algebras for any subfield K of $k(V)$.

PROOF. (1) A is an $M(A)$-invariant subspace of $A_{k(V)}$ and a faithful $M(A)$-module (since A is simple). Therefore, (6.9.1) follows from the Density Theorem and algebraic closedness of k, [J1, He]. Equality (6.9.2) follows from (6.9.1).

(2) Equality (6.9.3) follows from (6.9.2) and Corollary 3 of Theorem 8.

Let I_1, I_2 be nonzero ideals of $K \cdot B$. It follows from Theorem 8 that $\tilde{I}_j = k(V) \cdot I_j$ are nonzero ideals of $k(V)$-algebra $A_{k(V)}$. Since A is simple, this algebra is simple as well, [J2]. Hence $\tilde{I}_1 = \tilde{I}_2 = A_{k(V)}$. Therefore, it would follow from

$I_1I_2 = 0$ that A has zero multiplication. This contradiction shows that $K \cdot B$ is prime.

Using (6.9.3) and the simplicity of the $k(V)$-algebra $\operatorname{Mat}_n(k(V))$ one can prove in the same way that $K \cdot M(B)$ is a prime algebra. □

(6.10) We assume from now on that A is *simple* and *generated by* $\leq d$ *elements*.

It follows from (6.4.1) that $M(B)$ is a PI-algebra. It is prime due to Theorem 11. It follows from these properties that $M(B)$ has a nonzero center Z, see [J3] (in our case this fact also immediately follows from (6.9.3) and from existence of polylinear central polynomials for matrix rings, see [Ro] and subsection (6.13) below).

We assume that $k(V) \subseteq \operatorname{Mat}_n(k(V))$ identifying $f \in k(V)$ with fE_n. Then $k(V)$ is the center of $\operatorname{Mat}_n(k(V))$.

It follows from Theorem 11 (2), and description of $\operatorname{Tr} A^d$ given in (6.4) that

(6.10.1) $$Z \subset \operatorname{Tr} A^d.$$

(6.11) Now we shall show that $K := \operatorname{Frac}(Z) \subset k(V)$ has properties (a) and (b) indicated in (6.8).

Property (a) follows from (6.10.1), therefore, we need only to prove that (b) is fulfilled.

THEOREM 12. *If the algebra A is simple and generated by $\leq d$ elements and $K = \operatorname{Frac}(Z)$, then*:

(i) $K \cdot M(B)$ *is a finitely dimensional central simple K-algebra*,
(ii) $\dim_K K \cdot M(B) = n^2$,
(iii) $K \cdot B$ *is a simple K-algebra*,
(iv) $\dim_K K \cdot B = n$.

PROOF. Since $M(B)$ is a prime PI-algebra, (i) results from the Posner–Rowen Theorem, [J3].

By (i), there exists a finite field extension K'/K such that for some s one has an isomorphism of K'-algebras, [J1, He]:

(6.11.1) $$K \cdot M(B) \otimes_K K' \cong \operatorname{Mat}_s(K').$$

Hence

(6.11.2) $$\dim_K K \cdot M(B) = s^2.$$

It is well known, [J1, He], that if F is a field, then the standard polynomial $[x_1, \ldots, x_{2m}]$ is a polynomial identity of degree $2m$ of the F-algebra $\operatorname{Mat}_m(F)$ (the Amitsur–Levitski Theorem), and $\operatorname{Mat}_m(F)$ does not satisfy an identity of a smaller degree. Now (i) follows from here and (6.11.1), (6.9.3), and (6.11.2).

As a vector space over K, the algebra $A_{k(V)}$ is a $K \cdot M(B)$-module and $K \cdot B$ is a submodule. Submodules of $K \cdot M(B)$-module $K \cdot B$ are precisely ideals of the K-algebra $K \cdot B$. Assume that this algebra is not simple and let I be a proper ideal. It follows from (i) that any $K \cdot M(B)$-module is completely reducible, [He], therefore $K \cdot B = I \oplus J$ for some nonzero ideal J. Hence, $IJ \subset I \cap J = \{0\}$, contrary to primeness of $K \cdot B$ (cf. Theorem 11 (2)), whence (iii) follows.

Let $M(K \cdot B)$ and C be respectively the multiplication algebra and the centroid of the K-algebra $K \cdot B$. It follows from (iii), see [J1, He], that C is a field (containing K) and $M(K \cdot B)$ is a dense subring of the C-algebra $\mathrm{End}_C K \cdot B$ of all C-linear transformations of $K \cdot B$. It follows from (6.5.1) that the restriction of operators to $K \cdot B$ is an isomophism of K-algebras

(6.11.3) $$K \cdot M(B) \xrightarrow{\cong} M(K \cdot B).$$

It follows from (6.11.3) and (ii) that

(6.11.4) $$\dim_K M(K \cdot B) = n^2.$$

In particular, $M(K \cdot B)$ is finite dimensional over K. Since $K \subseteq C$, we obtain that $C \cdot M(K \cdot B)$ is finite dimensional over C. It follows from here and from the density property that

(6.11.5) $$\dim_C K \cdot B = s < \infty,$$

and $M(K \cdot B) = C \cdot M(K \cdot B) = \mathrm{End}_C K \cdot B$. Therefore, $M(K \cdot B)$ is isomorphic to $\mathrm{Mat}_s(C)$. Using the same arguments as in the proof of (ii), one gets from here and (6.11.3), (6.9.3) that

(6.11.6) $$s = n,$$

and hence

(6.11.7) $$\dim_C M(K \cdot B) = n^2.$$

It follows from (6.11.4) and (6.11.7) that

(6.11.8) $$C = K.$$

Now (iv) results from (6.11.5), (6.11.6), and (6.11.8). □

(6.12) REMARK. Since $k \cdot Z = Z$, one has $k \subset K$. Therefore, it is proved (see (6.8)) that $K = k(V)^G$.

(6.13) REMARK. There is a construction that gives "explicit" expressions of traces of elements from $M(B)$ as fractions of traces of elements from Z (this was used in the proof of Theorem 2 given in [I1]).

Namely, let $f = f(x_1, \ldots, x_s)$, $s \geqslant n^2$, be a central polynomial (its existence was proven in [Fo1, Ra1], cf. [Ro]), i.e. a polylinear skew-symmetric polynomial with integer coefficients in noncommutative variables $x_1, \ldots, x_s$ such that for any commutative ring R the following is fulfilled:

(a) $f(Q_1, \ldots, Q_s)$ is a scalar matrix for any $Q_i \in \mathrm{Mat}_n(R)$;
(b) $f(P_1, \ldots, P_s) \neq 0$ for some $P_i \in \mathrm{Mat}_n(R)$;
(c) $n \cdot \mathrm{tr}\, Q \cdot f(Q_1, \ldots, Q_s) = \sum_{i=1}^{n^2} f(Q_1, \ldots, Q_{i-1}, Q_i Q, Q_{i+1}, \ldots, Q_s)$ for any $Q_i, Q \in \mathrm{Mat}_n(R)$.

Let $R = k(V)$. It results from (a) that $f(Q_1, \ldots, Q_s) \in Z$ for any $Q_i \in M(B)$, and it follows from (6.9.3) that one can take $P_i \in M(B)$ in (b). Now it follows from (c) that for any $Q \in M(B)$ one has $\operatorname{tr} Q = z'/nz$, where $z = f(P_1, \ldots, P_s)$ and z' is the value of the right-hand side of (c) at $Q_i = P_i$.

References

[AG] M. Adamovich and E. O. Golovina, *Simple linear Lie group having a free algebra of invariants*, Voprosy Teorii Grupp i Gomologicheskoi Algebry **2** (1979), 3–41; English transl. in Selecta Math. Sovietica **3** (1984), no. 2.

[Ar] M. Artin, *On Azumaya algebras and finite-dimensional representations of rings*, J. Algebra **11** (1969), 532–563.

[BLB] C. Bessenrodt and L. Le Bruyn, *Stable rationality of certain* PGL_n-*quotients*, Invent. Math. **104** (1991), 179–199.

[Bo] N. Bourbaki, *Groupes et Algèbres de Lie*, 4-6, 7-8, Hermann, Paris, 1968, 1974.

[DG] M. Dugas and R. Göbel, *Automorphism groups of fields*, Manuscripta Math. **85,** (1994), no. 3-4, 227–242.

[Di] J. Dixmier, *Certaines algèbres non associatives simples définies par la trasvections du formes binaires*, J. Reine Angew. Math. **346** (1984), 110–128.

[Dy] E. B. Dynkin, *Maximal subgroups of classical groups*, Trudy Mosk. Mat. Obshch. **1** (1952), 39–166; English transl., Amer. Math. Soc. Transl. (2), vol. 6, Amer. Math. Soc., Providence, RI, 1957, pp. 245–378.

[E1] A. G. Elashvili, *Invariant algebras*, Lie Groups, Their Discrete Subgroups, and Invariant Theory, Advances in Soviet Math., vol. 8, Amer. Math. Soc., Providence, RI, 1992, pp. 57–64.

[E2] _____, *Algebras with an irreducible group of automorphisms*, Questions in Algebra, Proc. Tenth All Union Symposium on Group Theory, vol. 4, Universitetskoe, Minsk, 1989, pp. 152–162. (Russian)

[Fo1] E. Formanek, *Central polynomials for matrix rings*, J. Algebra **23** (1972), 129–132.

[Fo2] _____, *The center of the ring of* 3×3 *generic matrices*, Linear and Multilinear Algebra **7** (1979), 203–212.

[Fo3] _____, *The center of the ring of* 4×4 *generic matrices*, J. Algebra **62** (1980), 304–319.

[Ge] W. D. Geyer, *Jede endliche Gruppe ist Automorphismengruppe einer endlicher Erweiterung* $K/\mathbf{Q}$, Arch. Math. **41** (1983), 139–142.

[Gr] R. L. Griess, Jr., *The friendly Giant*, Invent. Math. **69** (1982), 1–102.

[Gu] G. B. Gurevich, *Foundations of the theory of algebraic invariants*, Moscow, 1948; English transl., Noordhoff, Groningen, 1964.

[GY] J. H. Grace and A.Young, *The algebra of invariants*, Cambridge Univ. Press, Cambridge, 1903.

[He] I. N. Herstein, *Noncommutative rings*, The Carus Math. Monographs, vol. 15, MAA, distributed by J. Wiley, New York, 1968.

[Hil] D. Hilbert, *Über die vollen Invariantensysteme*, Math. Ann. **42** (1893), 313–373.

[I1] A. V. Iltyakov, *Trace polynomials and Invariant theory*, Geometriae Dedicata (to appear).

[I2] _____, *On invariants of the group of automorphisms of Albert algebra*, Preprint (1994).

[IS] A. V. Iltyakov and I. P. Shestakov, *On invariants of* F_4 *and the center of Albert algebra*, Preprint (1994).

[J1] N. Jacobson, *Structure of rings*, Colloquium Publ., vol. XXXVII, Amer. Math. Soc., Providence, RI, 1956.

[J2] _____, *Lie algebras*, Interscience, New York, 1962.

[J3] _____, *P-I Algebras. An Introduction*, Lecture Notes in Math., vol. 441, Springer-Verlag, Heidelberg, 1975.

[Ka1] P. I. Katsylo, *Stable rationality of fields of invariants of linear representations of the groups* PSL_6 *and* PSL_{12}, Mat. Zametki **48** (1990), no. 1-2, 49–52; English transl. in Math. Notes **48** (1990).

[Ka2] _____, *Birational geometry of moduli varieties of vector bundles over* $\mathbf{P}^2$, Izv. Akad. Nauk SSSR, Ser. Mat. **55** (1991), 429–438; English transl. in Math. USSR-Izv. **38** (1992).

[Ko] B. Kostant, *Theorem of Frobenius, a theorem of Amitsur-Levitski and cohomology theory*, J. Math. Mech. **7** (1958), 237–264.

[Kr1] M. Krämer, *Eine Klassifikation bestimmter Untergruppen kompakter zusammenhängender Liegruppen*, Comm. Algebra **3** (1975), 691–737.

[Kr2] _____, *Über Untergruppen kompakter Liegruppen als Isotropiegruppen bei linearen Aktionen*, Math. Z. **147** (1976), 207–224.
[LB1] L. Le Bruyn, *Some remarks on rational matrix invariants*, J. Algebra **118** (1988), 487–493.
[LB2] _____, *Simultaneous equivalence of square matrices*, Lecture Notes in Math., vol. 1404, Springer-Verlag, Heidelberg, 1989, pp. 127–136.
[LBS] L. Le Bruyn and A. Schofield, *Rational invariants of quivers and the ring of matrix invariants*, Perspectives in Ring Theory, NATO ASI Series, Ser. C., vol. 233, Kluwer, Amsterdam, 1987, pp. 21–29.
[LBT] L. Le Bruyn and Y. Teranishi, *Matrix invariants and complete intersections*, Glasgow Math. J. **32** (1990), 227–229.
[MP] W. G. McKay and J. Patera, *Tables of dimensions, indices, and branching rules for representations of simple Lie algebras*, Marcel Dekker, New York and Basel, 1981.
[OV] A. L. Onishchik and E. B. Vinberg, Moscow (1988), "Nauka"; English transl. (1990), Springer-Verlag, Berlin and Heidelberg.
[**Pol**] S. V. Polikarpov, *Free affine Albert algebras*, Sibirsk. Mat. Zh. **32** (1991), no. 6, 131–141; English transl. in Siberian Math. J. **32** (1992).
[PS] S. V. Polikarpov and I. P. Shestakov, *Nonassociative affine algebras*, Algebra i Logika **29** (1990), 709–723; English transl. in Algebra and Logic **29** (1990).
[Pop] A. M. Popov, *Finite isotropy subgroups in general position in simple linear Lie groups*, Trudy Mosk. Mat. Obshch. **48** (1985), 7–59; English transl. in Trans. Moscow Math. Soc. **48** (1988).
[Po1] V. L. Popov, *Sections in invariant theory*, Proc. of the Sophus Lie Memorial Conference (Oslo, 1992), Scand. Univ. Press, Oslo, 1994, pp. 315–362.
[Po2] _____, *Analogues of M. Artin's conjecture on invariants for nonassociative algebras*, Notes of the talk given at the Advanced Workshop on Algebraic Geometry (August 15-26, 1994) August, ICTP, Tieste, Italy.
[**PV**] V. L. Popov and E. B. Vinberg, *Invariant theory*, Encyclopaedia of Math. Sci.: Algebraic Geometry IV, vol. 55, Springer-Verlag, Berlin, Heidelberg, and New York, 1994, pp. 123–284.
[Pr1] C. Procesi, *Invariant theory of $n \times n$ matrices*, Adv. in Math. **19** (1976), 306–381.
[Pr2] _____, *Non-commutative affine rings*, Atti Accad. Naz. Lincei VIII. Ser. **VIII, fo. 6** (1967), 239–255.
[Pr3] _____, *Relazioni tra geometria algebrica ed algebra non commutativa. Algebre cicliche e problema di Lüroth*, Boll. Un. Mat. Ital. A (5) **18** (1981), no. 1, 1–10.
[Ra1] Yu. P. Razmyslov, *On a problem of Kaplansky*, Izv. Akad. Nauk SSSR Ser. Mat. **37** (1973), 483–501; English transl. in Math. USSR-Izv. **7** (1973).
[Ra2] _____, *Trace identities of full matrix algebras over a field of charcteristic zero*, Izv. Akad. Nauk SSSR Ser. Mat. **38** (1974), 723–756; English transl. in Math. USSR Izv. **8** (1974).
[Re] Z. Reichstein, *On automorphisms of matrix invariants*, preprint MSRI 04708-91 (1991), Math. Sci. Research Inst., Berkeley, CA.
[Ro] L. H. Rowen, *Polynomial identities in ring theory*, Academic Press, New York, 1980.
[Ros] S. Rosset, *Generic matrices, K_2, and unirational fields*, Bull. Amer. Math. Soc. **81** (1975), no. 4, 707–708.
[Sa] D. J. Saltman, *Noether's problem over an algebraically closed field*, Invent. Math. **77** (1984), 71–84.
[Sc] A. Schofield, *Matrix invariants of composite size*, J. Algebra **147** (1992), 345–349.
[S1] G. W. Schwarz, *Representations of simple Lie groups with regular rings of invariants*, Invent. Math. **49** (1978), 167–191.
[S2] _____, *Invariant theory of G_2 and $Spin_7$*, Comment. Math. Helv. **63** (1988), 624–663.
[S3] _____, *On classical invariant theory and binary cubics*, Ann. Inst. Fourier **37** (1987), no. 3, 191–216.
[Sib] K. S. Sibirski, *Algebraic invariants for a set of matrices*, Sibirsk. Mat. Zh. **9** (1968), no. 1, 152–164; English transl. in Siberian Math. J. **9** (1968).
[Sim] V. T. Simoniya, *The First Fundamental Theorem in the theory of vector invariants of the exceptional Lie group G_2*, Soobshch. Akad. Nauk Gruz. SSR **24** (1960), no. 6, 641–648. (Russian)
[Sm] G. F. Smith, *A complete set of unitary invariants for N 3×3 complex matrices*, Tensor **21** (1970), 273–283.
[VdB] M. Van den Bergh, *The center of the generic division algebras*, J. Algebra **127** (1989), no. 1, 106–126.

[Vi] E. B. Vinberg, *On invariants of a set of matrices*, Preprint (1993).
[Vin] V. Vinnikov, *Complete description of determinantal representations of smooth irreducible curves*, Linear Algebra Appl. **125** (1989), 103–140.

DEPARTMENT OF MATHEMATICS, MOSCOW STATE UNIVERSITY MGIEM, BOL'SHOĬ TREKHSVYATITEL'SKIĬ PER., 3/12, 109028 MOSCOW, RUSSIA
  E-mail address: vladimir@popov.msk.su

On Reductive Algebraic Semigroups

E. B. VINBERG

§0. Introduction

An (affine) algebraic semigroup is an affine algebraic variety S with an associative multiplication

$$\mu\colon S \times S \to S,$$

which is a morphism of algebraic varieties. A zero of a semigroup S is such an element 0 (if it exists) that $0s = s0 = 0$ for any $s \in S$.

Any (affine) algebraic group is an algebraic semigroup. An important example of an algebraic semigroup which is not a group is the semigroup $\operatorname{End} V$ of endomorphisms of a (finite-dimensional) vector space V. Moreover, if $\dim V = n$, then for any $r = 1, \ldots, n$

$$\operatorname{End}_r V = \{A \in \operatorname{End} V : \operatorname{rk} A \leqslant r\}$$

is an algebraic semigroup with zero (but without unit, unless $r = n$).

It is well known that any algebraic group is isomorphic to a (Zariski) closed subgroup of the group $\operatorname{Aut} V = GL(V)$ of automorphisms of a suitable vector space V. A slight modification of the proof of this theorem allows us to prove that any algebraic semigroup S is isomorphic to a closed subsemigroup of $\operatorname{End} V$ for a suitable V. Moreover, if S has a unit, one may assume that it corresponds to the identity map of V under this isomorphism. (See [3] or [7] for details.) In this situation, an element of S is invertible if and only if it corresponds to an element of $GL(V)$. It follows that the group $G(S)$ of invertible elements (the unit group) of S is open in S and is an algebraic group. In particular, if S is a group, it is an algebraic group.

In what follows we assume that the base field k is algebraically closed, of characteristic 0, and the variety S is irreducible. An algebraic semigroup S is called (*geometrically*) *normal* if the variety S is normal.

For semigroups with units (monoids), we shall assume that their homomorphisms take the unit to the unit. Note that if $\varphi\colon S \to S'$ is a dominant homomorphism of algebraic semigroups with units, then $\varphi(G(S))$ is an open subgroup in $G(S')$ and hence $\varphi(G(S)) = G(S')$.

1991 *Mathematics Subject Classification.* Primary 20G15, 20M20.

©1995, American Mathematical Society

An algebraic semigroup S with unit is called *reductive*, if the group $G(S)$ is reductive. One can show (see [7, 8] and Proposition 1 below) that $G(S)$ cannot be semisimple, unless S is a group.

Reductive algebraic semigroups were studied by Putcha [2, 3] and Renner [4–7]. In particular, Renner classified the reductive semigroups S satisfying the following conditions:

(R1) the center of $G(S)$ is one-dimensional;
(R2) S has a zero;
(R3) S is normal.

Roughly speaking, Renner's result reduces to the assertion that such a semigroup S is uniquely determined by $G(S)$ and the closure of a maximal torus T of $G(S)$, which may be any affine embedding of T, equivariant with respect to the action of T on itself by multiplications and to the Weyl group.

An example of a reductive semigroup satisfying the conditions (R1) and (R2) is

$$S = \overline{k^* G_0} \subset \text{End } V,$$

where $G_0 \subset GL(V)$ is a connected semisimple linear group. In this example, $G(S) = k^* G_0$. In particular, if $G_0 = SL(V)$, then $S = \text{End } V$.

Apparently, condition (R1) is not really essential for Renner's method. However, in this article we propose another approach to the classification problem.

The commutative reductive semigroups were studied by Neeb [18].

Now we state the results of the article. Their proofs are given in §1–§9.

1. Let S be a reductive semigroup and $G = G(S)$. We define an action of $G \times G$ on S by

$$(g_1, g_2) \circ s = g_1 s g_2^{-1}.$$

The algebra $k[S]$ is a $(G \times G)$-invariant subalgebra of $k[G]$.

Let T be a Cartan subgroup of G and B a Borel subgroup containing T. We denote by $\mathfrak{X}$ the character group of T and by $\mathfrak{X}_+$ the semigroup of dominant characters with respect to B.

It is well known that

(1) $$k[G] = \bigoplus_{\Lambda \in \mathfrak{X}_+} k[G]_\Lambda,$$

where $k[G]_\Lambda$ denotes the linear space of the matrix entries of the irreducible linear representation $R^{(\Lambda)}$ of G with highest weight Λ. The summands of (1) are minimal $(G \times G)$-invariant subspaces, and the corresponding irreducible representations of $G \times G$ are mutually nonisomorphic. It follows that any $(G \times G)$-invariant subspace of $k[G]$ is the sum of some $k[G]_\Lambda$. In particular,

(2) $$k[S] = \bigoplus_{\Lambda \in \mathfrak{L}} k[G]_\Lambda,$$

where $\mathfrak{L} = \mathfrak{L}(S)$ is a subset of $\mathfrak{X}_+$.

The multiplication $\mu\colon G \times G \to G$ in the group G defines, and is defined by, the algebra homomorphism

$$\mu^*\colon k[G] \to k[G] \otimes k[G],$$

which is called the *comultiplication* in the algebra $k[G]$. It is given by the following formula: if $f_{ij}^{(\Lambda)}$ denotes the (i,j)th matrix entry of $R^{(\Lambda)}$, then

(3) $$\mu^* f_{ij}^{(\Lambda)} = \sum_k f_{ik}^{(\Lambda)} \otimes f_{kj}^{(\Lambda)}.$$

Obviously, the comultiplication in $k[S]$ is just the restriction of the comultiplication in $k[G]$.

Thus distinguishing S among all algebraic semigroups containing G as the unit group reduces to indicating $\mathfrak{L}$. We shall say that $\mathfrak{L}$ defines S, and denote $S = S(\mathfrak{L})$.

For any $\Lambda, M \in \mathfrak{X}_+$, we denote by $\mathfrak{X}(\Lambda, M)$ the set of the highest weights of irreducible components of the representation $R^{(\Lambda)} R^{(M)}$. It is known that $\mathfrak{X}(\Lambda, M) \ni \Lambda + M$. We have

(4) $$k[G]_\Lambda k[G]_M = \bigoplus_{N \in \mathfrak{X}(\Lambda, M)} k[G]_N.$$

It follows that $\mathfrak{L}(S) = \mathfrak{L}$ satisfies the condition

(5) $$\Lambda, M \in \mathfrak{L} \implies \mathfrak{X}(\Lambda, M) \subset \mathfrak{L}.$$

In particular, $\mathfrak{L}$ is a subsemigroup (containing 0) of $\mathfrak{X}_+$.

We call a subsemigroup $\mathfrak{L} \subset \mathfrak{X}_+$ *perfect* if it contains 0 and satisfies the condition (5).

THEOREM 1. *A subset $\mathfrak{L} \subset \mathfrak{X}_+$ defines an algebraic semigroup containing G as the unit group if and only if it is a perfect finitely generated subsemigroup generating the group $\mathfrak{X}$.*

2. If we require that S be normal, a more explicit description of $\mathfrak{L}(S)$ is available.

Let $\mathfrak{g}$ and $\mathfrak{t}$ be the tangent algebras of G and T, respectively. Identifying characters of T with their differentials[1], we put

$$\mathfrak{t}(\mathbb{Q}) = \{h \in \mathfrak{t} : \Lambda(h) \in \mathbb{Q}, \text{ for all } \Lambda \in \mathfrak{X}\},$$

so the dual space $\mathfrak{t}(\mathbb{Q})^*$ is identified with $\mathfrak{X} \otimes \mathbb{Q}$. Let $\alpha_1, \ldots, \alpha_n$ be the simple roots of G and $h_1, \ldots, h_n$ the corresponding dual roots. The Weyl chamber $C \subset \mathfrak{t}(\mathbb{Q})^*$ is defined by

$$C = \{\Lambda \in \mathfrak{t}(\mathbb{Q})^* : \Lambda(h_i) \geqslant 0,\ i = 1, \ldots, n\}.$$

The group G and the torus T decompose into the almost direct products $G = ZG_0$, $T = ZT_0$, where Z is the connected center and G_0 the commutator group of G, and $T_0 = T \cap G_0$ a Cartan subgroup of G_0.

---

[1] Since addition in the group $\mathfrak{X}$ corresponds to multiplication in the algebra $k[T]$, characters of T, when considered as elements of this algebra, are denoted as exponentials.

Let $\mathfrak{z}$, $\mathfrak{g}_0$, and $\mathfrak{t}_0$ denote the tangent algebras of Z, G_0, and T_0, respectively. Then

$$\mathfrak{g} = \mathfrak{z} \oplus \mathfrak{g}_0, \qquad \mathfrak{t} = \mathfrak{z} \oplus \mathfrak{t}_0.$$

If $\mathfrak{z}(\mathbb{Q}) = \mathfrak{z} \cap \mathfrak{t}(\mathbb{Q})$, $\mathfrak{t}_0(\mathbb{Q}) = \mathfrak{t}_0 \cap \mathfrak{t}(\mathbb{Q})$, then

(6) $$\mathfrak{t}(\mathbb{Q})^* = \mathfrak{z}(\mathbb{Q})^* \oplus \mathfrak{t}_0(\mathbb{Q})^*,$$
$$C = \mathfrak{z}(\mathbb{Q})^* + C_0,$$

where $C_0 \subset \mathfrak{t}_0(\mathbb{Q})^*$ is the Weyl chamber of G_0.

THEOREM 2. *A subset $\mathfrak{L} \subset \mathfrak{X}_+$ defines a normal algebraic semigroup, containing G as the unit group, if and only if $\mathfrak{L} = \mathfrak{X}_+ \cap K$, where K is a closed convex polyhedral cone in $\mathfrak{t}(\mathbb{Q})^*$ satisfying the conditions*

(1) $K \ni -\alpha_1, \ldots, -\alpha_n$;
(2) *the cone $K \cap C$ generates $\mathfrak{t}(\mathbb{Q})^*$.*

The semigroup $S(\mathfrak{L})$ has a zero if and only if

(3) *the cone $D = K \cap \mathfrak{z}(\mathbb{Q})^*$ is pointed*;
(4) $K \cap C_0 = \{0\}$.

We emphasize that any subset $\mathfrak{L} \subset \mathfrak{X}_+$ satisfying conditions (1) and (2) of the theorem automatically satisfies the conditions of Theorem 1.

REMARKS. 1. The projection of K on $\mathfrak{t}_0(\mathbb{Q})^*$ is a convex cone containing an interior point of C_0 and the negative simple roots. Hence it is the whole space $\mathfrak{t}_0(\mathbb{Q})^*$.

2. Since $C_0 \subset \operatorname{conv}\{\alpha_1, \ldots, \alpha_n\}$, the projection of $K \cap C$ on $\mathfrak{z}(\mathbb{Q})^*$ is contained in K (and coincides with D). It is a generating convex cone in $\mathfrak{z}(\mathbb{Q})^*$.

3. We may (and will) assume that the cone K is the greatest one among the convex cones having the same intersection with C. This means that any hyperplane bounding K bounds $K \cap C$. Under this condition, the cone K is uniquely determined by the semigroup.

COROLLARY. *Any normal reductive semigroup decomposes into an almost direct product of a reductive group and a (normal) reductive semigroup with zero.*

(For non-normal semigroups this is not true.)

"An almost direct product" means a quotient of the direct product with respect to a finite central subgroup. Note that if Γ is a finite central subgroup of an algebraic semigroup S, then the quotient semigroup S/Γ turns out to be an algebraic semigroup, being supplied with the structure of an affine algebraic variety as the invariant-theoretic quotient $S/\!/\Gamma$. (Since Γ is finite, the fibers of the canonical morphism $S \to S/\!/\Gamma$ are exactly the Γ-orbits: see, for example, [9].) We shall say that S is a covering semigroup of S/Γ.

3. Let S be a reductive semigroup with $G(S) = G$. The Borel subgroup of $G \times G$ has an open orbit in G, hence S is a spherical $(G \times G)$-variety and therefore contains only finitely many $(G \times G)$-orbits [11]. (More immediately, this follows from their description given below.)

Now consider the $(G_0 \times G_0)$-action on S. Let

(7) $$A = A(S) = S//(G_0 \times G_0)$$

be the invariant-theoretic quotient of S with respect to this action. By definition, A is the spectrum of the subalgebra $k[S]^{G_0 \times G_0} \subset k[S]$, consisting of the $(G_0 \times G_0)$-invariant polynomial functions on S.

Denote by $\mathfrak{X}_Z$ the subgroup of $\mathfrak{X}$ consisting of the characters vanishing on $\mathfrak{t}_0$. (These are the (highest) weights of the one-dimensional representations of G.) Then

(8) $$k[A] = k[S]^{G_0 \times G_0} = \bigoplus_{\Lambda \in \mathfrak{L}_Z} k[G]_\Lambda,$$

where $\mathfrak{L}_Z = \mathfrak{L} \cap \mathfrak{X}_Z$. The embedding $k[A] \subset k[S]$ defines the canonical morphism

(9) $$\pi \colon S \to A.$$

According to a general theorem of invariant theory (see, for example, [9]), π is surjective. If S is normal, A is also normal.

Since the subalgebra $k[A] \subset k[S]$ is $(G \times G)$-invariant, the action of the group $G \times G$ on S induces its action on A in such a way that the morphism π is equivariant. Obviously, the latter action reduces to an action of the torus $G/G_0 = Z/Z_0$, where $Z_0 = Z \cap G_0$.

Moreover, it follows from (3) that $\mu^* k[A] \subset k[A] \otimes k[A]$. Thereby A is endowed with the structure of a (commutative) algebraic semigroup in such a way that the morphism π is a semigroup homomorphism. The image of the unit of S is a unit of A. If S has a zero, its image is a zero of A.

DEFINITION 1. The algebraic semigroup $A = A(S)$, together with the homomorphism $\pi \colon S \to A$, is called the *abelization* of S.

According to the theory of toric varieties [13], the $G(A)$-orbits in A are in a one-to-one correspondence with the (closed) faces of the cone $D = \mathbb{Q}_+ \mathfrak{L}_Z$ in such a way that the ideal of (the closure of) an orbit is spanned by those subspaces $k[A]_\chi = k[G]_\chi$ for which χ does not belong to the corresponding face. This correspondence is monotone in the following sense: for two orbits O_1, O_2, corresponding to faces F_1, F_2, we have $O_1 \subset \overline{O}_2$ if and only if $F_1 \subset F_2$. The orbit O, corresponding to a face F, contains a (unique) idempotent e_F, defined by

(10) $$\chi(e_F) = \begin{cases} 1, & \chi \in F, \\ 0, & \chi \notin F. \end{cases}$$

In a similar way, the $(G \times G)$-orbits in S are in a monotone one-to-one correspondence with faces of the cone $\mathbb{Q}_+ \mathfrak{L}$, but in general not with all of them. The ideal of the orbit corresponding to a face F is spanned by those subspaces $k[G]_\Lambda$ for which $\Lambda \notin F$. In particular, there are only finitely many $(G \times G)$-orbits.

Denote by $\overline{Z}$ the closure of Z in S.

THEOREM 3. *Let S be a reductive semigroup. Then*
1) $\pi^{-1}(e) = G_0$;
2) $\pi(\overline{Z}) = A$;
3) *the closed* $(G_0 \times G_0)$-*orbits are exactly those meeting* $\overline{Z}$.

Moreover, if S is normal, then

4) π *induces an isomorphism* $\overline{Z}/Z_0 \simeq A$;
5) *the closure of any* $(G_0 \times G_0)$-*orbit is normal.*

It follows from 1) that

$$G(A) = G/G_0 = Z/Z_0$$

and the restriction of π on G is the canonical homomorphism $G \to G/G_0$. Moreover,

$$\pi^{-1}(G(A)) = G.$$

4. It is of special interest to distinguish the cases when the morphism π is flat. In these cases, the fibers of π are equidimensional and, if S has a zero, the triple (S, A, π) can be regarded as a multi-parameter contraction of the $(G_0 \times G_0)$-action on G_0 to that on $\pi^{-1}(0)$. We shall see that, under some restrictions, the result of this contraction does not depend on S. The action $G_0 \times G_0$ on $\pi^{-1}(0)$ is special, which means that the stabilizer of any point contains a maximal unipotent subgroup of $G_0 \times G_0$.

A canonical (one-parameter) contraction of any action of a reductive group on an affine variety to a special one was considered by Popov [10] (see also [11]). In the case of the action of $G_0 \times G_0$ on G_0, the result of our contraction is just the same.

DEFINITION 2. A normal reductive semigroup S is called *flat* if the morphism π is flat and its fibers are reduced (as schemas) and irreducible.

The morphism π is flat if and only if $k[S]$ is a free $k[A]$-module (Proposition 3). Even in this case, the fibers of π need not be reduced: see an example in 4.2.

According to the decomposition (6), we represent an element of $\mathfrak{t}(\mathbb{Q})^*$ as a pair (χ, λ), where $\chi \in \mathfrak{z}(\mathbb{Q})^*$, $\lambda \in \mathfrak{t}_0(\mathbb{Q})^*$.

THEOREM 4. *Let* $S = S(\mathfrak{L})$ *be a normal reductive semigroup. In the notation of Theorem 2, the semigroup S is flat if and only if there are such a convex polyhedral cone* $D \subset \mathfrak{z}(\mathbb{Q})^*$ *and a homomorphism* $\theta: Z \to T_0$ *that*

(11) $$\theta|_{Z_0} = \text{id}$$

and the cone $K = K(S)$ *has the form*

(12) $$K = \{(\chi, \lambda) \in \mathfrak{t}(\mathbb{Q})^* : \chi - \theta^*(\lambda) \in D\}.$$

For such a cone K, the conditions of Theorem 2 have the following form:

(1) $\theta^*(\alpha_i) \in D$ $(i = 1, \ldots, n)$;
(2) the cone D generates $\mathfrak{z}(\mathbb{Q})^*$;
(3) the cone D is pointed;
(4) $\theta^{*-1}(D) \cap (-C_0) = \{0\}$.

If S has a zero, the fiber $\pi^{-1}(0)$ is an ideal of S. As an algebraic semigroup with a $(G_0 \times G_0)$-action, it depends only on G_0, provided S is flat (see 4.4). Like the asymptotic cone of a hyperboloid, it reflects the behavior of G_0 at infinity. We call it the asymptotic semigroup of G_0 and denote by $\operatorname{As} G_0$. A separate paper [12] is devoted to its more detailed investigation.

In general, if S is flat, all the fibers of π are spherical $(G_0 \times G_0)$-varieties and $(G_0 \times G_0)$-orbits are just the intersections of $(G \times G)$-orbits with the fibers (Proposition 5). (For a reductive group L, an irreducible L-variety X is called *spherical* if the Borel subgroup of L has an open orbit in X. In this case, L has only finitely many orbits in X [11].)

5. Any homomorphism $\varphi \colon S' \to S$ of reductive algebraic semigroups gives rise to a homomorphism of their abelizations: $\varphi_{ab} \colon A' \to A$ in such a way that the diagram

$$\begin{array}{ccc} S' & \xrightarrow{\varphi} & S \\ \pi' \downarrow & & \downarrow \pi \\ A' & \xrightarrow{\varphi_{ab}} & A \end{array}$$

is commutative.

Consider the fiber product

$$\widehat{S} = A' \times_A S = \{(a', s) \in A' \times S : \varphi_{ab}(a') = \pi(s)\}.$$

It is a closed subsemigroup of $A' \times S$ and the canonical projections $\widehat{\pi} \colon \widehat{S} \to A'$, $\widehat{\varphi} \colon \widehat{S} \to S$ are semigroup homomorphisms. There is a (unique) homomorphism $\sigma \colon S' \to \widehat{S}$ such that $\widehat{\pi}\sigma = \pi$, $\widehat{\varphi}\sigma = \varphi$.

DEFINITION 3. The homomorphism φ is called *excellent* if σ is an isomorphism.

Note that the semigroup $\widehat{S}$ is reductive and φ maps the commutator group of $G(\widehat{S})$ isomorphically onto that of $G(S)$. So if the homomorphism φ is excellent, it maps the commutator group of $G(S')$ isomorphically onto that of $G(S)$. Moreover, if S is flat, so is S' (Proposition 6).

A standard consideration of commutative diagrams shows that the product of excellent homomorphisms is also excellent.

For a fixed connected semisimple group G_0, denote it by $\mathcal{FS}(G_0)$ the class of all flat reductive semigroups whose commutator group of the unit group is isomorphic to G_0. It turns out that there is a distinguished semigroup $S \in \mathcal{FS}(G_0)$ which is universal in a sense. In the statement of the following theorem, we identify the commutator group of $G(S)$ with G_0.

THEOREM 5. *There is a semigroup with zero $S \in \mathcal{FS}(G_0)$ satisfying the following condition:*

(∗) *For any semigroup $S' \in \mathcal{FS}(G_0)$ and any isomorphism φ_0 of the commutator group G_0' of $G(S')$ onto G_0, there is an excellent homomorphism $\varphi \colon S' \to S$, whose restriction to G_0' coincides with φ_0. Moreover, if S' has a zero, such a homomorphism is unique.*

It is clear that such a semigroup S is unique up to isomorphism. We call it the *enveloping semigroup* of G_0 and denote by $\operatorname{Env} G_0$.

In the terminology of Theorem 4, the semigroup $S = \mathrm{Env}\, G_0$ is described as follows:

(1) Z_0 is the whole center of G_0;
(2) θ is an isomorphism;
(3) the cone D is generated by the forms $\theta^*(\alpha_i)$, $i = 1, \ldots, n$.

For any $\lambda \in \mathfrak{t}_0(\mathbb{Q})^*$, we denote $\theta^*(\lambda)$ by $\bar\lambda$.

6. Now we describe the $(G \times G)$-orbit structure of $S = \mathrm{Env}\, G_0$.

The faces of the cone D are enumerated by the subsets of $\Omega = \{1, \ldots, n\}$ in such a way that to a subset I, there corresponds the face D_I spanned (as a convex cone) by α_i, $i \in I$. We denote by O_I the Z-orbit ($= G(A)$-orbit) in A, corresponding to D_I.

The cone $K \cap C$ is linearly, and hence combinatorially, isomorphic to the direct product $D \times C_0$ (so it is a simplicial cone). The cone C_0 is spanned by the fundamental weights $\omega_1, \ldots, \omega_n$ of $\mathfrak{g}_0$. For $J \subset \Omega$ let us denote by C_J its face spanned by ω_j, $j \in J$. In this notation, the faces of $K \cap C$ are

$$(13) \qquad F_{I,J} = \{(\chi, \lambda) \in \mathfrak{t}(\mathbb{Q})^* : \chi - \bar\lambda \in D_I,\ \lambda \in C_J\},$$

where $I, J \subset \Omega$.

Let Σ be the Dynkin diagram of $\mathfrak{g}$ and $v_1, \ldots, v_n$ its vertices enumerated in accordance with the enumeration of the simple roots. For $I \subset \Omega$, we denote by Σ_I the subdiagram of Σ constituted by the vertices v_i, $i \in I$. The subsets of I corresponding to the connected components of Σ_I will be called the *connected components* of I.

DEFINITION 4. A pair (I, J) and the corresponding face $F_{I,J}$ of $K \cap C$ are called *essential* if no connected component of the complement of J is entirely contained in I.

THEOREM 6. *The $(G \times G)$-orbits in $S = \mathrm{Env}\, G_0$ are in a monotone one-to-one correspondence with the essential faces of the cone $K \cap C$.*

We denote by $O_{I,J}$ the orbit corresponding to an (essential) face $F_{I,J}$. Its ideal is spanned by the subspaces $k[G]_\Lambda$ with $\Lambda \notin F_{I,J}$. Clearly,

$$(14) \qquad \pi(O_{I,J}) = O_I.$$

In particular, if $I = \Omega$, then the only possibility for J is to be equal to Ω as well. This means that $\pi^{-1}(G(A)) = G$, which also follows from Theorem 3. On the contrary, if $I = \varnothing$, then J may be an arbitrary subset of Ω, so $\pi^{-1}(0) = G_0$ decomposes into 2^n $(G \times G)$-orbits.

For any I, there is the least admissible J, namely, the union of the connected components of Ω entirely contained in I. The corresponding orbit $O_{I,J}$ is the unique orbit which is closed in $\pi^{-1}(O_I)$. At the same time, there is a greatest admissible J, namely, the whole set Ω. The corresponding orbit is the unique orbit which is open in $\pi^{-1}(O_I)$.

7. Let us describe the stabilizers of the $(G \times G)$-action on S.

According to general results of Putcha [1, 2] for reductive semigroups, each $(G \times G)$-orbit $O_{I,J}$ contains an idempotent defined up to conjugacy. It can be chosen in $\overline{T}$ and, under this condition, it is defined up to the action of the Weyl group. We denote such an idempotent by $e_{I,J}$ and will describe its stabilizer. An interpretation of $e_{I,J}$ is given in 7.3.

Let B be the Borel subgroup of G and $\mathfrak{b}$ its tangent algebra. For any subset $M \subset \Omega$, we denote by $P(M)$ the parabolic subgroup of G, whose tangent algebra is generated by $\mathfrak{b}$ and the root vectors corresponding to the roots $-\alpha_i$, $i \in M$ (so $B = P(\varnothing)$). We have

$$(15) \qquad P(M) = U(M)R(M),$$

where $U(M)$ is the unipotent radical and $R(M)$ a maximal reductive subgroup of $P(M)$. We shall assume that $R(M) \supset T$. Under this condition, $R(M)$ is uniquely defined. We denote by $G(M)$ its commutator group.

Let $P_-(M)$ be the parabolic subgroup which is opposite to $P(M)$ and $U_-(M)$ its unipotent radical. Then

$$(16) \qquad P_-(M) = U_-(M)R(M).$$

We denote by δ (respectively, δ_-) the projection of $P(M)$ (respectively, $P_-(M)$) onto $R(M)$ with respect to the decomposition (15) (respectively, (16)).

We call two elements of Ω *adjacent*, if such are the corresponding vertices of the Dynkin diagram. For a subset $M \subset \Omega$, we denote by $C(M)$ its complement and by M° its "interior", consisting of its elements that are not adjacent to any elements of $C(M)$.

Now let $O_{I,J}$ be a $(G \times G)$-orbit in $S = \operatorname{Env} G_0$. Put

$$(17) \qquad M = (I \cap J^\circ) \cup C(J)$$

and define a torus $T_{I,J} \subset T$ by

$$(18) \qquad T_{I,J} = \{t \in T : \Lambda(t) = 1 \text{ for } \Lambda \in F_{I,J}\}.$$

Note that $T_{I,J}$ contains $T \cap G(C(J))$ and $G(C(J))T_{I,J}$ is a normal subgroup of $R(M)$.

THEOREM 7. *Under a suitable choice of the idempotent $e_{I,J} \in O_{I,J} \cap \overline{T}$, its stabilizer $H_{I,J}$ is the subgroup of $P(M) \times P_-(M)$, consisting of the pairs (g, g_-), satisfying the condition*

$$(19) \qquad \delta(g) \equiv \delta_-(g_-) \pmod{G(C(J))T_{I,J}}.$$

In other words, $H_{I,J}$ is the (semidirect) product of $U(M) \times U_-(M)$, the diagonal in $R(M) \times R(M)$, and the group $G(C(J))T_{I,J} \times \{e\}$.

In particular, $H_{I,J}$ is reductive if and only if $M = \Omega$, which means that $I \supset J = J^\circ$. This is just the case when $O_{I,J}$ is closed in $\pi^{-1}(O_I)$. Another characterization of this case is that $e_{I,J} \in \overline{Z}$.

On the contrary, for the orbit $O_{I,\Omega}$, which is open in $\pi^{-1}(O_I)$, we have

$$J = J^\circ = \Omega, \qquad M = I \cap J^\circ = I,$$

so $H_{I,\Omega}$ is the product of $U(I) \times U_-(I)$, the diagonal in $R(I) \times R(I)$, and the torus $T_I \times \{e\}$, where

$$(20) \qquad T_I = T_{I,\Omega} = \{z\theta(z)^{-1} : z \in Z, \ \overline{\alpha}_i(z) = 1 \ (i \in I)\}.$$

The idempotents $e_{I,J}$ chosen as in Theorem 7 subject to the relations

$$(21) \qquad e_{I_1,J_1} e_{I_2,J_2} = e_{I_1 \cap I_2, J_1 \cap J_2}.$$

8. The idempotents $e_{I,\Omega}$ are just those lying in the closure of the diagonal torus

$$(22) \qquad T_\varnothing = \{z\theta(z)^{-1} : z \in Z\},$$

so

$$(23) \qquad G\overline{T}_\varnothing G = \bigcup_I O_{I,\Omega}.$$

Clearly, this is an open subvariety of S.

It is easy to see that $\overline{T}_\varnothing \simeq k^n$. This fact gives rise to the following theorem.

THEOREM 8. *The variety*

$$(24) \qquad S^{\mathrm{pr}} = \bigcup_I O_{I,\Omega}$$

is smooth and there is a geometric quotient S^{pr}/Z which is a smooth projective variety.

For the definition of the geometric quotient see, for example, in [9].

The variety S^{pr}/Z inherits the $(G_0 \times G_0)$-action and contains the adjoint group G_0/Z_0 as an open orbit of this action. One can show that it is nothing else than the "wonderful" equivariant completion of G_0/Z_0 constructed by DeConcini and Procesi [14].

The subset $S^{\mathrm{pr}} \subset S$ is not a subsemigroup, so there is no natural semigroup structure on S^{pr}/Z. On the other hand, the set-theoretic quotient S/Z is a semigroup but not an algebraic variety. It seems that, for classical groups G_0, the semigroup S/Z is close, if not identical, to that constructed by Neretin [15]. Neretin's results are not used in this work, but his ideology influenced me to a certain extent.

9. If the group G_0 acts on an affine variety X_0, we can be interested in the extension of this action to an action of the semigroup $S = \mathrm{Env}\, G_0$ on an affine variety E containing X_0.

More generally, we can consider G_0-equivariant morphisms $\varphi : X_0 \to E$, where E is an affine variety with an action of S on it. (We assume that the unit of S acts as the identity map.) Let us call such an S-variety E, together with the morphism φ, an enveloping S-variety of X_0, if for any pair (E', φ') of the same kind, there is a unique S-equivariant morphism $\psi : E \to E'$ such that the diagram

$$\begin{array}{ccc} X_0 & \xrightarrow{\varphi} & E \\ {\scriptstyle \varphi'} \searrow & & \swarrow {\scriptstyle \psi} \\ & E' & \end{array}$$

is commutative. It is clear that an enveloping S-variety, if it exists, is unique in a natural sense.

THEOREM 9. *For any affine G_0-variety X_0, there exists an enveloping S-variety E. The corresponding morphism φ is an isomorphism of X_0 onto a closed subvariety of E.*

We shall denote the enveloping S-variety of X_0 by $\operatorname{Env} X_0$ and identify X_0 with its image in $\operatorname{Env} X_0$.

For any affine S-variety E, we can consider the invariant-theoretic quotient $E/\!/G_0$. It inherits the action of S, which reduces to an action of the abelization $A = S/\!/(G_0 \times G_0)$ of S.

For $E = \operatorname{Env} X_0$, we have

$$(25) \qquad E/\!/G_0 = A \times X_0/\!/G_0.$$

Moreover, $X_0/\!/G_0$ is embedded into E as the subvariety of fixed points of S (see 9.6).

10. The basic results of this work were obtained during my visit to the Institut des Hautes Études Scientifiques in August of 1993. A preliminary version of the research was reported at the meeting on "Invariant ordering in geometry and algebra" at Mathematisches Forschungsinstitut Oberwolfach in October of 1993 and at the international meeting organized by Sondervorschungsbereiche 343 "Diskrete Strukturen in der Mathematik" in Bielefeld in November of 1993. I thank all these institutions for their hospitality. I also thank Yu. A. Neretin for fruitful discussions.

Accomplishing this research was made possible in part by Grant MQZ000 from the International Science Foundation.

§1. Proof of Theorem 1

1. In the subsequent proofs, we make use of the following results which are apparently due to Khadzhiev [16], Vust [20], and Popov [10]. (See [10] for details.)

Let A be a commutative associative algebra with unit, and let a reductive group R act on A by automorphisms. We assume that any element of A is contained in a finite-dimensional R-invariant subspace and, for any such subspace V, the induced linear representation $R \to GL(V)$ is algebraic. Let U be a maximal unipotent subgroup of R and A^U the subalgebra of U-invariant elements of A.

Consider the following properties of an algebra:

  (a) it is finitely generated;
  (b) it has no nilpotent elements;
  (c) it has no zero divisors;
  (d) it is normal.

THEOREM ([16, 20, 10]). *Let (P) be any of the properties* (a)–(d). *The algebra A has the property (P) if and only if the algebra A^U has this property.*

2. Let $\mathfrak{L}$ be a perfect subsemigroup of $\mathfrak{X}_+$. To prove Theorem 1, we must find out, under what conditions the algebra

$$(26) \qquad k[G]_{\mathfrak{L}} = \bigoplus_{\Lambda \in \mathfrak{L}} k[G]_\Lambda$$

is finitely generated and generates the field $k(G)$.

Consider the action of $G \times G$ on $k[G]$. Denote by U the unipotent radical of the Borel subgroup B of G and by U_- the unipotent radical of the opposite Borel subgroup B_-. Then $U_- \times U$ is a maximal unipotent subgroup of $G \times G$.

For each irreducible linear representation $R^{(\Lambda)}: G \to GL(V^{(\Lambda)})$ choose a basis in $V^{(\Lambda)}$ consisting of weight vectors, the highest vector being the first of them. Then the algebra $k[G]^{U_- \times U}$ is spanned by the functions $\delta^{(\Lambda)} = f_{11}^{(\Lambda)}$. Since the highest vector of $V^{(\Lambda+M)}$ is the tensor product of those of $V^{(\Lambda)}$ and $V^{(M)}$, we have $\delta^{(\Lambda+M)} = \delta^{(\Lambda)}\delta^{(M)}$, so the algebra $k[G]^{U_- \times U}$ is isomorphic to the semigroup algebra of $\mathfrak{X}_+$.

In the same way, the algebra $k[G]_{\mathfrak{L}}^{U_- \times U}$ is isomorphic to the semigroup algebra of $\mathfrak{L}$. It follows that the algebra $k[G]_{\mathfrak{L}}$ is finitely generated if and only if such is the semigroup $\mathfrak{L}$.

3. Let Quot A denote the field of quotients of an algebra A.

We have Quot $k[G]_{\mathfrak{L}} = k(G)$ if and only if the functions of $k[G]_{\mathfrak{L}}$ separate points of G, i.e., if the intersection of the kernels of the representations $R^{(\Lambda)}$, $\Lambda \in \mathfrak{L}$, is trivial. Let us denote this intersection by G_1.

If $G_1 \neq \{e\}$, then $T_1 = G_1 \cap T \neq \{e\}$, so $\mathfrak{L}$ belongs to the proper subgroup

$$\mathfrak{X}_1 = \{\Lambda \in \mathfrak{X} : \Lambda|_{T_1} = 1\} \subset \mathfrak{X}.$$

To prove the converse, we need the following

LEMMA. *Let* $\Lambda \in \mathfrak{X}_+$ *and* $i \in \Omega$ *be such that* $(\Lambda, \alpha_i) > 0$. *Then* $2\Lambda - \alpha_i \in \mathfrak{X}(\Lambda, \Lambda)$.

This lemma follows from Proposition 9 below and the decomposition rule for products of irreducible representations of SL_2.

So, if $\Lambda \in \mathfrak{L}$ and $(\Lambda, \alpha_i) > 0$, then $2\Lambda - \alpha_i \in \mathfrak{L}$, and $\alpha_i \in \mathfrak{L} - \mathfrak{L}$. Moreover, if $(\alpha_i, \alpha_j) < 0$, then $(2\Lambda - \alpha_i, \alpha_j) > 0$. It follows that the set

$$\Omega_1 = \{i \in \Omega : \alpha_i \in \mathfrak{L} - \mathfrak{L}\}$$

is a union of connected components of Ω.

Now let $G_1 = \{e\}$. Then, for each connected component of Ω, there is a $\Lambda \in \mathfrak{L}$ such that $(\Lambda, \alpha_i) > 0$ for some i of this component. Hence $\Omega_1 = \Omega$, so the group $\mathfrak{L} - \mathfrak{L}$ contains the root lattice $\mathfrak{R}$ of $\mathfrak{g}$.

The quotient group $\mathfrak{X}/\mathfrak{R}$ is naturally isomorphic to the character group of the center of G, and if $\mathfrak{L} - \mathfrak{L} \neq \mathfrak{X}$, there is an element $z \neq e$ of the center such that $\Lambda(z) = 1$ for all $\Lambda \in \mathfrak{L}$, and hence $z \in G_1$. This contradicts our assumption.

§2. Proof of Theorem 2

1. First let $\mathfrak{L} = \mathfrak{X}_+ \cap K$, where $K \subset \mathfrak{t}(\mathbb{Q})^*$ is a convex polyhedral cone satisfying the conditions 1) and 2) of the theorem.

It is known (and easy to show) that the intersection of a lattice in $\mathbb{Q}^m$ with a convex polyhedral cone is a finitely generated semigroup. Hence the semigroup $\mathfrak{L} = \mathfrak{X} \cap (C \cap K)$ is finitely generated. Since the cone $C \cap K$ generates the space $\mathfrak{t}(\mathbb{Q})^*$, the semigroup $\mathfrak{L}$ generates the group $\mathfrak{X}$.

Since any $N \in \mathfrak{X}(\Lambda, M)$ has the from

$$N = \Lambda + M - \sum_i k_i \alpha_i, \qquad k_i \geqslant 0,$$

condition 1) of the theorem guarantees that the semigroup $\mathfrak{L}$ is perfect. So it defines an algebraic semigroup S with $G(S) = G$.

Now let S be an algebraic semigroup with $G(S) = G$ defined by a semigroup $\mathfrak{L}$.

In view of the Theorem stated in 1.1, the algebra $k[S]$ defined by (2) is normal if and only if such is the algebra $k[S]^{U_- \times U}$. The last algebra is isomorphic to the semigroup algebra of $\mathfrak{L}$, which is normal if and only if

$$\mathfrak{L} = \mathfrak{X} \cap \mathbb{Q}_+ \mathfrak{L}$$

(see, for example, [13] or [17]).

The cone $\mathbb{Q}_+\mathfrak{L}$ is a convex polyhedral cone contained in the Weyl chamber C. Let $H_1, \ldots, H_s$ be its walls distinct from the walls of C and $H_1^+, \ldots, H_s^+$ the half-spaces bounded by $H_1, \ldots, H_s$ respectively and containing $\mathbb{Q}_+\mathfrak{L}$. We put

$$K = H_1^+ \cap \cdots \cap H_s^+$$

so that

$$\mathfrak{L} = \mathfrak{X} \cap (K \cap C) = \mathfrak{X}_+ \cap K.$$

Let us prove that $K \ni -\alpha_1, \ldots, \alpha_n$.

Suppose $-\alpha_i \notin H_j^+$ for some i, j and take an interior point Λ of the face $H_j \cap (K \cap C)$ of the cone $K \cap C$. Multiplying Λ by an integer, we may assume that $\Lambda \in \mathfrak{L}$. Then $2\Lambda - \alpha_i \in \mathfrak{X}(\Lambda, \Lambda)$ but $2\Lambda - \alpha_i \notin \mathfrak{L}$, which is a contradiction.

A maximal ideal of the algebra $k[S]$ defines a zero of S if and only if it is $(G \times G)$-invariant. If such an ideal exists, it must be equal to

$$k[S]_+ = \bigoplus_{\Lambda \in \mathfrak{L} \setminus \{0\}} k[G]_\Lambda.$$

The subspace $k[S]_+$ is really an ideal if and only if

(27) $\qquad 0 \notin \mathfrak{X}(\Lambda, M), \qquad \Lambda, M \in \mathfrak{L} \setminus \{0\}.$

It follows from condition 1) of the theorem that the projection of $K \cap C$ on $\mathfrak{z}(\mathbb{Q})^*$ is contained in $K \cap C$. So if the condition 3) is satisfied, $0 \in \mathfrak{X}(\Lambda, M)$ implies $\Lambda, M \in C_0$ and, moreover, if the condition 4) is satisfied, $\Lambda = M = 0$.

Conversely, if condition 3) is violated, there are such $\Lambda, M \in (\mathfrak{L} \setminus \{0\}) \cap \mathfrak{z}(\mathbb{Q})^*$ that $\Lambda + M = 0$. If condition 4) is violated, there exists $\Lambda \in (\mathfrak{L} \setminus \{0\}) \cap C_0$. Let m be the dimension of the representation $R^{(\Lambda)}$. We have $\det R(g) = 1$ for $g \in G$, which implies that the mth (tensor) power of $R^{(\Lambda)}$ contains the trivial representation. It follows that the condition (27) is violated.

2. For example, consider the case $G = k^* \times SL_2$. We identify the space $\mathfrak{t}(\mathbb{Q})^*$ with $\mathbb{Q}^2$ in such a way that $\mathfrak{t}_0(\mathbb{Q})^*$ is identified with the x-axis, $\mathfrak{z}(\mathbb{Q})^*$ with the y-axis,

the group $\mathfrak{X}$ with $\mathbb{Z}^2$, and the only simple root α with $(2, 0)$. Then the Weyl chamber C has the form

$$C = \{(x, y) \in \mathbb{Q}^2 : x \geq 0\}.$$

A normal algebraic semigroup S with $G(S) = G$ is defined, in terms of Theorem 2, by a convex cone $K \subset \mathbb{Q}^2$ with the following properties: $K \ni (-1, 0)$ and K^0 meets the right half-plane C. Moreover, since only the intersection $K \cap C$ is essential, we may assume that each side of K meets C^0, so K is either the whole plane or a half-plane distinct from $-C$. Since the above identification of $\mathfrak{t}(\mathbb{Q})^*$ with $\mathbb{Q}^2$ is defined up to multiplication y by -1, we also may assume that $K \ni (0, 1)$.

If K is the whole plane, we have $\mathfrak{L} = \mathfrak{X}_+$ and $S = G$. If K is the upper half-plane, $\mathfrak{L} = \mathbb{Z}_+^2$ and $S = k \times SL_2$.

In all other cases, K has the form

$$K = \{(x, y) \in \mathbb{Q}^2 : y \geq ax\}, \qquad a \in \mathbb{Q}, \ a > 0$$

(see Fig. 1), so the conditions 3) and 4) of Theorem 2 are satisfied and S is a semigroup with zero. For further citation, we denote this semigroup by S_a.

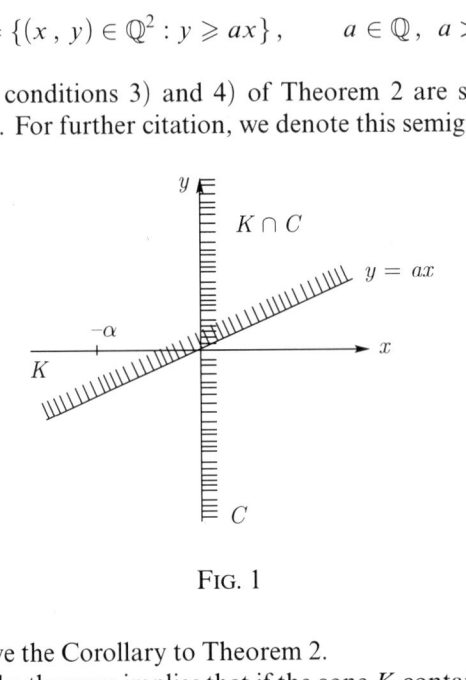

FIG. 1

3. Now we prove the Corollary to Theorem 2.

Condition 1) of the theorem implies that if the cone K contains an interior point of some face of C_0, it contains the whole face. So $F = K \cap C_0$ is a face of C_0. Moreover, if $\alpha_i|_F \neq 0$ and $(\alpha_i, \alpha_j) < 0$, then $\alpha_j|_F \neq 0$, too. Hence F is spanned by the fundamental weights of some ideal of $\mathfrak{g}_0$, say, $\mathfrak{g}_0^{(1)}$. Let $\mathfrak{g}_0^{(2)}$ be the complementary ideal, $G_0^{(1)}$ and $G_0^{(2)}$ connected subgroups of G_0 whose tangent algebras are $\mathfrak{g}_0^{(1)}$ and $\mathfrak{g}_0^{(2)}$, respectively.

Further let $Z^{(1)}$ and $Z^{(2)}$ be almost complementary subtori of Z such that $\mathfrak{z}^{(2)}$, the tangent algebra of $Z^{(2)}$, is the annihilator of $K \cap (-K) \cap \mathfrak{z}(\mathbb{Q})^*$.

Put

$$G^{(1)} = Z^{(1)} G_0^{(1)}, \qquad G^{(2)} = Z^{(2)} G_0^{(2)}.$$

Passing to a suitable covering semigroup, we may assume that $G = G^{(1)} \times G^{(2)}$. Then $\mathfrak{X} = \mathfrak{X}^{(1)} \oplus \mathfrak{X}^{(2)}$, where $\mathfrak{X}^{(1)}$ and $\mathfrak{X}^{(2)}$ are the character groups of $T^{(1)} = T \cap G^{(1)}$

and $T^{(2)} = T \cap G^{(2)}$, respectively. With respect to this decomposition,

$$\mathfrak{L} = \mathfrak{X}_+^{(1)} \oplus (\mathfrak{X}_+^{(2)} \cap K^{(2)}),$$

where the cone $K^{(2)} = K \cap \mathfrak{z}^{(2)}(\mathbb{Q})^*$ satisfies conditions 3) and 4) of the theorem. It follows that $S = G^{(1)} \times S^{(2)}$, where $S^{(2)}$ is an algebraic semigroup with $G(S^{(2)}) = G^{(2)}$ defined by the semigroup $\mathfrak{L}^{(2)} = \mathfrak{X}_+^{(2)} \cap K^{(2)}$. According to the theorem, $S^{(2)}$ has a zero.

§3. Proof of Theorem 3

1. The first assertion of Theorem 3 is implied by the following well-known fact [8].

PROPOSITION 1. *Any reductive algebraic semigroup whose unit group is semisimple is a group.*

Indeed, the proposition implies that $\overline{G}_0 = G_0$. It follows that G_0 is the unique closed $(G_0 \times G_0)$-orbit in the fiber $\pi^{-1}(e)$. All other orbits in $\pi^{-1}(e)$, if they exist, must be higher-dimensional. But each coset gG_0 ($g \in G$) is a $(G_0 \times G_0)$-orbit, and, consequently, these orbits have the highest dimension. Hence $\pi^{-1}(e) = G_0$.

For the sake of completeness, we give a proof of Proposition 1 below. It is based on the following

PROPOSITION 2. *Let $H \subset GL(V)$ be a connected semisimple algebraic linear group. Then any irreducible linear representation of H is realized in a suitable tensor power of V.*

PROOF. We have $H \subset SL(V)$, so any irreducible linear representation of H is contained in the restriction to G of some irreducible linear representation of $SL(V)$. But any irreducible linear representation of $SL(V)$ is realized in a suitable tensor power of V. □

PROOF OF PROPOSITION 1. Let S be an algebraic semigroup, whose unit group $G(S) = G$ is semisimple, and $\mathfrak{L}$ the corresponding subsemigroup of $\mathfrak{X}_+$ (see the formula (2)). Since $\mathfrak{L}$ generates the group $\mathfrak{X}$, it contains such characters $\Lambda_1, \ldots, \Lambda_m$ that the representation $R = R^{(\Lambda_1)} + \cdots + R^{(\Lambda_m)}$ of the group G is faithful. By Proposition 2, any irreducible representation of G is contained in some product of $R^{(\Lambda_1)}, \ldots, R^{(\Lambda_m)}$. Hence $\mathfrak{L} = \mathfrak{X}_+$ and $S = G$. □

2. Since $G_0 \overline{Z} \supset G$, the restriction homomorphism

$$\rho: k[A] = k[S]^{G_0 \times G_0} \to k[\overline{Z}]$$

is injective. We shall prove that $k[\overline{Z}]$ is integral over $\rho(k[A])$, which will imply that the corresponding morphism $\rho^* = \pi|_{\overline{Z}}: \overline{Z} \to A$ is surjective.

The algebra $k[\overline{Z}] \subset k[Z]$ is spanned by the characters $e^\Lambda|_Z$, $\Lambda \in \mathfrak{L}$, of Z. For $\Lambda \in \mathfrak{L}$, let $m = \dim R^{(\Lambda)}$. Then the mth exterior power of $R^{(\Lambda)}$ is a one-dimensional representation of G. Its (highest) weight $M \in \mathfrak{L}_Z$ is such that $e^M|_Z = (e^\Lambda)^m|_Z$, so $e^\Lambda|_Z$ is integral over $\rho(k[A])$.

3. Since $p(k[A]) \subset k[\overline{Z}]^{Z_0}$, we have the commutative diagram

$$\overline{Z} \xrightarrow{\pi} A$$
$$p \searrow \quad \swarrow \sigma$$
$$\overline{Z}/Z_0$$

where p is the canonical homomorphism. Moreover, since the fibers of $\pi|_{\overline{Z}}$ are just the cosets of Z_0, σ is a birational morphism. Since $\pi|_{\overline{Z}}$ is surjective, σ is also surjective. But any surjective birational morphism onto a normal variety is an isomorphism. This implies the fourth assertion of the theorem.

4. The $(G_0 \times G_0)$-orbit of an element $u \in \overline{Z}$ is $G_0 u$. Let us prove that it is closed. Since Zu contains an idempotent, we may assume that u is an idempotent. In this case, the morphism $G_0 \to G_0 u$, $g \mapsto gu$, is a semigroup homomorphism. It defines an isomorphism $G_0/N \simeq G_0 u$, where $N = \{g \in G_0 : gu = u\}$. Hence $\overline{G_0 u}$ is an algebraic semigroup, whose unit group is semisimple. In virtue of Proposition 1, this implies that $\overline{G_0 u} = G_0 u$.

Now let O be any closed $(G_0 \times G_0)$-orbit. Since $\pi(\overline{Z}) = A$, the fiber $\pi^{-1}(\pi(O))$ meets $\overline{Z}$. But any fiber contains only one closed orbit. Hence O meets $\overline{Z}$. This proves the third assertion of the theorem.

5. Let S be normal and O be the $(G \times G)$-orbit in S corresponding to a face F of the cone $K \cap C$ (in the notation of Theorem 2). We know (see 1.1) that the normality of $k[\overline{O}]$ is equivalent to the normality of $k[\overline{O}]^{U_- \times U}$. In 1.2, we saw that the algebra $k[S]^{U_- \times U}$ is naturally isomorphic to the semigroup algebra of $\mathfrak{L}$. Correspondingly, the algebra $k[\overline{O}]^{U_- \times U}$ is isomorphic to the semigroup algebra of

$$\mathfrak{L}_F = \mathfrak{L} \cap F = \mathfrak{X} \cap F$$

and hence normal (see 2.1).

§4. Proof of Theorem 4

1. Let $S = S(\mathfrak{L})$ be a normal reductive semigroup with $G(S) = G$ and A its abelization.

We introduce a preorder on $\mathfrak{L}$:

$$\Lambda_1 \geqslant \Lambda_2, \quad \text{if } \Lambda_1 - \Lambda_2 \in \mathfrak{L}_Z.$$

If $\Lambda_1 \geqslant \Lambda_2$ and $\Lambda_2 \geqslant \Lambda_1$, we shall call Λ_1 and Λ_2 equivalent and write $\Lambda_1 \sim \Lambda_2$. More explicitly, $\Lambda_1 \sim \Lambda_2$, if $\Lambda_1 - \Lambda_2 \in \mathfrak{M}_0$, where

$$\mathfrak{M}_0 = \mathfrak{L}_Z \cap (-\mathfrak{L}_Z)$$

is the greatest subgroup contained in $\mathfrak{L}_Z$ (and in $\mathfrak{L}$). An element $M \in \mathfrak{L}$ will be called *minimal*, if $\Lambda \leqslant M$ implies $\Lambda \sim M$. Let $\mathfrak{M}$ denote the set of all minimal elements of $\mathfrak{L}$. It is evident that

$$\mathfrak{L} = \mathfrak{M} + \mathfrak{L}_Z.$$

PROPOSITION 3. *The following conditions are equivalent*:

(1) $k[S]$ *is a flat* $k[A]$*-module*;

(2) $k[S]$ *is a free* $k[A]$*-module*;

(3) *if* $M_1 + \chi_1 = M_2 + \chi_2$ ($M_1, M_2 \in \mathfrak{M}$, $\chi_1, \chi_2 \in \mathfrak{L}_Z$), *then* $M_1 \sim M_2$ (*and* $\chi_1 \sim \chi_2$).

(4) $k[S]$ *decomposes as a vector space into the tensor product*

$$k[S] = k[A] \otimes k[G]_{\mathfrak{M}_1}, \tag{28}$$

where $\mathfrak{M}_1$ *is a set of representatives of the cosets of* $\mathfrak{M}_0$ *in* $\mathfrak{M}$ *and*

$$k[G]_{\mathfrak{M}_1} = \bigoplus_{M \in \mathfrak{M}_1} k[G]_M. \tag{29}$$

In the case when S has a zero $\mathfrak{M}_0 = \{0\}$ and $\mathfrak{M}_1 = \mathfrak{M}$.

PROOF. Obviously, 3) $\implies$ 4) $\implies$ 2) $\implies$ 1), so we only must prove the implication 1) $\implies$ 3).

Let $k[S]$ be a flat $k[A]$-module. It is easy to see that the subalgebra $k[S]^{U_- \times U}$ (see 1.2) is a direct summand of $k[S]$ as a $k[A]$-module. Consequently, it is also a flat $k[A]$-module.

For a semigroup $\mathfrak{S}$, we denote by $k\mathfrak{S}$ its semigroup algebra over k. We saw in 1.2 that $k[S]^{U_- \times U} \simeq k\mathfrak{L}$. Under this isomorphism, the subalgebra $k[A]$ corresponds to $k\mathfrak{L}_Z$. Thus $k\mathfrak{L}$ is a flat $k\mathfrak{L}_Z$-module. It follows that for any ideal $\mathfrak{J}$ of $\mathfrak{L}_Z$ the natural homomorphism

$$k\mathfrak{J} \otimes_{k\mathfrak{L}_Z} k\mathfrak{L} \to k\mathfrak{L} \tag{30}$$

is injective.

Let us call two pairs $(\Lambda_1, \chi_1), (\Lambda_2, \chi_2) \in \mathfrak{L} \times \mathfrak{J}$ *adjacent*, if $\Lambda_1 + \chi_1 = \Lambda_2 + \chi_2$ and $\Lambda_1 \geq \Lambda_2$ or $\Lambda_2 \geq \Lambda_1$. Extending this relation by transitivity, we obtain an equivalence relation on $\mathfrak{L} \times \mathfrak{J}$, which we shall call the $\mathfrak{J}$-*equivalence*. The injectivity of the homomorphism (30) means that any two pairs $(\Lambda_1, \chi_1), (\Lambda_2, \chi_2) \in \mathfrak{L} \times \mathfrak{J}$ such that $\Lambda_1 + \chi_1 = \Lambda_2 + \chi_2$ are $\mathfrak{J}$-equivalent.

Now we prove that if $M_1, M_2 \in \mathfrak{M}$ satisfy $M_1 - M_2 \in \mathfrak{L}_Z - \mathfrak{L}_Z$, then $M_1 \sim M_2$, which is equivalent to condition 3) of the proposition.

Consider the ideal

$$\mathfrak{J} = (M_1 - M_2 + \mathfrak{L}_Z) \cap \mathfrak{L}_Z.$$

Let χ_2 be a minimal element of $\mathfrak{J}$ and $\chi_1 \in \mathfrak{L}_Z$ such that $M_1 + \chi_1 = M_2 + \chi_2$. Since M_2 is minimal in $\mathfrak{L}$ and χ_2 is minimal in $\mathfrak{J}$, the only pairs adjacent to (M_2, χ_2) are $(M_2 + \chi, \chi_2 - \chi)$, where $\chi \in \mathfrak{M}_0$. It follows that $M_1 \sim M_2$. $\square$

2. Now assume that $k[S]$ is a flat $k[A]$-module, i.e., the morphism $\pi \colon S \to A$ is flat.

We keep the notation $\mathfrak{M}$ for the set of minimal elements of $\mathfrak{L}$.

PROPOSITION 4. *The fibers of π are reduced and irreducible if and only if $\mathfrak{M}$ is a subsemigroup of $\mathfrak{L}$.*

PROOF. Let e_0 be the idempotent of A defined by

$$\chi(e_0) = \begin{cases} 1 & \text{for } \chi \in \mathfrak{M}_0, \\ 0 & \text{for } \chi \in \mathfrak{L}_Z \setminus \mathfrak{M}_0. \end{cases} \tag{31}$$

It is easy to see that any neighborhood of e_0 contains representatives of all $G(A)$-orbits in A. Since the set of points $a \in A$ for which the fiber $\pi^{-1}(a)$ is reduced and irreducible, is open [19] and $G(A)$-invariant, we may restrict ourselves to the investigation of the fiber $\pi^{-1}(e_0)$.

Let $\mathfrak{p}_0$ denote the ideal of $k[S]$ generated by the maximal ideal of $k[A]$ corresponding to e_0. Obviously, it is spanned by the subspaces $k[G]_\Lambda$ with $\Lambda \in \mathfrak{L} \setminus \mathfrak{M}$ and $(e^\chi - 1)k[G]_M$ with $M \in \mathfrak{M}$ and $\chi \in \mathfrak{M}_0$, so the subspace (28) is complementary to $\mathfrak{p}_0$.

"The fiber $\pi^{-1}(e_0)$ is reduced and irreducible" means that the quotient algebra $k[S]/\mathfrak{p}_0$ has no zero divisors. If $M_1, M_2 \in \mathfrak{M}$ but $M_1 + M_2 \notin \mathfrak{M}$, then

$$k[G]_{M_1}, k[G]_{M_2} \not\subset \mathfrak{p}_0, \quad \text{but } k[G]_{M_1+M_2} \subset \mathfrak{p}_0,$$

so the above condition is not fulfilled.

Now let $\mathfrak{M}$ be a subsemigroup. Then the algebra $(k[S]/\mathfrak{p}_0)^{U_- \times U}$ is isomorphic to the semigroup algebra of $\mathfrak{M}/\mathfrak{M}_0$ and consequently has no zero divisors. According to the theorem stated in 1.1, $k[S]/\mathfrak{p}_0$ has no zero divisors as well.

EXAMPLE. For the semigroup $S = S_a$ defined in §2, the morphism π is always flat and its fibers are always irreducible, but they are reduced if and only if $a \in \mathbb{N}$. The case $a = 1/2$ is depicted in Fig. 2. The elements of $\mathfrak{L}$ are represented by dots, the minimal ones being distinguished by small circle. Those of them lying above the line $y = x/2$ correspond to nilpotent elements of the algebra $k[S]/I_0$.

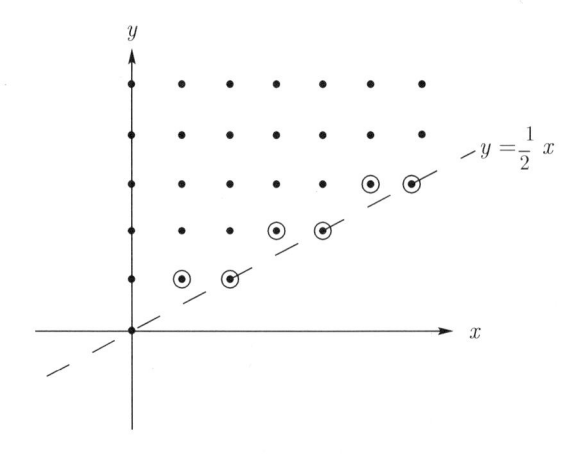

FIG. 2

3. Let us represent the above results in terms of Theorem 4.

We denote by $\mathfrak{X}(T_0)$ (respectively, $\mathfrak{X}(Z)$) the character group of T_0 (respectively, Z) and by $\mathfrak{X}_+(T_0)$ the semigroup of dominant characters of T_0.

If the cone K has the form (12), then

$$\mathfrak{M}_0 = \mathfrak{X}_Z \cap (D \cap (-D)), \tag{32}$$

$$\mathfrak{M} = \{(\theta^*(\lambda), \lambda) : \lambda \in \mathfrak{X}(T_0)\} + \mathfrak{M}_0, \tag{33}$$

and the conditions of Proposition 3 and 4 are satisfied.

Conversely, if these conditions are satisfied, then the projection on $\mathfrak{t}_0(\mathbb{Q})^*$ defines a semigroup epimorphism $p: \mathfrak{M} \to \mathfrak{X}_+(T_0)$ such that $p(M_1) = p(M_2)$ if and only if $M_1 \sim M_2$. It can be extended to a group epimorphism $p: \mathfrak{M} - \mathfrak{M} \to \mathfrak{X}(T_0)$ with the same property. Let $q: \mathfrak{X}(T_0) \to \mathfrak{M} - \mathfrak{M}$ be a homomorphism satisfying $pq = \text{id}$. We have

$$q(\lambda) = (r(\lambda), \lambda) \in \mathfrak{M} - \mathfrak{M} \subset \mathfrak{X},$$

where $r: \mathfrak{X}(T_0) \to \mathfrak{X}(Z)$ is a group homomorphism such that the restrictions of λ and $r(\lambda)$ to Z_0 coincide for any $\lambda \in \mathfrak{X}(T_0)$. Hence $r = \theta^*$, where $\theta: Z \to T_0$ is a homomorphism satisfying (11).

We have

$$\mathfrak{L} = \{(\chi, \lambda) \in \mathfrak{X}(Z) \times \mathfrak{X}_+(T_0) : \chi - \theta^*(\lambda) \in \mathfrak{L}_Z\}.$$

Hence the cone K has the form (12) with $D = K \cap \mathfrak{z}(\mathbb{Q})^* = \mathbb{Q}\mathfrak{L}_Z$.

REMARKS. 1. The cone D and, if S has a zero, the homomorphism θ are determined uniquely by S.

2. As a set of representatives of the cosets of $\mathfrak{M}_0$ in $\mathfrak{M}$ (see Proposition 3), we can choose

$$\mathfrak{M}_1 = \{(\theta^*(\lambda), \lambda) : \lambda \in \mathfrak{X}(T_0)\}. \tag{34}$$

4. The algebra $k[G_0]$ decomposes (as a vector space) into the direct sum

$$k[G_0] = \bigoplus_{\lambda \in \mathfrak{X}_+(T_0)} k[G_0]_\lambda, \tag{35}$$

where $K[G_0]_\lambda$ is the linear span of the matrix entries of the irreducible linear representation $R^{(\lambda)}$ of G_0 with highest weight λ. The multiplication in $k[G_0]$ has the form

$$fg = \sum_\nu p_\nu(f, g) \quad (f \in k[G_0]_\lambda, \; g \in k[G_0]_\mu, \; p_\nu(f, g) \in k[G_0]_\nu), \tag{36}$$

where ν runs over the highest weights of the irreducible components of $R^{(\lambda)} R^{(\mu)}$. The comultiplication has the form

$$\mu^*(f_{ij}^{(\lambda)}) = \sum_k f_{ik}^{(\lambda)} \otimes f_{kj}^{(\lambda)}, \tag{37}$$

where $f_{ij}^{(\lambda)}$ denotes the (i, j)th matrix entry of $R^{(\lambda)}$.

If S has a zero, then $e_0 = 0$, $\mathfrak{M}_0 = \{0\}$ and the algebra $k[\pi^{-1}(0)] = k[S]/I_0$ is naturally identified as a $(G_0 \times G_0)$-module with the subspace

$$k[S]_{\mathfrak{M}} = \bigoplus_{M \in \mathfrak{M}} k[G]_M.$$

On the other hand, by (33) we can identify $k[G_0]$ with $k[S]_{\mathfrak{M}}$ by means of the mapping $f \mapsto e^{\theta^*\lambda} f$ ($f \in k[G_0]_\lambda$). Thereby $k[\pi^{-1}(0)]$ is identified with $k[G_0]$.

To the multiplication in $k[\pi^{-1}(0)]$, there corresponds the $*$-multiplication in $k[G_0]$, defined by

$$f * g = p_{\lambda+\mu}(f, g) \qquad (f \in k[G_0]_\lambda, \; g \in k[G_0]_\mu).$$

To the comultiplication in $k[\pi^{-1}(0)]$, there corresponds the comultiplication (37) in $k[G_0]$.

So we see that the semigroup $\pi^{-1}(0)$, together with the action of $G_0 \times G_0$, depends only on G_0, provided S is a flat semigroup with zero.

5. For a flat reductive semigroup S, the structure of $(G \times G)$-orbits in S is closely connected with the structure of $(G_0 \times G_0)$-orbits in the fibers of the morphism π.

PROPOSITION 5. *Let S be a flat reductive semigroup. Then any fiber $\pi^{-1}(a)$ ($a \in A$) of the morphism π is a spherical $(G_0 \times G_0)$-variety and*

$$GsG \cap \pi^{-1}(a) = G_0 s G_0$$

for any $s \in \pi^{-1}(a)$.

PROOF. For a reductive group L, an irreducible affine L-variety X is spherical if and only if $k[X]$ is a multiplicity free L-module [23].

Since the morphism π is flat and its fibers are reduced, all the $(G_0 \times G_0)$-modules $k[\pi^{-1}(a)]$, $a \in A$, are isomorphic. Since $k[\pi^{-1}(e)] = k[G_0]$ is multiplicity free, such are all of them.

Thus, any fiber $\pi^{-1}(a)$ ($a \in A$) is a spherical $(G_0 \times G_0)$-variety and, in particular, contains only finitely many $(G_0 \times G_0)$-orbits. Denote by Z_a the stabilizer of a in Z. Obviously,

$$GsG \cap \pi^{-1}(a) = Z_a G_0 s G_0$$

for any $s \in \pi^{-1}(a)$. The group Z_a, acting in $\pi^{-1}(a)$, can only permute $(G_0 \times G_0)$-orbits. We must prove that in fact it leaves each of them invariant. This will follow from the connectedness of Z_a/Z_0, which is proved below.

We have $\mathfrak{X}(Z/Z_0) = \mathfrak{X}_Z$ and $\mathfrak{X}_Z \cap D = \mathfrak{L}_Z$. There is a face F of the cone D such that, for $\chi \in \mathfrak{L}_Z$,

$$\chi(a) \begin{cases} \neq 0, & \chi \in F, \\ = 0, & \chi \notin F. \end{cases}$$

The subgroup $Z_a \subset Z$ is defined by the equations

$$\chi(z) = 1, \qquad \chi \in \mathfrak{L}_Z \cap F.$$

Since $\mathfrak{L}_Z \cap F = \mathfrak{X}_Z \cap F$, the subgroup of $\mathfrak{X}_Z$ generated by $\mathfrak{L}_Z \cap F$ is primitive (i.e., the quotient group is torsion-free). This means that the group Z_a/Z_0 is connected. □

§5. Proof of Theorem 5

1. Let S and S' be reductive semigroups and $\varphi\colon S' \to S$ be a homomorphism. We denote all the objects associated to S' by the same letters, as those associated to S, but with a prime.

The homomorphism φ is excellent if and only if

$$(38) \qquad k[S'] \simeq k[A'] \otimes_{k[A]} k[S],$$

the structure of a $k[A]$-module on $k[A']$ being defined by means of the homomorphism $\varphi_{\mathrm{ab}}^*\colon k[A] \to k[A']$ and the isomorphism being realized by means of the map

$$\pi'^* \times \varphi^*\colon k[A'] \times k[S] \to k[S'].$$

Note that φ_{ab}^* is nothing else than the restriction of φ^* to $k[A]$ and π'^* is the identity embedding of $k[A']$ into $k[S']$.

PROPOSITION 6. *Let the semigroup S be flat and the homomorphism φ be excellent. Then the semigroup S' is also flat.*

PROOF. Since $k[S]$ is a flat $k[A]$-module, it follows from (38) that $k[S']$ is a flat $k[A']$-module [21].

Now let $\mathfrak{m}'$ be a maximal ideal of $k[A']$ and $\mathfrak{m}$ be its pullback in $k[A]$. Then

$$k[S']/\mathfrak{m}'k[S'] \simeq (k[A']/\mathfrak{m}') \otimes_{k[A]/\mathfrak{m}} (k[S]/\mathfrak{m}k[S]) = k[S]/\mathfrak{m}k[S],$$

so $k[S']/\mathfrak{m}'k[S']$ has no zero divisors. □

PROPOSITION 7. *Let the semigroups S and S' be flat. The homomorphism φ is excellent if and only if $\varphi^*(\mathfrak{M}_1)$ is a set of representatives of the cosets of $\mathfrak{M}'_0$ in $\mathfrak{M}'$.*

PROOF. By (28), the right-hand side of (38) can be represented in the form

$$k[A'] \otimes_{k[A]} k[S] = k[A'] \otimes k[G]_{\mathfrak{M}_1}.$$

So the homomorphism φ is excellent if and only if

$$k[S'] = k[A'] \otimes k[G']_{\varphi^*(\mathfrak{M}_1)},$$

which is equivalent to the property stated in the proposition. □

2. For a connected semisimple group G_0, we construct a reductive semigroup $S = \mathrm{Env}\, G_0$ as described in the Introduction. Obviously, it has a zero. We must prove the property $(*)$.

Let S' be another reductive semigroup from the class $\mathcal{FS}(G_0)$ and $\varphi_0\colon G'_0 \to G_0$ an isomorphism. To simplify the notation, let us identify G'_0 with G_0 by means of this isomorphism.

Since θ is an isomorphism, there is a unique homomorphism $\varphi\colon Z' \to Z$ such that the diagram

$$(39) \qquad \begin{array}{ccc} Z' & \xrightarrow{\varphi} & Z \\ {\scriptstyle \theta'} \searrow & & \swarrow {\scriptstyle \theta} \\ & T_0 & \end{array}$$

is commutative. Since $Z'_0 \subset Z_0$, we have $\varphi|_{Z'_0} = \mathrm{id}$. It follows that, being combined with the identity map of G_0, the map φ gives rise to a homomorphism $G' \to G$. We also denote it by φ.

For any $\lambda \in \mathfrak{t}_0(\mathbb{Q})^*$ we have

$$\varphi^*(\bar{\lambda}) = \varphi^*\theta^*(\lambda) = \theta'^*(\lambda).$$

Since the cone $D \subset \mathfrak{z}(\mathbb{Q})^*$ is generated by $\bar{\alpha}_1, \ldots, \bar{\alpha}_n$, while the cone $D' \subset \mathfrak{z}'(\mathbb{Q})^*$ contains $\theta'^*(\alpha_1), \ldots, \theta'^*(\alpha_n)$, we obtain $\varphi^*(D) \subset D'$, $\varphi^*(K) \subset K'$. Hence

(40) $$\varphi^*(\mathfrak{L}) \subset \mathfrak{L}',$$

and, moreover,

(41) $$\varphi^*(\mathfrak{L}_Z) \subset \mathfrak{L}'_Z, \qquad \varphi^*(\mathfrak{M}) = \mathfrak{M}'_1 \subset \mathfrak{M}',$$

where $\mathfrak{M}'_1 = \{(\theta'^*(\lambda), \lambda) : \lambda \in \mathfrak{X}(T_0)\}$.

It follows from (40) that, for any $\Lambda \in \mathfrak{L}$,

$$\varphi^*(k[G]_\Lambda) = k[G']_{\varphi^*(\Lambda)} \subset k[S'].$$

This means that φ is extended to a homomorphism $S' \to S$. We still denote it by φ.

According to Proposition 7, the property (41) implies that the homomorphism φ is excellent.

Conversely, if the semigroup S' has a zero, any excellent homomorphism $\varphi \colon S' \to S$ which is the identity map on $G'_0 = G_0$ must satisfy the condition $\varphi^*(\mathfrak{M}) = \mathfrak{M}'$. This condition is equivalent to commutativity of the diagram (39) and hence defines φ uniquely.

§6. Proof of Theorem 6

1. We need some known facts from representation theory. For the reader's convenience, we give their proofs (cf. [22]).

For any $I \subset \Omega$ and $\Lambda \in \mathfrak{X}_+$, we use the following notation:

Π_I — the linear span of $\{\alpha_i : i \in I\}$ in $\mathfrak{t}(\mathbb{Q})^*$,

G_I — the connected (reductive) algebraic subgroup of G, whose tangent algebra is spanned by $\mathfrak{t}$ and the root subspaces corresponding to the roots lying in Π_I,

$R_I^{(\Lambda)}$ — the irreducible linear representation of G_I with highest weight Λ,

$V_I^{(\Lambda)}$ — the subspace of $V^{(\Lambda)}$ spanned by the weight subspaces, corresponding to the weights lying in $\Lambda + \Pi_I$.

Evidently, $V_I^{(\Lambda)}$ is G_I-invariant.

PROPOSITION 8. *The representation of G_I in $V_I^{(\Lambda)}$ is irreducible (and hence isomorphic to $R_I^{(\Lambda)}$).*

PROOF. Every weight M of $R^{(\Lambda)}$ is (uniquely) represented in the form

$$M = \Lambda - \sum_i k_i \alpha_i \qquad (k_i \in \mathbb{Z}_+).$$

We have $M \in \Lambda + \Pi_I$ if and only if $k_j = 0$ for $j \notin I$. So if $M \in \Lambda + \Pi_I$ and $j \notin I$, then $M + \alpha_j$ is not a weight of $R^{(\Lambda)}$. It follows that every highest vector for G_I in $V_I^{(\Lambda)}$ is a highest vector for G. Consequently, such a vector is unique up to proportionality. □

Now let $R: G \to GL(V)$ be a (not necessarily irreducible) linear representation of G and $\overline{\Lambda} \in \mathfrak{t}(\mathbb{Q})^*$ be such that any weight M of R can be represented in the form

$$M = \overline{\Lambda} - \sum_i k_i \alpha_i \qquad (k_i \in \mathbb{Z}_+).$$

Denote by V_I the subspace of V spanned by the weight subspaces corresponding to the weights lying in $\overline{\Lambda} + \Pi_I$.

PROPOSITION 9. *The representation of G_I in V_I is isomorphic to $\sum_\Lambda m_\Lambda R_I^{(\Lambda)}$, where Λ runs over $\mathfrak{X}_+ \cap (\overline{\Lambda} + \Pi_I)$ and m_Λ denote the multiplicity of $R^{(\Lambda)}$ in R.*

PROOF. In the same way as in the preceding proof, we can see that any highest vector for G_I in V_I is a highest vector for G. □

PROPOSITION 10. *For Λ, $M \in \mathfrak{X}_+$ and*

$$N = \Lambda + M - \sum_{i \in I} k_i \alpha_i \in \mathfrak{X}_+ \qquad (k_i \in \mathbb{Z}_+),$$

the multiplicity of $R^{(N)}$ in $R^{(\Lambda)} R^{(M)}$ is equal to that of $R_I^{(N)}$ in $R_I^{(\Lambda)} R_I^{(M)}$.

PROOF. Apply Proposition 9 to $R = R^{(\Lambda)} R^{(M)}$, taking $\overline{\Lambda} = \Lambda + M$. Note that in this case the representation of G_I in V_I is isomorphic to $R_I^{(\Lambda)} R_I^{(M)}$. □

2. Let G_0 be a connected semisimple group and $S = \text{Env}\, G_0$. Any prime $(G \times G)$-invariant ideal $\mathfrak{p}$ of $k[S]$ has the form

(42) $$\mathfrak{p} = \bigoplus_{\Lambda \in \mathfrak{J}} k[G]_\Lambda,$$

where $\mathfrak{J}$ is an ideal of the semigroup $\mathfrak{L}$ such that its complement $\mathfrak{L} \setminus \mathfrak{J}$ is a subsemigroup. In its turn, any such ideal $\mathfrak{J}$ of $\mathfrak{L}$ has the form

(43) $$\mathfrak{J} = \mathfrak{L} \setminus F,$$

where F is a (closed) face of $\mathbb{Q}\mathfrak{L} = K \cap C$ [13].

Conversely, let $\mathfrak{J}$ be an ideal of $\mathfrak{L}$ of the form (43). Assume that the subspace $\mathfrak{p}$ defined by (42) is an ideal of $k[S]$. Then

$$(k[S]/\mathfrak{p})^{U_- \times U} \simeq k(\mathfrak{L} \setminus \mathfrak{J}),$$

and, by the theorem stated in 1.1, $k[S]/\mathfrak{p}$ has no zero divisors, i.e., $\mathfrak{p}$ is a prime ideal.

3. Let us now find out for what faces F of $K \cap C$ the subspace $\mathfrak{p}$ defined by (42) is an ideal of $k[S]$, that is

(44) $$\Lambda \in \mathfrak{L},\; M \in \mathfrak{J} \implies \mathfrak{X}(\Lambda, M) \subset \mathfrak{J}.$$

Let $F = F_{I,J}$ in the notation of the introduction.

Assume that a connected component $\tilde{I}$ of the complement $C(J)$ of J is entirely contained in I. Let
$$\alpha = \sum_{i \in \tilde{I}} m_i \alpha_i = \sum_{i=1}^{n} l_i \omega_i$$
be the highest root of $G_{\tilde{I}}$. We have $m_i > 0$ for $i \in \tilde{I}$, while
$$l_i \begin{cases} \geq 0 & \text{for } i \in \tilde{I}, \\ \leq 0 & \text{for } i \in J, \\ = 0 & \text{for } i \notin \tilde{I} \cup J; \end{cases}$$
moreover, there is an $i \in \tilde{I}$ such that $l_i > 0$. So
$$\lambda = \sum_{i \in \tilde{I}} l_i \omega_i \in C_0 \setminus C_J, \qquad \Lambda = (\bar{\lambda}, \lambda) \in (K \cap C) \setminus F.$$

Take an integer $k > 0$ such that $k\Lambda \in \mathfrak{X}$. Then $k\Lambda \in \mathfrak{J}$.

Consider the representation $R^{(k\Lambda)}$ of G. Its restriction to $(G_{\tilde{I}}, G_{\tilde{I}})$ is the irreducible representation with the highest weight $k\alpha$. It is self-dual, which implies that its square contains the trivial representation. Hence $(R_{\tilde{I}}^{(k\Lambda)})^2 \supset R_{\tilde{I}}^{(2k(\Lambda-\alpha))}$ and, by Proposition 10, $(R^{(k\Lambda)})^2 \supset R^{(2k(\Lambda-\alpha))}$. However
$$2k(\Lambda - \alpha) = 2k\left(\bar{\lambda}, -\sum_{j \in J} l_j \omega_j\right) \in F,$$
so $2k(\Lambda - \alpha) \notin \mathfrak{J}$. Consequently, $\mathfrak{p}$ is not an ideal of $k[S]$.

4. Conversely, let no connected component of $C(J)$ be entirely contained in I. Let
$$\Lambda = (\chi, \lambda) \in \mathfrak{L}, \qquad M = (\psi, \mu) \in \mathfrak{J}.$$
Then any $N \in \mathfrak{X}(\Lambda, M)$ has the form $N = (\chi + \psi, \nu)$, where
$$\nu = \lambda + \mu - \sum_{i=1}^{n} k_i \alpha_i \qquad (k_i \geq 0).$$

We have either $\psi - \bar{\mu} \notin D_I$, or $\mu \notin C_J$. In the first case
$$\chi + \psi - \bar{\nu} = (\chi - \bar{\lambda}) + (\psi - \bar{\mu}) + \sum_{i=1}^{n} k_i \bar{\alpha}_i \notin D_I,$$
so $N \in \mathfrak{J}$. The same result is obtained if $k_s > 0$ for some $s \notin I$.

Now let $\mu \notin C_J$, so $\lambda + \mu \notin C_J$ as well, and
$$(45) \qquad \nu = \lambda + \mu - \sum_{i \in I} k_i \alpha_i \qquad (k_i \geq 0).$$

Suppose $\nu \in C_J$ and let $I' = \{i \in I : k_i > 0\}$. If j is adjacent to some $i \in I'$, then it follows from (45) that $\nu(h_j) > 0$ and hence $j \in J$. But then our assumption about I and J implies that $I' \subset J$. So
$$\nu(h_s) = (\lambda + \mu)(h_s) \quad \text{for } s \notin J,$$
which makes it impossible for ν to belong to C_J.

§7. Proof of Theorem 7

1. Let S be a reductive semigroup with $G(S) = G$ defined by a subsemigroup $\mathfrak{L} \subset \mathfrak{X}_+$. For any $\Lambda \in \mathfrak{L}$, the representation $R^{(\Lambda)}$ of G is extended to a representation of S, which will be denoted in the same way. The sum of all these representations will be denoted by $\mathcal{R}$. Obviously, $\mathcal{R}$ is a faithful (infinite-dimensional) representation of S.

For any $s \in S$, $\mathcal{R}(s)$ can be represented as the set $\{R^{(\Lambda)}(s) : \Lambda \in \mathfrak{L}\}$, where $R^{(\Lambda)}(s) \in \operatorname{End} V^{(\Lambda)}$.

PROPOSITION 11. *A set* $\{\mathcal{A}^{(\Lambda)} : \Lambda \in \mathfrak{L}\}$, *where* $\mathcal{A}^{(\Lambda)} \in \operatorname{End} V^{(\Lambda)}$, *belongs to* $\mathcal{R}(S)$ *if and only if for any* Λ, M, $N \in \mathfrak{L}$ *and any G-equivariant linear map*

$$\varphi \colon V^{(\Lambda)} \otimes V^{(M)} \to V^{(N)} \tag{46}$$

the diagram

$$\begin{array}{ccc} V^{(\Lambda)} \otimes V^{(M)} & \xrightarrow{\varphi} & V^{(N)} \\ \mathcal{A}^{(\Lambda)} \otimes \mathcal{A}^{(M)} \downarrow & & \downarrow \mathcal{A}^{(N)} \\ V^{(\Lambda)} \otimes V^{(M)} & \xrightarrow{\varphi} & V^{(N)} \end{array} \tag{47}$$

is commutative.

PROOF. The maps (46) contain all the information about the decomposition rule of the tensor products of the G-modules $V^{(\Lambda)}$, $\Lambda \in \mathfrak{L}$, and thereby about the multiplication law of the matrix entries of the representations $R^{(\Lambda)}$, $\Lambda \in \mathfrak{L}$. The commutativity of the diagrams (47) means that the matrix entries of $\mathcal{A}^{(\Lambda)}$'s are multiplied in the same way as the matrix entries of the representations $R^{(\Lambda)}$. □

REMARK. One can easily get a finite-dimensional faithful representation of S. Namely, if $\Lambda_1, \ldots, \Lambda_m$ generate the semigroup $\mathfrak{L}$, then $R = R^{(\Lambda_1)} + \cdots + R^{(\Lambda_m)}$ is such a representation. Moreover, $R(S)$ is a closed subsemigroup of $\operatorname{End} V^{(\Lambda_1)} \times \cdots \times \operatorname{End} V^{(\Lambda_m)}$. However, to describe $R(S)$ explicitly is a difficult task.

2. Now let G_0 be a connected semisimple group, $S = \operatorname{Env} G_0$, and A the abelization of S.

The characters $\overline{\alpha}_i = \theta^* \alpha_i$ $(i = 1, \ldots, n)$ of Z, considered as elements of the algebra $k[A] \subset k[S]$, will be denoted by π_i. We have $k[A] = k[\pi_1, \ldots, \pi_n]$, so $A = k^n$ and the homomorphism $\pi \colon S \to A$ is given by $\pi(s) = (\pi_1(s), \ldots, \pi_n(s))$.

For any $I \subset \Omega$, let e_I be the corresponding idempotent of A, defined by

$$\pi_i(e_I) = \begin{cases} 1, & i \in I, \\ 0, & i \in I. \end{cases} \tag{48}$$

We shall describe the subsemigroup

$$S_I = \pi^{-1}(e_I) \subset S \tag{49}$$

in terms of the representation $\mathcal{R}$.

For $\lambda \in \mathfrak{X}_+(T_0)$, let

(50) $$R^{(\lambda)}: G_0 \to GL(V^{(\lambda)})$$

be the irreducible representation of G_0 with highest weight λ.

Any $\Lambda = (\chi, \lambda) \in \mathfrak{L}$ is represented in the form

(51) $$\Lambda = \left(\bar{\lambda} + \sum_i k_i \bar{\alpha}_i, \lambda\right) \qquad (k_i \in \mathbb{Z}_+).$$

The space of the representation $R^{(\Lambda)}$ can be identified with $V^{(\lambda)}$ and if we put

(52) $$\overline{R}^{(\lambda)} = R^{(\bar{\lambda}, \lambda)},$$

then

$$R^{(\Lambda)} = \left(\prod_i \pi_i^{k_i}\right) \overline{R}^{(\lambda)}.$$

In particular, for $s \in S_I$ we have

(53) $$R^{(\Lambda)}(s) = \begin{cases} \overline{R}^{(\lambda)}(s), & \text{if } \chi - \bar{\lambda} \in D_I, \\ 0, & \text{otherwise.} \end{cases}$$

It follows that

(54) $$\overline{\mathcal{R}} = \bigoplus_{\lambda \in \mathfrak{X}_+(T_0)} \overline{R}^{(\lambda)}$$

is a faithful representation of S_I.

Proposition 11 together with (53) implies the following description of $\overline{\mathcal{R}}(S_I)$.

PROPOSITION 12. *A set* $\{\mathcal{A}^{(\lambda)}: \lambda \in \mathfrak{X}_+(T_0)\}$, *where* $\mathcal{A}^{(\lambda)} \in \text{End } V^{(\lambda)}$, *belongs to* $\overline{\mathcal{R}}(S_I)$ *if and only if for any* $\lambda, \mu, \nu \in \mathfrak{X}_+(T_0)$ *and any G_0-equivariant linear map*

(55) $$\varphi: V_0^{(\lambda)} \otimes V_0^{(\mu)} \to V_0^{(\nu)},$$

one has

(56) $$\varphi \circ (\mathcal{A}^{(\lambda)} \otimes \mathcal{A}^{(\mu)}) = \begin{cases} \mathcal{A}^{(\nu)} \circ \varphi, & \text{if } \nu \in \lambda + \mu + \Pi_I, \\ 0, & \text{otherwise.} \end{cases}$$

(Here Π_I denotes the linear span of $\{\alpha_i : i \in I\}$ in $\mathfrak{t}_0(\mathbb{Q})^*$.)

3. For $I \subset \Omega$ and $\lambda \in \mathfrak{X}_+(T_0)$ let $V_I^{(\lambda)}$ denote the subspace of $V^{(\lambda)}$ spanned by the weight subspaces, corresponding to the weights lying in the plane $\lambda + \Pi_I$, and $\mathcal{P}_I^{(\lambda)}$ the (unique) T_0-equivariant projection on $V_I^{(\lambda)}$.

Now let $I, J \subset \Omega$ constitute an essential pair.

LEMMA 1. *If* $\lambda \in \mathfrak{X}_+(T_0)$ *and* $\lambda - \sum_{i \in I} k_i \alpha_i \in C_J$ ($k_i \in \mathbb{Z}_+$), *then*

1) $\lambda \in C_J$;
2) $k_i = 0$ *for* $i \notin J^\circ$.

(See the notation in 0.6 and 0.7.)

PROOF. Let $I' = \{i \in I : k_i \neq 0\}$. Suppose $i_0 \in I' \setminus J^\circ$. Let m be an element of $C(J)$ such that $\alpha_{i_0}(h_m) < 0$, and let M be the connected component of $C(J)$ containing m. Since $I' \not\supset M$, we may assume that $m \notin I'$. Then

$$\left(\lambda - \sum_i k_i \alpha_i\right)(h_m) = \lambda(h_m) - \sum_i k_i \alpha_i(h_m) > 0,$$

which is a contradiction. Hence $I' \subset J^\circ$ and for any $m \in C(J)$

$$\lambda(h_m) = \left(\lambda - \sum k_i \alpha_i\right)(h_m) = 0,$$

so $\lambda \in C_J$. □

For any $\lambda \in \mathfrak{X}_+(T_0)$ let us define

(57) $$\mathcal{P}_{I,J}^{(\lambda)} = \begin{cases} \mathcal{P}_{I \cap J^\circ}^{(\lambda)}, & \text{if } \lambda \in C_J, \\ 0, & \text{otherwise}, \end{cases}$$

and prove that the set $\{\mathcal{P}_{I,J}^{(\lambda)}\}$ satisfies the condition of Proposition 12.

Consider any nontrivial G_0-equivariant linear map (55). Note that $\mathcal{P}_{I,J}^{(\lambda)} \otimes \mathcal{P}_{I,J}^{(\mu)}$ is either 0, or, if $\lambda, \mu \in C_J$, the T_0-equivariant projection of $V^{(\lambda)} \otimes V^{(\mu)}$ on the sum of the weight subspaces, corresponding to the weights, lying in the plane $\lambda + \mu + \Pi_{I \cap J^\circ}$.

If $\nu \notin \lambda + \mu + \Pi_I$, then $R^{(\nu)}$ has no weights in the plane $\lambda + \mu + \Pi_I$, so

$$\varphi \circ (\mathcal{P}_{I,J}^{(\lambda)} \otimes \mathcal{P}_{I,J}^{(\mu)}) = 0.$$

Now let $\nu \in \lambda + \mu + \Pi_I$. Under this condition, we must prove that

(58) $$\varphi \circ (\mathcal{P}_{I,J}^{(\lambda)} \otimes \mathcal{P}_{I,J}^{(\mu)}) = \mathcal{P}_{I,J}^{(\nu)} \circ \varphi.$$

If $\lambda + \mu \notin C_J$, then by Lemma 1 $\nu \notin C_J$ and both sides of (58) vanish.

If $\lambda + \mu \in C_J$, but $\nu \notin C_J$, then $\nu \notin \lambda + \mu + \Pi_{I \cap J^\circ}$ and hence $R^{(\nu)}$ has no weights in this plane. In this case, both sides of (58) still vanish.

Finally, if $\lambda, \mu, \nu \in C_J$, then by Lemma 1 $\nu \in \lambda + \mu + \Pi_{I \cap J^\circ}$, so

$$\nu + \Pi_{I \cap J^\circ} = \lambda + \mu + \Pi_{I \cap J^\circ}.$$

It follows that

$$\varphi \circ (\mathcal{P}_{I \cap J^\circ}^{(\lambda)} \otimes \mathcal{P}_{I \cap J^\circ}^{(\mu)}) = \mathcal{P}_{I \cap J^\circ}^{(\nu)} \circ \varphi,$$

which is just the equality (58) in this case.

Thus, there is an idempotent $e_{I,J} \in S_I$ such that

(59) $$\overline{\mathcal{R}}(e_{I,J}) = \{\mathcal{P}_{I,J}^{(\lambda)}\}.$$

The definition of $\mathcal{P}_{I,J}^{(\lambda)}$ implies that $e_{I,J} \in O_{I,J}$.

4. In this subsection we prove that $e_{I,J} \in \overline{T}$. This can also be regarded as an independent proof of the existence of an element $e_{I,J} \in S_I$ satisfying (59).

LEMMA 2. *Let Δ be an indecomposable root system and Π its base (whose elements will be called simple roots). For any $\alpha \in \Pi$ there exists a positive linear combination of simple roots whose scalar products with all of them but α are negative.*

PROOF. Let $\Pi_1, \ldots, \Pi_s$ be the indecomposable components of $\Pi \setminus \{\alpha\}$ and α_i the root of Π_i adjacent to α. Proceeding by induction on rk Δ, we may assume that for each i there exists a positive linear combination β_i of the roots of Π_i whose scalar products with all of them but α_i are negative. The sum $\sum_{i=1}^{s} \beta_i + c\alpha$ meets the requirement for sufficiently large positive c. □

LEMMA 3. *For any essential pair (I, J) there exists an element $h \in \mathfrak{t}(\mathbb{Q})$ such that*

$$\overline{\alpha}_i(h) = \begin{cases} 0, & i \in I, \\ a_i > 0, & i \notin I, \end{cases} \tag{60}$$

$$(\overline{\omega}_j + \omega_j)(h) = \begin{cases} 0, & j \in J, \\ b_j > 0, & j \notin J, \end{cases} \tag{61}$$

$$\alpha_i(h) = \begin{cases} 0, & i \in I \cap J^\circ, \\ c_i < 0, & i \notin I \cap J^\circ. \end{cases} \tag{62}$$

PROOF. For any positive rational a_i ($i \notin I$) and b_j ($j \notin J$) there exists a unique $h \in \mathfrak{t}(\mathbb{Q})$ satisfying (60) and (61). We must show that a_i and b_j can be chosen in such a way that (62) be also satisfied.

For any connected component M of $C(J)$ choose $m \in M \setminus I$ and, using Lemma 2, take a positive linear combination h_M of the corresponding dual roots satisfying the condition $\alpha_i(h_M) < 0$ for all $i \in M \setminus \{m\}$. The sum $h_0 = \sum_M h_M$ satisfies the conditions

$$\omega_j(h_0) = \begin{cases} 0, & j \in J, \\ b_j > 0, & j \notin J, \end{cases} \tag{63}$$

$$\alpha_i(h_0) = \begin{cases} 0, & i \in J^\circ, \\ d_i < 0, & i \in I \setminus J^\circ. \end{cases} \tag{64}$$

Since $\overline{\alpha}_i + \alpha_i$ are expressed in terms of $\overline{\omega}_j + \omega_j$ in the same way as α_i are expressed in terms of ω_j, any $h \in \mathfrak{t}(\mathbb{Q})$ satisfying (61) automatically satisfies the conditions

$$(\overline{\alpha}_i + \alpha_i)(h) = \begin{cases} 0, & i \in J^\circ, \\ d_i < 0, & i \in I \setminus J^\circ. \end{cases} \tag{65}$$

Since

$$\alpha_i(h) = (\overline{\alpha}_i + \alpha_i)(h) - \overline{\alpha}_i(h),$$

(65) and (60) implies (62), provided a_i are sufficiently large. □

PROPOSITION 13. *If $h \in \mathfrak{t}(\mathbb{Q})$ satisfies the conditions of Lemma 3, then*

$$\lim_{t \to -\infty} \exp th = e_{I,J}. \tag{66}$$

PROOF. To prove (66) it suffices to prove that all the eigenvalues of $d\mathcal{R}(h)$ are nonnegative and

(67) $$\operatorname{Ker} d\mathcal{R}(h) = \operatorname{Im} \mathcal{R}(e_{I,J}).$$

Let $\Lambda \in \mathfrak{L}$ have the form (51). Then all the weights of $R^{(\Lambda)}$ have the form

(68) $$M = \sum_i k_i \overline{\alpha}_i + (\overline{\lambda} + \lambda) - \sum_i l_i \alpha_i \qquad (k_i, l_i \geq 0),$$

and it follows from (60)–(62) that $M(h) \geq 0$. Moreover, $M(h) = 0$ if and only if
1) $k_i = 0$ for $i \notin I$;
2) $\lambda \in C_J$;
3) $l_i = 0$ for $i \notin I \cap J^\circ$.

This gives (67). □

5. Now we find the stabilizer of $e_{I,J}$ in $G \times G$. Obviously, $g_1 e_{I,J} g_2^{-1} = e_{I,J}$ ($g_1, g_2 \in G$) if and only if

(69) $$R^{(\Lambda)}(g_1) \mathcal{P}_{I\cap J^\circ}^{(\lambda)} R^{(\Lambda)}(g_2)^{-1} = \mathcal{P}_{I\cap J^\circ}^{(\lambda)}$$

for any $\Lambda = (\chi, \lambda) \in \mathfrak{X} \cap F_{I,J}$.

Denote by $U_{I\cap J^\circ}^{(\lambda)}$ the T_0-invariant complementary subspace of $V_{I\cap J^\circ}^{(\lambda)}$ in $V^{(\lambda)}$. The condition (69) is equivalent to the three following ones:

(S1) $V_{I\cap J^\circ}^{(\lambda)}$ is invariant under $R^{(\Lambda)}(g_1)$;
(S2) $U_{I\cap J^\circ}^{(\lambda)}$ is invariant under $R^{(\Lambda)}(g_2)$;
(S3) if we identify the spaces $V_{I\cap J^\circ}^{(\lambda)}$ and $V^{(\lambda)}/U_{I\cap J^\circ}^{(\lambda)}$ in the natural way, then their endomorphisms induced by $R^{(\Lambda)}(g_1)$ and $R^{(\Lambda)}(g_2)$, respectively, coincide.

The condition (S1) is satisfied if and only if $g_1 \in P(M)$, where $M = (I \cap J^\circ) \cup C(J)$. The kernel of the representation of $P(M)$ in $V_{I\cap J^\circ}^{(\lambda)}$ is $U(M)G(C(J))T_{I,J}$ (see the notation in 0.7).

In an analogous way, the condition (S2) is satisfied if and only if $g_2 \in P_-(M)$. The kernel of the representation of P_- in $V_{I\cap J^\circ}^{(\lambda)} = V^{(\lambda)}/U_{I\cap J^\circ}^{(\lambda)}$ is $U_-(M)G(C(J))T_{I,J}$.

This gives Theorem 7.

6. Formula (21) follows from the formula

$$\mathcal{P}_{M_1}^{(\lambda)} \mathcal{P}_{M_2}^{(\lambda)} = \mathcal{P}_{M_1 \cap M_2}^{(\lambda)} \qquad (M_1, M_2 \subset \Omega, \lambda \in \mathfrak{X}_+(T_0)),$$

and the obvious fact that the interior of $J_1 \cap J_2$ is $J_1^\circ \cap J_2^\circ$.

§8. Proof of Theorem 8

1. Define a character $\widetilde{\alpha}_i$ of the torus $T_\varnothing$ by

(70) $$\widetilde{\alpha}_i(z\theta(z)^{-1}) = \overline{\alpha}_i(z) = \alpha_i(\theta(z)).$$

Obviously, the group $\mathfrak{X}(T_\varnothing)$ is freely generated by $\widetilde{\alpha}_1, \ldots, \widetilde{\alpha}_n$. The algebra $k[\overline{T}_\varnothing]$ is spanned by the restrictions to $T_\varnothing$ of the weights of the representations $R^{(\Lambda)}$, $\Lambda \in \mathfrak{L}$. Since any such weight is represented in the form (68) and

$$(\overline{\lambda} + \lambda)(z\theta(z)^{-1}) = 1 \qquad (\lambda \in \mathfrak{X}_+(T_0), \ z \in Z),$$

we obtain

(71) $$k[\overline{T}_\varnothing] = k[e^{\widetilde{\alpha}_1}, \ldots, e^{\widetilde{\alpha}_n}].$$

It follows that the semigroup $\overline{T}_\varnothing$ is isomorphic to k^n. It contains 2^n idempotents, enumerated by the subsets of $\Omega = \{1, \ldots, n\}$. To a subset $I \subset \Omega$ there corresponds an idempotent $\widetilde{e}_I \in \overline{T}_\varnothing \subset \overline{T}$ with the following properties:

$$\pi_i(\widetilde{e}_I) = \widetilde{\alpha}_i(\widetilde{e}_I) = \begin{cases} 1, & i \in I, \\ 0, & i \notin I, \end{cases}$$
$$\overline{R}^{(\lambda)}(\widetilde{e}_I) = \mathcal{P}_I^{(\lambda)} \qquad (\lambda \in \mathfrak{X}_+(T_0))$$

(see the notation in §7). Comparing this with results of 7.3, we obtain $\widetilde{e}_I = e_{I,\Omega}$. It follows that

(72) $$G\overline{T}_\varnothing G = S^{\mathrm{pr}}.$$

It is easy to see that $T = Z \times T_\varnothing$.

2. The following proposition describes "the big cell" of S.

PROPOSITION 14. *The map*

$$\varphi\colon U_- \times Z \times \overline{T}_\varnothing \times U \to S, \qquad (u_-, z, t, u) \mapsto u_- z t u,$$

is an open embedding. Its image is contained in S^{pr} and contains representatives of all $(G \times G)$-orbits in S^{pr}.

PROOF. Two last assertions follow from (72). To prove the first one, we have only to check that φ is dominant and injective, because every injective dominant morphism of normal irreducible algebraic varieties is an open embedding.

Since $ZT_\varnothing = T$ and the big cell $\mathrm{BC}(G) = U_- T U$ is dense in G, φ is dominant. To prove injectivity, we must show that

(73) $$u_- z t_1 e_{I,\Omega} u = t_2 e_{I,\Omega} \qquad (u_- \in U_-, \ u \in U, \ z \in Z, \ t_1, t_2 \in T_\varnothing)$$

implies

(74) $$u_- = u = z = e.$$

We can rewrite (73) as follows:

(75) $$(t_2^{-1} u_- t_2) z t e_{I,\Omega} u = e_{I,\Omega},$$

where $t = t_1 t_2^{-1} \in T_\varnothing$. According to Theorem 7, the stabilizer of $e_{I,\Omega}$ in $G \times G$ is the semidirect product of $U(I) \times U_-(I)$, the diagonal in $R(I) \times R(I)$, and the torus $T_I \times \{e\}$. Note that $T_I \subset T_\varnothing$. Therefore (75) implies that

$$u_-, u \in R(I), \qquad (t_2^{-1} u_- t_2) z t u \in T_I,$$

which, in turn, implies (74). $\square$

COROLLARY. *The variety S^{pr} is smooth.*

The affine chart $U_-Z\overline{T}_\varnothing U$ will be called the *big cell* of S^{pr} and denoted by $\mathrm{BC}(S^{\mathrm{pr}})$.

3. The study S^{pr}/Z can be reduced to the case when G_0 is simply connected. Namely, let $\widetilde{G}_0$ be the simply connected covering group of G_0 and $\widetilde{S} = \mathrm{Env}\,\widetilde{G}_0$. Let $\widetilde{G} = G(\widetilde{S})$ and $\widetilde{Z}$ be the connected center of $\widetilde{G}$. Then $S = \widetilde{S}/\Gamma$, where Γ is a finite central subgroup of $\widetilde{G}_0$ (contained in $\widetilde{Z}$). It is easy to see that $S^{\mathrm{pr}} = \widetilde{S}^{\mathrm{pr}}/\Gamma$. If there exists a geometric quotient $\widetilde{S}^{\mathrm{pr}}/\widetilde{Z} = X$, it can be regarded as a geometric quotient S^{pr}/Z via the commutative diagram

$$\begin{array}{ccc} \widetilde{S}^{\mathrm{pr}} & \longrightarrow & S^{\mathrm{pr}} \\ & \searrow \quad \swarrow & \\ & X & \end{array}$$

In what follows we assume that G_0 is simply connected.

4. Let $\omega_1, \ldots, \omega_n$ be the fundamental weights of $\mathfrak{g}_0$. Put

$$V^{(i)} = V^{(\omega_i)}, \quad R^{(i)} = \overline{R}^{(\omega_i)}(= R^{(\overline{\omega}_i,\,\omega_i)}), \quad \delta^{(i)} = \delta^{(\overline{\omega}_i,\,\omega_i)}$$

(see the notation in 7.2 and 1.2) and

$$V = V^{(1)} \oplus \cdots \oplus V^{(n)}, \qquad R = R^{(1)} + \cdots + R^{(n)}.$$

Further, put

$$\mathrm{End}'\,V^{(i)} = (\mathrm{End}\,V^{(i)}) \setminus \{0\}, \qquad \mathrm{End}'\,V = \mathrm{End}'\,V^{(1)} \times \cdots \times \mathrm{End}'\,V^{(n)}.$$

It is clear that

(76) $$S^{\mathrm{pr}} = R^{-1}(\mathrm{End}'\,V).$$

PROPOSITION 15. *The representation R maps the variety S^{pr} onto a closed subvariety of $\mathrm{End}'\,V$ isomorphically.*

The first step in proving the proposition is to study the representation of the big cell. In each of the spaces $V^{(1)}, \ldots, V^{(n)}$ we choose a basis as in 1.2 and put

$$\mathrm{End}''\,V^{(i)} = \{\mathcal{A} \in \mathrm{End}\,V^{(i)} : a_{11} \neq 0\},$$

where (a_{kl}) is the matrix of $\mathcal{A}$ in the chosen basis, and

$$\mathrm{End}''\,V = \mathrm{End}''\,V^{(1)} \times \cdots \times \mathrm{End}''\,V^{(n)}.$$

PROPOSITION 16. $\mathrm{BC}(S^{\mathrm{pr}}) = R^{-1}(\mathrm{End}''(V))$.

In other words, the complement of $\mathrm{BC}(S^{\mathrm{pr}})$ in S is the union of the divisors defined by the equations

(77) $$\delta^{(i)} = 0 \quad (i = 1, \ldots, n).$$

As we shall see, these divisors are prime.

PROOF. Since $\mathrm{BC}(S^{\mathrm{pr}})$ is an affine variety, its complement in S is a divisor, say, D. A straightforward verification shows that $\delta^{(i)}(s) \neq 0$ for $s \in \mathrm{BC}(S^{\mathrm{pr}})$. This means that D contains all the divisors (77).

On the other hand, we have

$$\mathrm{BC}(S^{\mathrm{pr}}) \cap G = \mathrm{BC}(G).$$

It is well known that the complement of $\mathrm{BC}(G)$ in G is the union of n prime divisors defined in G by the equations (77) (the center components of the highest weights do not matter here). Consequently, D is the union of the closures of these divisors and, maybe, some divisor that does not meet G.

Obviously, any prime divisor beyond G is $(G \times G)$-invariant and hence the closure of a $(G \times G)$-orbit. Since a pair of the form (Ω, J) is not essential for $J \neq \Omega$, the only $(G \times G)$-orbits of codimension 1 are $O_{I,\Omega}$, where $I = \Omega \setminus \{i\}$. But they all are represented in $\mathrm{BC}(S^{\mathrm{pr}})$. Hence the divisors (77) are prime and exhaust D. □

PROPOSITION 17. *The representation R maps the variety $\mathrm{BC}(S^{\mathrm{pr}})$ onto a closed subvariety of $\mathrm{End}'' V$ isomorphically.*

PROOF. In algebraic terms, the assertion means that the algebra $k[\mathrm{BC}(S^{\mathrm{pr}})]$ is generated by the matrix entries of $R^{(1)}, \ldots, R^{(n)}$ and the function $(\delta^{(1)})^{-1}, \ldots, (\delta^{(n)})^{-1}$. By Proposition 16,

$$k[\mathrm{BC}(S^{\mathrm{pr}})] = k[S][(\delta^{(1)})^{-1}, \ldots, (\delta^{(n)})^{-1}].$$

Since the algebra $k[S]$ is generated by the matrix entries of $R^{(1)}, \ldots, R^{(n)}$ and the functions $\pi_1, \ldots, \pi_n$ (see 7.2), we have only to check that the latter functions can be expressed as polynomials in the matrix entries of $R^{(1)}, \ldots, R^{(n)}$ and $(\delta^{(1)})^{-1}, \ldots, (\delta^{(n)})^{-1}$.

For any i, the square of the representation $R^{(i)}$ contains the representation

$$R^{(2\overline{\omega}_i, 2\omega_i - \alpha_i)} = \pi_i R^{(2\overline{\omega}_i - \overline{\alpha}_i, 2\omega_i - \alpha_i)} = \pi_i \overline{R}^{(2\omega_i - \alpha_i)}$$

(see Proposition 10). In particular, the function $\pi_i \delta^{(2\overline{\omega}_i - \overline{\alpha}_i, 2\omega_i - \alpha_i)}$ can be expressed as a sum of products of matrix entries of $R^{(i)}$. On the other hand, if $2\omega_i - \alpha_i = \sum_j k_j \omega_j$ ($k_j \in \mathbb{Z}_+$), then

$$\delta^{(2\overline{\omega}_1 - \overline{\alpha}_1, 2\omega_1 - \alpha_1)} = \prod_j (\delta^{(j)})^{k_j}.$$

Hence π_i can be represented in the desired form. □

PROOF OF PROPOSITION 15. Since $S^{pr} = G \cdot BC(S^{pr}) \cdot G$, it follows from Proposition 16 and 17 that R maps isomorphically S^{pr} onto a closed subvariety of $R(G) \cdot \text{End}'' V \cdot R(G)$. We will prove that

$$R(G) \cdot \text{End}'' V \cdot R(G) = \text{End}' V, \tag{78}$$

which implies the proposition.

If we identify in the canonical way $\text{End}\, V^{(i)}$ with $V^{(i)} \otimes (V^{(i)})^*$, the matrix entry a_{11} as a linear form on $\text{End}\, V^{(i)}$ is identified with some nonzero element $u \in V^{(i)*} \otimes V^{(i)}$. Since the representation $R^{(i)*} \otimes R^{(i)}$ of $G \times G$ is irreducible, the $(G \times G)$-orbit of u spans the space $V^{(i)*} \otimes V^{(i)}$. This means that for any $\mathcal{A} \in \text{End}'\, V^{(i)}$, there exist such $g_1, g_2 \in G$ that

$$\langle \mathcal{A}, (R^{(i)*}(g_1) \otimes R^{(i)}(g_2))u \rangle = \langle R^{(i)}(g_1)\mathcal{A} R^{(i)}(g_2)^{-1}, u \rangle \neq 0.$$

Thus

$$R(G) \cdot \text{End}'' V^{(i)} \cdot R(G) = \text{End}'\, V^{(i)}. \tag{79}$$

For any set $(\mathcal{A}_1, \ldots, \mathcal{A}_n) \in \text{End}'\, V$, let

$$M^{(i)} = \{(g_1, g_2) \in G \times G : R^{(i)}(g_1)\mathcal{A}_i R^{(i)}(g_2)^{-1} \in \text{End}''\, V^{(i)}\}.$$

It is clear that $M^{(i)}$ is open in $G \times G$ and it follows from (79) that $M^{(i)}$ is not empty. Hence $\bigcap_i M^{(i)} \neq \varnothing$, which means that

$$(\mathcal{A}_1, \ldots, \mathcal{A}_n) \in R(G) \cdot \text{End}''\, V \cdot R(G). \qquad \square$$

5. For $z \in Z$, we have $R^{(i)}(z) = \varpi_i(z)\mathcal{E}$, where $\mathcal{E}$ denotes the identity operator. It follows that the action of Z on S^{pr} is induced via the representation R by the action of $(k^*)^n$ on $\text{End}'\, V$, defined by

$$(t_1, \ldots, t_n) \circ (\mathcal{A}_1, \ldots, \mathcal{A}_n) = (t_1\mathcal{A}_1, \ldots, t_n\mathcal{A}_n).$$

The latter action has the standard geometric quotient

$$p \colon \text{End}'\, V \to P(\text{End}\, V^{(1)}) \times \cdots \times P(\text{End}\, V^{(n)}),$$

where $P(U)$ denotes the projective space associated with the vector space U. The restriction of p to $R(S^{pr})$ defines a geometric quotient of S^{pr}, which is a closed subvariety of $P(\text{End}\, V^{(1)}) \times \cdots \times P(\text{End}\, V^{(n)})$ and hence a projective variety. Moreover, since $R(S^{pr})$ is a smooth variety, so is $p(R(S^{pr}))$. Thus, Theorem 8 is proved.

6. For a centerless connected semisimple group H, the wonderful $(H \times H)$-equivariant embedding of H constructed by DeConcini and Procesi [14] can be characterized by the following properties:

1) it is complete;
2) it is simple, i.e., contains only one closed orbit;
3) it is toroidal, i.e., the closed orbit is not contained in the closure of the complement of the big cell in H.

The embedding $\text{Ad}(G_0) = G_0/Z_0 \subset S^{pr}/Z$ is $(G_0 \times G_0)$-equivariant and projective. It follows from Proposition 14 that S^{pr}/Z decomposes into 2^n orbits, ordered as the subsets of Ω, so only one of them is closed. It is not contained in the closure of the complement of the big cell in G_0/Z_0, because this complement lies beyond $BC(S^{pr})$, while all the orbits meet $BC(S^{pr})$. Thus S^{pr}/Z coincides with the wonderful embedding of $\text{Ad}(G_0)$.

§9. Proof of Theorem 9

1. Let G be a connected reductive group, and let us use the notation of 0.1 for the objects associated to G.

Let G act on an affine variety X. For each $\Lambda \in \mathfrak{X}_+$, denote by $k[X]_\Lambda$ the isotypic component of the G-module $k[X]$ corresponding to the irreducible representation, dual to $R^{(\Lambda)}$. Then

$$(80) \qquad k[X] = \bigoplus_{\Lambda \in \mathfrak{X}_+} k[X]_\Lambda$$

and one can choose a basis $\{\varphi_{ip}\}$ of $k[X]_\Lambda$ such that

$$(81) \qquad \varphi_{ip}(gx) = \sum_j f_{ij}^{(\Lambda)}(g) \varphi_{jp}(x) \qquad (g \in G, \ x \in X).$$

The space $k[X]_\Lambda$ need not be finite-dimensional, but it is known (see, for example, [9]) that it is a finitely generated module over $k[X]^G = k[X]_0$.

Obviously,

$$(82) \qquad k[X]_\Lambda k[X]_M \subset \bigoplus_{N \in \mathfrak{X}(\Lambda, M)} k[X]_N.$$

Considering the products of the highest vectors, we see that if X is irreducible, then the set

$$(83) \qquad \mathfrak{L}(X) = \{\Lambda \in \mathfrak{X}_+ : k[X]_\Lambda \neq 0\}$$

is a subsemigroup of $\mathfrak{X}_+$.

2. Now let S be an algebraic semigroup with $G(S) = G$, defined by a (perfect) subsemigroup $\mathfrak{L} \subset \mathfrak{X}_+$. The G-action on X is extended to an S-action if and only if $\mathfrak{L}(X) \subset \mathfrak{L}$, the extension being given by the formulas (81), in which $g \in G$ is replaced by $s \in S$.

In the general case, the subspace

$$k[X]_\mathfrak{L} = \bigoplus_{\Lambda \in \mathfrak{L}} k[X]_\Lambda$$

is a subalgebra of $k[X]$. If it is finitely generated, we can consider the variety

$$(84) \qquad E = \operatorname{Spec} k[X]_\mathfrak{L}.$$

The semigroup S acts in a natural way on E, and the morphism $\varphi \colon X \to E$, defined by the embedding $k[X]_\mathfrak{L} \subset k[X]$, is G-equivariant.

Moreover, it is easy to see that for each affine S-variety E' and G-equivariant morphism $\varphi' \colon X \to E'$ there is a unique S-equivariant morphism $\psi \colon E \to E'$ such that the diagram

$$\begin{array}{ccc} X & \xrightarrow{\varphi} & E \\ {\scriptstyle \varphi'} \searrow & & \swarrow {\scriptstyle \psi} \\ & E' & \end{array}$$

is commutative.

3. If the commutator group G_0 of G acts on an affine variety X_0, then the formula $(z_1, g) \circ (z_2, x) = (z_1 z_2, gx)$ defines an action of $Z \times G_0$ on $Z \times X_0$ and thereby an action of $G = (Z \times G_0)/Z_0$ on $X = (Z \times X_0)/Z_0$, where $Z_0 = Z \cap G_0$ is assumed to be embedded in $Z \times G_0$ by means of the map $z_0 \mapsto (z_0, z_0^{-1})$. The formula $\varphi(x) = (e, x) \mod Z$ defines a G_0-equivariant closed embedding $\varphi: X_0 \to X$. Obviously, $X = Z\varphi(X_0)$.

Now let X' be an affine G-variety and $\varphi': X_0 \to X'$ a G_0-equivariant morphism. Then there is a unique G_0-equivariant morphism $\psi: X \to X'$ such that the diagram

$$\begin{array}{ccc} X_0 & \xrightarrow{\varphi} & X \\ & \searrow^{\varphi'} \swarrow^{\psi} & \\ & X' & \end{array}$$

is commutative. It is defined by $\psi(z\varphi(x)) = z\varphi'(x)$.

4. The combination of the preceding constructions permits us to obtain in a canonical way an affine S-variety E, starting from an affine G_0-variety X_0, provided the algebra $k[X]_{\mathfrak{L}}$ is finitely generated, Moreover, there is a canonical G_0-equivariant morphism $\varphi: X_0 \to E$ and, for any affine S-variety E' and G_0-equivariant morphism $\varphi': X_0 \to E'$, there is a unique S-equivariant morphism $\psi: E \to E'$ such that the diagram

$$\begin{array}{ccc} X_0 & \xrightarrow{\varphi} & E \\ & \searrow^{\varphi'} \swarrow^{\psi} & \\ & E' & \end{array}$$

is commutative.

The first part of Theorem 9 will follow if we prove that in the case $S = \text{Env } G_0$ the algebra $k[X]_{\mathfrak{L}}$ is always finitely generated. We will prove a more general result.

PROPOSITION 18. *If the semigroup S is flat, then the algebra $k[X]_{\mathfrak{L}}$ is finitely generated.*

PROOF. We use the theorem stated in 1.1. Since $k[X_0]_\lambda^U k[X_0]_\mu^U \subset k[X_0]_{\lambda+\mu}^U$, the subspace

$$\bigoplus_{\lambda \in \mathfrak{X}_+(T_0)} k[X]^U_{(\theta^*\lambda, \lambda)}$$

is a subalgebra isomorphic to $k[X_0]^U$. Since every $\Lambda \in \mathfrak{L}$ is uniquely represented in the form $\Lambda = \chi + (\theta^*\lambda, \lambda)$ ($\chi \in \mathfrak{L}_Z$, $\lambda \in \mathfrak{X}_+(T_0)$) and

$$\bigoplus_{\chi \in \mathfrak{L}_Z} k[X]_\chi = k[A],$$

we obtain $k[X]_{\mathfrak{L}}^U \simeq k[A] \otimes k[X_0]^U$. It follows that the algebra $k[X]_{\mathfrak{L}}^U$, and hence the algebra $k[X]_{\mathfrak{L}}$, is finitely generated. $\square$

5. The second part of Theorem 9 is also valid in a more general situation.

PROPOSITION 19. *If the semigroup S is normal, then the morphism $\varphi \colon X_0 \to E$ is a closed embedding.*

PROOF. In algebraic terms, the assertion means that the corresponding algebra homomorphism
$$\varphi^* \colon k[E] = k[X]_{\mathfrak{L}} \to k[X_0]$$
is surjective. It is clear that $\varphi^* k[X]_{(\chi, \lambda)} = k[X_0]_\lambda$, so we must prove that for any $\lambda \in \mathfrak{L}(X)$, there exists such $\chi \in \mathfrak{z}(\mathbb{Q})^*$ that $(\chi, \lambda) \in \mathfrak{L}$. Actually, we shall prove this for any $\lambda \in \mathfrak{X}_+(T_0)$.

In the notation of Theorem 2, there is $\chi_0 \in \mathfrak{z}(\mathbb{Q})^*$ such that $(\chi_0, \lambda) \in K$ (see Remark 1 to the theorem). Then $(\chi_0, \lambda) + D \subset K$. Since the cone D generates $\mathfrak{z}(\mathbb{Q})^*$ (see Remark 2 to the theorem), we can find $\chi \in \chi_0 + D$ such that $(\chi, \lambda) \in \mathfrak{X}$. Then $(\chi, \lambda) \in \mathfrak{L}$. □

REMARK. I think that the proposition is true for any S.

6. For any S, we have in the preceding notation
$$k[E] = k[X]_{\mathfrak{L}} = (k[Z] \otimes k[X_0])_{\mathfrak{L}}$$
and hence
$$k[E]^{G_0} = (k[Z] \otimes k[X_0]^{G_0})_{\mathfrak{L}} = k[Z]_{\mathfrak{L}} \otimes k[X_0]^{G_0} = k[A] \otimes k[X_0]^{G_0},$$
or, in geometric terms,

(85) $$E /\!/ G_0 = A \times (X_0 /\!/ G_0).$$

If S contains a zero, then (27) holds and hence
$$k[E]_+ = \bigoplus_{\Lambda \in \mathfrak{L} \setminus \{0\}} k[E]_\Lambda$$
is an ideal of E. The quotient algebra $k[E]/k[E]_+$ is naturally isomorphic to
$$k[E]_0 = k[E]^G = k[X_0]^{G_0}.$$

This defines a closed embedding of $X_0 /\!/ G_0$ into E, which is a section of the decomposition (85). Its image is nothing else than the set of fixed points of S.

§10. An example

We denote by L_m the semigroup of all the matrices of order m.

It is easy to see that $\operatorname{Env} SL_2 = L_2$. However, $\operatorname{Env} SL_3$ is more complicated than L_3. It can be described in terms of its faithful linear representation
$$R^{(1)} + R^{(2)} + \pi_1 + \pi_2$$
(see the notation in 8.4 and 7.2).

The restriction of $R^{(1)}$ to the group SL_3 is its tautological representation, while the restriction of $R^{(2)}$ is the dual one or, which is the same, the exterior square of $R^{(1)}$. So, if

(86) $$R^{(1)}(g) = A_1, \quad R^{(2)}(g) = A_2 \quad (g \in SL_3),$$

then

(87) $$A_1 A_2^\top = A_1^\top A_2 = E,$$
(88) $$\wedge^2 A_1 = A_2, \quad \wedge^2 A_2 = A_1,$$

where $\wedge^2 A$ denotes the matrix formed by the algebraic complements of the entries of A.

The center Z of $G(\text{Env } SL_3)$ is represented by quadruples

(89) $$(\lambda_1 E, \lambda_2 E, \lambda_1^2 \lambda_2^{-1}, \lambda_2^2 \lambda_1^{-1}) \quad (\lambda_1, \lambda_2 \in k^*).$$

It follows from (87)–(89) that any quadruple $(A_1, A_2, t_1, t_2) \in \text{Env } SL_3$ satisfies the relations

(90) $$A_1 A_2^\top = A_1^\top A_2 = t_1 t_2 E,$$
(91) $$\wedge^2 A_1 = t_1 A_2, \quad \wedge^2 A_2 = t_2 A_1.$$

One can show that these relations define Env SL_3.

The orbital decomposition of Env SL_3 is given by the following table.

I	t_1	t_2	J	rk A_1	rk A_2	dim $O_{I,J}$
$\{1,2\}$	$\neq 0$	$\neq 0$	$\{1,2\}$	3	3	10
$\{1\}$	$\neq 0$	0	$\{1,2\}$	2	1	9
			$\{1\}$	1	0	6
			$\varnothing$	0	0	1
$\{2\}$	0	$\neq 0$	$\{1,2\}$	1	2	9
			$\{2\}$	0	1	6
			$\varnothing$	0	0	1
$\varnothing$	0	0	$\{1,2\}$	1	1	8
			$\{1\}$	1	0	5
			$\{2\}$	0	1	5
			$\varnothing$	0	0	0

References

1. M. S. Putcha, *Green's relation on a connected algebraic monoid*, Linear and Multilinear Algebra **12** (1982), 205–214.
2. _____, *Reductive groups and regular semigroups*, Semigroup Forum **30** (1984), 253–261.
3. _____, *Linear algebraic monoids*, London Math. Soc. Lecture Notes, vol. 133, Cambridge Univ. Press, Cambridge, 1988.

4. L. Renner, *Reductive monoids are von Neumann regular*, J. Algebra **93** (1985), 237–245.
5. _____, *Classification of semisimple algebraic monoids*, Trans. Amer. Math Soc. **292** (1985), 193–223.
6. _____, *Classification of semisimple algebraic varieties*, J. Algebra **122** (1989), 275–287.
7. _____, *Algebraic varieties and semigroups*, The analytical and topological theory of semigroups (K. H. Hofmann, J. D. Lawson, and J. S. Pym, eds.), de Gruyter, Berlin, 1990, pp. 81–101.
8. W. C. Waterhouse, *The unit groups of affine algebraic monoids*, Proc. Amer. Math. Soc. **85** (1982), 506–508.
9. V. L. Popov and E. B. Vinberg, *Invariant theory*, Encyclopaedia of Math. Sci., vol. 55, Springer-Verlag, Berlin, Heidelberg, and New York, 1994.
10. V. L. Popov, *A contraction of actions of reductive algebraic groups*, Mat. Sb. **130** (1986), 310–334; English transl. in Math. USSR-Sb.
11. E. B. Vinberg, *Complexity of actions of reductive groups*, Funktsional. Anal. i Prilozhen. **20** (1986), 1–13; English transl. in Functional Anal. Appl. **20** (1986).
12. _____, *The asymptotic semigroup of a semisimple Lie group*, Invariant Ordering in Geometry and Algebra (K. H. Hofmann, J. D. Lawson, and E. B. Vinberg, eds.), de Gruyter (to appear).
13. G. Kempf, F. Knudsen, D. Mumford, and B. Saint-Donat, *Toroidal embeddings*, I, Lecture Notes in Math., vol. 339, Springer-Verlag, Heidelberg, 1973.
14. C. DeConcini and C. Procesi, *Complete symmetric varieties*, Lecture Notes in Math., vol. 996, Springer-Verlag, Heidelberg and Berlin, 1983, pp. 1–44.
15. Yu. A. Neretin, *Universal completions of the complex classical groups*, Funktsional. Anal. i Prilozhen. **26** (1992), 30–44; English transl. in Functional Anal. Appl. **26** (1992).
16. D. Khadzhiev, *Some questions on the theory of vector invariants*, Mat. Sb **72** (1967), 420–435; English transl. in Math. USSR-Sb.
17. E. B. Vinberg and V. L. Popov, *On a class of quasihomogeneous affine varieties*, Izv. Akad. Nauk SSSR Ser. Mat. **36** (1972), 749–763; English transl. in Math. USSR-Izv.
18. K.-H. Neeb, *Toric varieties and algebraic monoids*, Seminar Sophus Lie **2** (1992), 159–188.
19. A. Grothendieck and J. Dieudonné, *Eléments de géométrie Algébrique*, IV, Inst. Hates Études Sci. Publ. **28** (1966).
20. Th. Vust, *Sur la théorie des invariants des groupes classiques*, Ann. Inst. Fourier (Grenoble) **26** (1976), 1–31.
21. N. Bourbaki, *Algèbre commutative*, Ch. 1–2, Hermann, Paris, 1961.
22. M. Krämer, *Some tips on the decomposition of tensor product representations of compact connected Lie groups*, Rep. Math. Phys. **13** (1978), 295–304.
23. E. B. Vinberg and B. N. Kimel'feld, *Homogeneous domains on flag manifolds and spherical subgroups of semisimple Lie groups*, Funktsional. Anal. i Prilozhen. **12** (1978), 12–19; English transl. in Functional Anal. Appl. **12** (1978).

Translated by THE AUTHOR

Crystal Bases and the Problem of Reduction in Classical and Quantum Modules

D. P. ZHELOBENKO

ABSTRACT. The paper contains a study of crystal bases, originally proposed by M. Kashiwara for semisimple (or reductive) finite-dimensional Lie algebras $\mathfrak{g}$, in terms of Schubert filtrations of finite-dimensional $\mathfrak{g}$-modules. The main result (Theorem 4.5) is a description of all solutions of the chain of extremal equations in the corresponding crystal modules. An application to the reduction problem $\mathfrak{g} \downarrow \mathfrak{g}_0$ is given, where $\mathfrak{g}_0$ is a simple embedded reductive subalgebra of the algebra $\mathfrak{g}$.

The subject of this paper is a development of new "quantum" methods in the classical theory of representations of reductive Lie algebras. This approach became possible in 1985, after the discovering of the quantum envelope $U_q(\mathfrak{g})$ of an algebra $\mathfrak{g}$ in the works of V. Drinfeld [D] and M. Jimbo [J]. As was shown in subsequent works (see, for example, [Z2]), the theory of finite-dimensional $\mathfrak{g}$-modules has a natural analog for the algebra $U_q(\mathfrak{g})$.

The introduction of the quantum parameter q supplies the corresponding quantum modules with a new degree of freedom, which allows one to unify some aspects of the theory of representations, from the classical case ($q = 1$) to the theory of representations of Chevalley algebras over finite fields ($q^n = 1$). Of special interest is the specialization at the point $q = 0$. This map (called crystallization by M. Kashiwara) supplies any quantum module V (dim $V < \infty$) with the corresponding "crystallic picture", given in terms of "crystal bases" [Ka]. By applying similar methods, G. Lusztig [Lu] and M. Kashiwara [Ka] solved the old problem of constructing so-called "canonical" bases for finite-dimensional $\mathfrak{g}$-modules.

In this paper we present a further study of crystal bases in terms of Schubert filtrations of quantum modules, introduced as an analog of the classical case $q = 1$ (see, for example, [Z2]). This method allows us to give a complete solution of the chain of extremal equations in the corresponding crystal modules (Theorem 4.5). In §6, we point out the relationship of this results with the well-known conjecture of P. Littelmann [Li].

As a simplest application of these constructions, we consider in §5 the reduction problem from the algebra $\mathfrak{g}$ to a subalgebra $\mathfrak{g}_0$ in classical and quantum modules. The solution of this problem given in Theorem 5.7 in the classical case $q = 1$ is

---

1991 *Mathematics Subject Classification.* Primary 17B37, 22E47.
This work is partially supported by the International Scientific Foundation, grant MSU000.

formulated in quantum terms, namely in terms of the crystal bases of G. Lusztig and M. Kashiwara.

As to technical aspects, this paper is based, apart from general constructions of [Lu] and [Ka], on results of [Z4] describing Schubert filtrations in the algebra $U_q(\mathfrak{g})$. It is essential that this descriptions provides synthesis between the methods of G. Lusztig and M. Kashiwara [Z4] (see also [Z2]). Everywhere in this paper we shall use appropriate notations and results of [Ka, Z4].

The results of this article have been announced in [Z3, Z5]. A preliminary version of Theorem 4.5 given in these papers is corrected in the English translation of the Russian paper [Z3].

§1. The algebra $U_q(\mathfrak{g})$

1.1. We recall [D, J] (see also [Ka, Z2]) that the quantum functor U_q is defined in the class of symmetrizable Kac–Moody algebras $\mathfrak{g} = \mathfrak{g}(a)$, where $a = (a_{ij})$ is a generalized symmetrizable Cartan matrix:

$$(1.1) \qquad a_{ij} = \langle h_i, \alpha_j \rangle = 2(\alpha_i, \alpha_j)/(\alpha_i, \alpha_i).$$

Here we set $i, j \in I$ (finite set of indices), $h_i \in \mathfrak{h}$, $\alpha_j \in \mathfrak{h}^*$ (simple roots of the algebra $\mathfrak{g}$), $\mathfrak{h}$ is a Cartan subalgebra of $\mathfrak{g}$, $\langle \cdot, \cdot \rangle$ is the canonical bilinear form on $\mathfrak{h} \times \mathfrak{h}^*$, $(\cdot, \cdot)$ is an associated symmetric bilinear from in the dual space $\mathfrak{h}^*$.

Moreover, we have $(\alpha_i, \alpha_i) > 0$ for any $i \in I$, $(\alpha_i, \alpha_j) \leqslant 0$ for $i \neq j$, and we may assume (by normalizing the last bilinear form) that $(\alpha_i, \alpha_j) \in \mathbb{Z}$ for any $i, j \in I$.

1.2. According to [K], the algebra $\mathfrak{g} = \mathfrak{g}(a)$ is defined as a Lie algebra over the field $\mathbb{C}$ with generators e_i, f_i $(i \in I)$, $h \in \mathfrak{h}$ and with a special system of defining relations. In particular, we have

$$(1.2) \qquad [h, h'] = 0, \qquad [e_i, f_j] = \delta_{ij} h_i,$$

where $h, h', h_i \in \mathfrak{h}$ and δ_{ij} is the Kronecker symbol. The remaining relations are weight conditions and Serre conditions for the elements e_i, f_i $(i \in I)$.

The algebra $U_q(\mathfrak{g})$ possesses an involution $x \mapsto x'$ (the Chevalley involutive antiautomorphism) defined on the generators by the rule $e_i' = f_i$, $f_i' = e_i$, $h' = h$ $(i \in I, h \in \mathfrak{h})$.

The condition $\dim \mathfrak{g} < \infty$ is valid if and only if the matrix a is a classical Cartan matrix (i.e., a is nondegenerate and positive definite over the field $\mathbb{R}$). The family $\mathfrak{g} = \mathfrak{g}(a)$ for $\dim \mathfrak{g} < \infty$ exhausts, up to isomorphism, the class of all finite-dimensional semisimple Lie algebras over $\mathbb{C}$.

REMARK. In [K] the following condition is assumed: $\dim \mathfrak{h} = 2n - r$, where $n = \operatorname{card} I$, $r = \operatorname{rank} a$. Omitting this condition, we may include the class of all finite-dimensional reductive Lie algebras in the family $\mathfrak{g}(a)$.

1.3. Let $\mathbb{C}(q)$ be the field of complex rational functions of an independent variable q. We shall also consider the formal symbols q^h, where $h \in \mathfrak{h} \oplus \mathbb{C}$, are defined by the following relations: $q^h q^{h'} = q^{h+h'}$, $q^0 = 1$ for $h, h' \in \mathfrak{h} \oplus \mathbb{C}$. We shall use everywhere the following notations:

$$q_i = q^{(\alpha_i, \alpha_i)}, \quad t_i = q^{(\alpha_i, \alpha_i)h_i}, \quad \theta_i = q_i - q_i^{-1}, \quad [x]_i = \theta_i^{-1}(q_i^x - q_i^{-x}),$$
$$[n]_i! = [1]_i \cdots [n]_i \quad \text{for } n = 0, 1, 2, \ldots,$$
$$x_i^{(n)} = x_i^n/[n]_i!, \quad x_i^{(n)} = 0 \quad \text{for } n < 0.$$

Here x_i ($i \in I$) are elements of some associative algebra over the field $\mathbb{C}(q)$. The elements $[n]_i \in \mathbb{Z}[q, q^{-1}]$ for $n \in \mathbb{Z}$ are called quantum integers. We note that $\lim_{q \to 1} [n]_i = n$. Correspondingly, $[n]_i!$ is the quantum factorial. We shall omit the index i in the case $(\alpha_i, \alpha_i) = 1$, i.e., $q_i = q$. For any set of elements $x = (x_i)_{i \in I}$, we set

$$\tau_{ij}(x) = \sum_n (-1)^n x_i^{(b-n)} x_j x_i^{(n)},$$

where $i \neq j$, $b = 1 - a_{ij}$. It is clear that this polynomial contains $b + 1$ members. For example, $\tau_{ij}(x) = [x_i, x_j]$ for $a_{ij} = 0$.

1.4. DEFINITION. Let $U_q(\mathfrak{g})$ be an associative algebra with unit over the field $\mathbb{C}(q)$ generated by the elements e_i, f_i ($i \in I$), q^h ($h \in \mathfrak{h}$) and by the following fundamental conditions:

(1.3a) $$q^h q^{h'} = q^{h+h'}, \quad q^0 = 1,$$

(1.3b) $$q^h e_i q^{-h} = q^{\langle h, \alpha_i \rangle} e_i, \quad q^h f_i q^{-h} = q^{-\langle h, \alpha_i \rangle} f_i,$$

(1.3c) $$[e_i, f_j] = \delta_{ij} \theta_i^{-1}(t_i - t_i^{-1}),$$

(1.3d) $$\tau_{ij}(e) = \tau_{ij}(f) = 0 \quad \text{for } i \neq j,$$

where $h, h' \in \mathfrak{h}$, $i, j \in I$, $e = (e_i)_{i \in I}$, $f = (f_i)_{i \in I}$. The relations (1.3a), (1.3b) may be regarded as quantum analogs of the classical conditions (1.2). Correspondingly, (1.3c) are the *weight conditions*, and the relations (1.3d) are called the *quantum Serre conditions*.

The symbol $U_{q_0}(\mathfrak{g})$ at points $q_0 \in \mathbb{C}$ in the general position may be considered also as an algebra over $\mathbb{C}$ (a specialization of the algebra $U_q(\mathfrak{g})$ at the point $q = q_0$). The specialization map at the singular point $q = 1$ is described in [Z2] (see also [CK]). In this sense the algebra $U_1(\mathfrak{g})$ is isomorphic to the universal enveloping algebra $U(\mathfrak{g})$.

The algebra $U_q(\mathfrak{g})$ is usually called the *quantum envelope* of the algebra $\mathfrak{g}$ (or the *quantum deformation* of the algebra $U(\mathfrak{g})$).

1.5. It should be noted that the following analog of the triangular decomposition of the algebra $U(\mathfrak{g})$ is valid:

(1.4) $$U_q(\mathfrak{g}) = XHX^+ \approx X \otimes H \otimes X^+,$$

where H (respectively, X, X^+) is the subalgebra with the unit generated by the elements q^h (respectively, f_i, e_i).

The commutative subalgebra H is said to be the *Cartan subalgebra* of $U_q(\mathfrak{g})$. The subalgebras X, X^+ are conjugated with respect to the involution $x^+ = \overline{x}'$, where $x \mapsto x'$ is the Chevalley involution and $x \mapsto \overline{x}$ is the automorphism of the algebra $U_q(\mathfrak{g})$; these operations are defined on the generators as follows:

$$(q^h)' = q^h, \quad e_i' = f_i, \quad f_i' = e_i, \quad q' = q,$$
$$(q^h)^- = q^{-h}, \quad \overline{e}_i = e_i, \quad \overline{f}_i = f_i, \quad \overline{q} = q^{-1}.$$

The algebra $U_q(\mathfrak{g})$ possesses a $\mathbb{Z}$-grading defined on the generators by the rule $\deg e_i = -\deg f_i = 1$, $\deg q^h = 0$. The algebra $U_q(\mathfrak{g})$ with respect to this $\mathbb{Z}$-grading, the Chevalley involution, and the triangular decomposition (1.4) is a Cartan type algebra in the terminology of [Z6].

Correspondingly, we shall use (see §6) some general results of [Z6] concerning Cartan type algebras.

1.6. Let us fix a lattice $P \subset \mathfrak{h}^*$ satisfying the conditions $\alpha_i \in P$, $h_i \in P^*$ for any $i \in I$, where $P^* \subset \mathfrak{h}$ is the *dual lattice*, i.e.,

(1.5) $\qquad P^* = \{h \in \mathfrak{h} \mid \langle h, P \rangle \subset \mathbb{Z}\}.$

Moreover we shall assume that the lattice P is *autodual*, i.e., $P^{**} = P$. Also we set

(1.6) $\qquad P_+ = \{\lambda \in P \mid \langle h_i, \lambda \rangle \subset \mathbb{Z}_+ \text{ for any } i \in I\}.$

The lattice P is called the *weight lattice*, and elements $\lambda \in P_+$ are called *dominant weights* of the algebra $U_q(\mathfrak{g})$.

A module V over the algebra $U_q(\mathfrak{g})$ is called *P-graded* if U coincides with the (direct) sum of its weight subspaces

(1.7) $\qquad V_\mu = \{v \in V \mid q^h v = q^{\langle h, \mu \rangle} v \text{ for any } h \in \mathfrak{h}\},$

where $\mu \in P$. A module V is said to be a *module of category* $\mathcal{O}$ if V is P-graded and locally nilpotent with respect to the family of operators e_i ($i \in I$); this means that for any $v \in V$ there exists an $n \geqslant 1$ such that

(1.8) $\qquad e_{i_1} \ldots e_{i_n} v = 0 \quad \text{for any } i_1, \ldots, i_n \in I.$

We note that condition (1.8) is equivalent to that of locally finiteness under X^+, namely $\dim X^+ v < \infty$ for any $v \in V$.

1.7. A vector $v \in V$ is called *extremal with respect to X^+* if $v \neq 0$, $e_i v = 0$ for any $i \in I$ (i.e., $X^+ v = \mathbb{C}(q)v$). The subspace V^{ext} spanned by extremal vectors is called the *extremal subspace* of the module V.

It is evident that V^{ext} is an H-submodule of the module V. If $V \neq 0$ is a module of the category $\mathcal{O}$, then $V^{\text{ext}} \neq 0$.

In particular, for any $\lambda \in P$ there exists a unique (up to isomorphism) simple module $D(\lambda)$ generated by the external vector 1_λ of weight λ. According to the triangular decomposition (1.4), this module is of the form

(1.9) $\qquad D(\lambda) = X 1_\lambda \approx X/J(\lambda),$

where $J(\lambda)$ is the annulator of 1_λ in the algebra X. The family of modules $D(\lambda)$ for $\lambda \in P$ exhausts (up to isomorphism) the class of simple objects of the category $\mathcal{O}$. See, for example, [Z2, Z6].

A module V is called *integrable* if V is P-graded and locally nilpotent with respect to every operators e_i, f_i ($i \in I$), i.e., $e_i^n v = f_i^n v = 0$ for any $v \in V$ and $n \geqslant n_0(v)$. According to [Z2], the module $D(\lambda)$ is integrable if and only if $\lambda \in P_+$.

REMARK. The family $J(\lambda)$, $\lambda \in P_+$, has zero intersection. Hence we may regard X as a projective limit of the spaces $D(\lambda)$ for $\lambda \in P_+$.

1.8. Let $\mathcal{O}_{\text{int}}$ be the subcategory of the category $\mathcal{O}$ consisting of integrable modules. As is known [Z2], the category $\mathcal{O}_{\text{int}}$ is *semisimple*, i.e., any module V of $\mathcal{O}_{\text{int}}$ is a direct sum of modules $D(\lambda)$ for $\lambda \in P_+$.

The general definitions given above are nothing but the quantum analogs of the corresponding constructions for the algebra $\mathfrak{g} = \mathfrak{g}(a)$ (i.e., for the classical case $q = 1$). Moreover, the quantum module $D(\lambda)$ may be considered as the "quantum hull" of the corresponding classical module $D_1(\lambda)$. Namely, the following relation (i.e., an isomorphism of $\mathfrak{g}$-modules) is valid:

$$(1.10) \qquad D_1(\lambda) \approx D_N(\lambda)/(q-1)D_N(\lambda),$$

where $D_N(\lambda) \subset D(\lambda)$ is a lattice over the ring $N = \mathbb{Z}[q, q^{-1}]$ and $D(\lambda) \approx \mathbb{C}(q) \otimes_N D_N(\lambda)$. The quotient map (1.10) may be considered as the limiting process as $q \to 1$ in the lattice $D_N(\lambda)$. For a more detail exposition, see, for example, [CK, Z2].

For $\dim \mathfrak{g} < \infty$, the category $\mathcal{O}_{\text{int}}$ over the algebra $\mathfrak{g}$ coincides with the category $\mathcal{F}$ of finite-dimensional $\mathfrak{g}$-modules [Z2]. Correspondingly, the functor U_q defines an embedding of the category $\mathcal{F}$ into the category $\mathcal{O}_{\text{int}}$ over the algebra $U_q(\mathfrak{g})$.

§2. Crystallization method

2.1. We shall use the crystallization method proposed by M. Kashiwara [Ka]. Essentially, this method consists in passing to the limit as $q \to 0$ while pressing important information about the structure of quantum modules.

Let U_i be the subalgebra with unit of the algebra $U_q(\mathfrak{g})$ generated by the elements e_i, f_i, t_i, t_i^{-1} (for fixed i). Using the theory of finite-dimensional representations of the algebra $U_i \approx U_q(sl_2)$, for any module U of the category $\mathcal{O}_{\text{int}}$ we find

$$(2.1) \qquad V = \bigoplus_{0 \leqslant n \leqslant \mu_i} f_i^{(n)}(U_\mu \cap \ker e_i),$$

where $\mu_i = \langle h_i, \mu_i \rangle$ for any $\mu \in P$. The decomposition (2.1) allows one to define the following modification of the operators e_i, f_i in the module V:

$$(2.2) \qquad \tilde{e}_i f_i^{(n)} x = f_i^{(n-1)} x, \qquad \tilde{f}_i f_i^{(n)} x = f_i^{(n+1)} x,$$

where $x \in \ker e_i$.

The operators (2.2) are convenient for giving a description of the crystallization method.

2.2. Let $A \subset \mathbb{C}(q)$ be the subalgebra of functions regular at the point $q = 0$. A subset $L \subset V$ is called an *A-lattice* if (1) L is a free A-module, (2) $V \approx \mathbb{C}(q) \otimes_A L$.

DEFINITION [Ka]. A pair (L, B) is called a *crystal basis* of the module V over $U_q(\mathfrak{g})$ if the following conditions are fulfilled.

(α) L is an A-lattice in the space V.
(β) B is a basis of the space $K = L/qL$.
(γ) The pair (L, B) inherits the P-grading of the module V, namely
$$L = \bigoplus_\mu L_\mu, \qquad B = \bigsqcup_\mu B_\mu,$$
where $L_\mu = L \cap V_\mu$ and $B_\mu = B \cap (L_\mu/qL_\mu)$.
(δ) The lattice L is invariant under $\tilde{e}_i, \tilde{f}_i$ for any $i \in I$. Correspondingly, the operators $\tilde{e}_i, \tilde{f}_i$ are defined in the space K.
(ε) The set $B \cup \{0\}$ is invariant under $\tilde{e}_i, \tilde{f}_i$ for any $i \in I$. The equality $a = \tilde{e}_i b$ for $a, b \in B$ is equivalent to $b = \tilde{f}_i a$.

The set L in this definition is said to be a *crystal lattice* of the module V. The set B is equipped with the structure of an oriented graph by the way: a pair $a, b \in B$ is joined by an arrow passing from a to b if $b = \tilde{f}_i a$ for some $i \in I$. Correspondingly, the set B is called the *crystal graph* of the module V.

According to general results of Kashiwara [Ka], a crystal basis exists for any module V of the category $\mathcal{O}_{\text{int}}$. In particular, for any $\lambda \in P_+$, the complementary condition $1_\lambda \in L(\lambda)$ defines a unique crystal basis $(L(\lambda), B(\lambda))$ of the module $D(\lambda)$. In view of (δ), (ε), this basis is generated by the elements
$$(2.3) \qquad f_{i_1 \ldots i_n}(\lambda) = \tilde{f}_{i_1} \ldots \tilde{f}_{i_n} 1_\lambda,$$
where $i_1, \ldots, i_n \in I$ for $n \geq 0$, including the element $f_0(\lambda) = 1_\lambda$. More exactly, let $F(\lambda)$ be the set of elements (2.3). Then we have
$$L(\lambda) = AF(\lambda), \qquad B(\lambda) = F(\lambda)^0 \setminus \{0\},$$
where N^0 is the canonical image of the set $N \subset L(\lambda)$ in the quotient space $K(\lambda) = L(\lambda)/qL(\lambda)$. It is evident from (2.3) that the graph $B(\lambda)$ is connected. Moreover, a crystal graph B of a module V is connected if and only if V is simple, i.e., $V \approx D(\lambda)$ for some $\lambda \in P_+$.

Hence the decomposition problem of a module V into simple submodules is equivalent to that of the corresponding crystal graph B into its connected components.

2.3. The uniqueness theorem for the basis $(L(\lambda), B(\lambda))$ has an interesting interpretation in terms of "canonical" bilinear form in the space $D(\lambda)$. This form $(\cdot, \cdot)$ is uniquely determined by the normalization $(1_\lambda, 1_\lambda) = 1$ and by the following contravariance condition:
$$(ax, y) = (x, a^* y)$$
for any $a \in U_q(\mathfrak{g})$, $x, y \in D(\lambda)$, where $a \mapsto a^*$ is the involution of the algebra $U_q(\mathfrak{g})$ defined on the generators as follows:
$$f_i^* = \bar{q}_i t_i e_i, \qquad e_i^* = q_i f_i \bar{t}_i, \qquad (q^n)^* = q^n.$$

Below we shall use the following notation. For any subset N of a vector space (over $\mathbb{C}$) let $[N]$ be the set of elements $\pm x$, where $x \in N$.

2.4. PROPOSITION [Ka]. *Let $(\cdot, \cdot)$ be the canonical bilinear from defined in 2.3. Then*

(i) *The crystal lattice $L(\lambda)$ coincides with the set of all $x \in D(\lambda)$ for which $(x, x) \in A$. Moreover, $(x, y) \in A$ for any $x, y \in L(\lambda)$. Hence, in the space $K(\lambda)$ the following bilinear form is defined*:

$$(2.4) \qquad (x, y)_0 = (x, y)|_{q=0}.$$

(ii) *The basis $B(\lambda)$ is orthonormal with respect to the form (2.4). Moreover, the set $[B(\lambda)]$ coincides with the set of all $x \in \mathbb{Z}B(\lambda)$ for which $(x, x)_0 = 1$.*

(iii) *The operators $\tilde{e}_i$, $\tilde{f}_i$ in the space $K(\lambda)$ are conjugated with respect to the form (2.4).*

REMARK. The Hermitian form $(x, y) = (x, \tilde{y})_0$, where $x \mapsto \tilde{x}$ is complex conjugation under $B(\lambda)$, is a scalar product in the space $K(\lambda)$.

2.5. The construction of crystal bases may be carried over the space X (see 1.7) by changing the operators $e_i \in \mathrm{End}\, D(\lambda)$ to their asymptotical version $a_i \in \mathrm{End}\, X$ defined by the following way. According to the structural relations of the algebra $U_q(\mathfrak{g})$ (see 1.4), we have for any pair $i \in I$, $x \in X$:

$$(2.5) \qquad (\mathrm{ad}\, e_i)\, x = \theta_i^{-1}(t_i \overline{a}_i - \overline{t}_i a_i)\, x,$$

where $(\mathrm{ad}\, x)\, y = [x, y]$ and the operators $a_i, \overline{a}_i \in \mathrm{End}\, X$ are conjugated with respect to the involution $a \mapsto \overline{a}$ of the algebra $U_q(\mathfrak{g})$. Moreover, these operators satisfy the following condition:

$$(2.6) \qquad a_i \overline{a}_j = q_{ij} \overline{a}_j a_i,$$

where $q_{ij} = q^{2(\alpha_i, \alpha_j)}$. The operators a_i, f_j are related by certain contragradience conditions (where f_j is regarded as the left multiplication operator in the algebra X). Moreover, for any $i \in I$ the following analog of (2.1) is valid:

$$(2.7) \qquad X = \bigoplus_n f_i^{(n)} \ker a_i.$$

Correspondingly, by changing (2.1) to (2.7), one may repeat the definition of crystal bases given in 2.2. The crystal basis (L, B) of the space X is defined uniquely by the additional condition $1 \in L$. This basis (L, B) is generated by the set F consisting of the elements

$$f_{i_1 \ldots i_n} = \tilde{f}_{i_1} \ldots \tilde{f}_{i_n} 1,$$

including the element $f_0 = 1$.

The crystal basis (L, B) possesses the following universal property in the category $\mathcal{O}_{\mathrm{int}}$:

$$L(\lambda) = L 1_\lambda, \qquad B(\lambda) = B 1_\lambda \setminus \{0\}$$

for any $\lambda \in P_+$.

2.6. For the pair (L, B) in the space X the corresponding analog of Proposition 2.4 is valid. In this case the canonical bilinear form $(\cdot, \cdot)$ in the space X is uniquely determined by the normalization $(1, 1) = 1$ and by the following contravariance condition:

$$(2.8) \qquad (a_i x, y) = (x, f_i y)$$

for any $i \in I$, $x, y \in X$.

2.7. As is known (see, for example, [Ka]), the algebra $U_q(\mathfrak{g})$ possesses a comultiplication Δ defined on the generators as follows:

$$(2.9\text{a}) \qquad \Delta(q^h) = q^h \otimes q^h,$$
$$(2.9\text{b}) \qquad \Delta(f_i) = f_i \otimes 1 + t_i \otimes f_i, \qquad \Delta(e_i) = e_i \otimes \bar{t}_i + 1 \otimes e_i.$$

Correspondingly, for any pair of $U_q(\mathfrak{g})$-modules V_1, V_2, we may define the tensor product $V_1 \otimes V_2$.

It is essential that the tensor product $\otimes$ preserves the category $\mathcal{O}_{\text{int}}$. In particular, for any pair $\lambda, \mu \in P_+$ the module $D(\lambda) \otimes D(\mu)$ possesses the following crystal basis:

$$(L(\lambda) \otimes L(\mu), B(\lambda) \otimes B(\mu)).$$

We remark that the set $B(\lambda) \otimes B(\mu)$ is orthonormal with respect to the following bilinear form in the space $D(\lambda) \otimes D(\mu)$:

$$(x \otimes y, x' \otimes y') = (x, x')(y, y'),$$

where $x, x' \in P(\lambda)$ and $y, y' \in D(\mu)$.

REMARK. The bialgebra $U_q(\mathfrak{g})$ also possesses a Hopf algebra structure. See, for example, [Z2].

2.8. Let $U_N(\mathfrak{g})$ be the subring of $U_q(\mathfrak{g})$ generated over $\mathbb{Z}$ by the elements f_i, e_i, q^x and by the elements $\varphi_n = (1 - q^{2n})^{-1}$, where $i \in I$, $n \in \mathbb{Z}_+$, $x \in P^* \oplus \mathbb{Z}$. As is known (see, for example, [Z2]), the ring $U_N(\mathfrak{g})$ inherits the triangular decomposition of the algebra $U_q(\mathfrak{g})$. We set

$$D_N(\lambda) = U_N(\mathfrak{g}) 1_\lambda = X_N 1_\lambda,$$

where $\lambda \in P_+$ and $X_N = X \cap U_N(\mathfrak{g})$ is the subring generated by the elements f_i, φ_n and by elements of the ring $N = \mathbb{Z}[q, q^{-1}]$. Also we set $L_N(\lambda) = L(\lambda) \cap D_N(\lambda)$ and

$$K_N(\lambda) = L_N(\lambda)/qL_N(\lambda).$$

According to [Ka] (see the remark at the end of 6.1.2), $K_N(\lambda)$ is a free $\mathbb{Z}$-module with basis $B(\lambda)$. Moreover (see 2.4) the set $[B(\lambda)]$ coincides with the set of all $x \in K_N(\lambda)$ for which $(x, x)_0 = 1$.

§3. The group $W_q(\mathfrak{g})$

3.1. Everywhere in the rest of this paper we set $\dim \mathfrak{g} < \infty$. As is known [Z2], in this case another construction of crystal bases exists, based on the works of Lusztig [Lu] and connected with the action of the quantum Weyl group in the algebra $U_q(\mathfrak{g})$.

Recall that the Weyl group W of the algebra $\mathfrak{g}$ is isomorphic to Γ/Γ_0, where Γ is the associated braid group with generators r_i $(i \in I)$ and Γ_0 is the normal subgroup generated by the elements r_i^2 $(i \in I)$. The element r_i acts in the space $\mathfrak{h}^*$ as the reflection with respect to the hyperplane orthogonal to the root α_i. Correspondingly, the group W acts in the space $\mathfrak{h}$ (dual action).

On the other hand, in [Lu] Lusztig defines the action of Γ by the automorphisms of the algebra $U_q(\mathfrak{g})$. Namely, this action is defined on the generators by the following rule:

(3.1a) $$r_i e_i = -f_i t_i, \quad r_i f_i = -\bar{t}_i e_i, \quad r_i q^h = q^{r_i h},$$

(3.1b) $$r_i e_j = \sum_n (-1)^{c-n} q_i^{n-c} e_i^{(n)} e_j e_i^{(c-n)},$$

(3.1c) $$r_i f_j = \sum_n (-1)^n q_i^{c-n} f_i^{(c-n)} f_j f_i^{(n)},$$

where $i \neq j$, $c = |a_{ij}|$ and the action $h \mapsto r_i h$ is defined with respect to the action of W in the space $\mathfrak{h}$. The image of Γ in $\operatorname{Aut} U_q(\mathfrak{g})$ is denoted by $W_q(\mathfrak{g})$ and is called the *quantum Weyl group* associated to the algebra $U_q(\mathfrak{g})$.

We note that the group W may be embedded (as a subset) in $W_q(\mathfrak{g})$ by identifying every reduced decomposition $w = r_{i_1} \ldots r_{i_n}$ in W with the corresponding element of the group $W_q(\mathfrak{g})$.

3.2. Let Δ be the system of roots of the algebra $\mathfrak{g}$ with respect to $\mathfrak{h}$, and let Δ_+ be the subsystem of positive roots generated by $\Pi = \{\alpha_i\}_{i \in I}$. We recall that the group W preserves the system Δ. On the other hand, for any $w \in W$ we set

(3.2) $$\Delta_w = \{\alpha \in \Delta_+ \mid w^{-1}\alpha < 0\}.$$

Then we have $\operatorname{card} \Delta_w = l(w)$, where $l(w)$ is the length of the element w defined as the length of every reduced decomposition $w = r_{i_1} \ldots r_{i_n}$. The relation $u \leqslant w$ for $\Delta_u \subset \Delta_w$ defines in W the ordering with minimal element 1 and (unique) maximal element $w_{\max}$ satisfying the condition $l(w_{\max}) = \operatorname{card} \Delta_+$.

An ordering $\Delta_+ = \{\alpha(1), \ldots, \alpha(m)\}$ is called *normal* [Z2], if any root of the form $\alpha(j) = \alpha(i) + \alpha(k)$ is arranged between its summands, i.e., we have $i < j < k$ or $k < j < i$. The set $NO(\Delta_+)$ of all normal orderings in the system Δ_+ can be mapped bijectively to the set of all reduced decompositions $w_{\max} = r_{i_1} \ldots r_{i_m}$. The bijection is defined in the following way:

(3.3) $$\alpha(s) = w_{s-1}\alpha_{i_s}, \quad \text{for } s = 1, \ldots, m,$$

where $w_s = r_{i_1} \ldots r_{i_s}$ $(w_0 = 1)$.

Let us fix an ordering $\tau \in NO(\Delta_+)$. We shall write the defining relation (3.3) in the form $\alpha = w\alpha_i$. Setting

(3.4) $$f_\alpha = wf_i, \quad e_\alpha = we_i,$$

it is easy to verify [Z2] that $f_\alpha \in X$, $e_\alpha \in X^+$ for any $\alpha \in \Delta_+$. The elements (3.4) are called *generalized root vectors* of the algebra $U_q(\mathfrak{g})$.

The elements (3.4) at the classical limit $q = 1$ do not depend on the choice of $\tau \in NO(\Delta_+)$ and coincide with the corresponding root vectors of the algebra $\mathfrak{g}$ such that e_α (respectively, f_α) is a vector of weight α (respectively, $-\alpha$).

3.3. According to the general scheme of §2, we may restrict ourselves to considering the algebra X only. Let us fix $\tau \in NO(\Delta_+)$, and let F_τ be the set of monomials of the following form:

$$(3.5) \qquad f(k) = f_{\alpha(1)}^{(k_1)} \cdots f_{\alpha(m)}^{(k_m)},$$

where $k = (k_1, \ldots, k_m) \in \mathbb{Z}_+^m$. As is known [Z2] (see also [Lu]), the set F_τ is a basis of the vector space X. Following [Z2], we call any set F_τ the *Lusztig basis* of the space X.

For any two subsets $\mathcal{M}, \mathcal{N}$ of a vector space, we shall write $\mathcal{M} \sim \mathcal{N}$ if there exists a bijection between $\mathcal{M}$ and $\mathcal{N}$ mapping any element $x \in \mathcal{M}$ to one of elements $\pm x \in \mathcal{N}$,

PROPOSITION [Z2]. *Let* $\dim \mathfrak{g} < \infty$. *Then for any* $\tau \in NO(\Delta_+)$ *we have* $L = AF_\tau$, $B \sim F_\tau^0$. *In particular, F_τ is a basis of the free A-module L, and F_τ^0 is an orthonormal basis of the space $K = L/qL$.*

REMARK. According to the general criterion 2.4, the proof given in [Z2] may be reduced to verifying the orthonormality of the set F_τ^0.

COROLLARY. *For any pair $\lambda \in P_+$, $\tau \in NO(\Delta_+)$ we have*

$$(3.6) \qquad L(\lambda) = AF_\tau(\lambda), \qquad B(\lambda) \sim F_\tau(\lambda)^0,$$

where $F_\tau(\lambda) = F_\tau 1_\lambda$.

3.4. Changing Π to $w\Pi$ for $w \in W$ and applying the corresponding automorphism $w \in W_q(\mathfrak{g})$, we may consider a new system $\hat{e}_i = we_i$, $\hat{f}_i = wf_i$, q^h ($h \in \mathfrak{h}$) of generators in the algebra $U_q(\mathfrak{g})$ satisfying the structural relation of 1.2.

According to the theory of crystal bases, for any pair $\lambda \in P_+$, $w \in W$ there exists a vector $1_{w\lambda} \in L_N(\lambda)$ of weight $w\lambda$, unique mod $qL_N(\lambda)$. It is clear that this vector is extremal with respect to the system $w\Pi$, i.e., $\hat{e}_i 1_{w\lambda} = 0$ for any $i \in I$.

3.5. THEOREM. *Let $(L'(\lambda), B'(\lambda))$ be the crystal basis of the module $D(\lambda)$ defined by the system $w\Pi$ and normalized by the condition $1_{w\lambda} \in L'(\lambda)$. Let $(\cdot, \cdot)'$ be the canonical bilinear form of the module $D(\lambda)$ defined by the system $w\Pi$ and normalized by the condition $(1_{w\lambda}, 1_{w\lambda})' = (1_{w\lambda}, 1_{w\lambda})$. Then we have:*

(i) $(x, y)' = (x, y)$ *for any* $x, y \in D(\lambda)$,
(ii) $L'(\lambda) = L(\lambda)$ *for any* $\lambda \in P_+$,
(iii) $B'(\lambda) \sim B(\lambda)$ *for any* $\lambda \in P_+$.

PROOF. It is easy to verify that the involution $a \mapsto a^*$ defined in 2.3 commutes with the action of the group $W_q(\mathfrak{g})$ (see 3.1). From this we have

$$(3.7) \qquad (w(a)x, y) = (x, w(a^*)y)$$

for any $a \in U_q(\mathfrak{g})$ and x, $y \in D(\lambda)$. This relation is nothing but the contravariance property with respect to the system $w\Pi$. Using the uniqueness of such a form, we obtain (i). Also, using the uniqueness criterion 2.4, we obtain (ii).

On the other hand, let $L'_N(\lambda) = wU_N(\mathfrak{g})1_{w\lambda}$ be the corresponding analog of $L_N(\lambda)$ for the system $w\Pi$. Using the equality $wU_N(\mathfrak{g}) = U_N(\mathfrak{g})$ (see 3.1) and the condition $1_{w\lambda} \subset D_N(\lambda)$, we find $L'_N(\lambda) \subset L_N(\lambda)$. Hence, $B'(\lambda) \subset \mathbb{Z}B(\lambda)$. At the same time, we have $(x, x)_0 = 1$ for any $x \in B'(\lambda)$. Using the criterion 2.4 again, we obtain (iii). $\square$

3.6. DEFINITION. Let us fix $\tau \in NO(\Delta_+)$ and $\alpha \in \Delta_+$. Changing in 2.1 the subalgebra U_i to the subalgebra U_α generated by the elements e_α, f_α, t_α, t_α^{-1}, where $t_\alpha = wt_i$ (in the notation of 3.2), we define the following analogs of the operators $\tilde{e}_i$, $\tilde{f}_i$:

$$(3.8) \qquad \tilde{e}_\alpha f_\alpha^{(n)} x = f_\alpha^{(n-1)} x, \qquad \tilde{f}_\alpha f_\alpha^{(n)} x = f_\alpha^{(n+1)} x,$$

where $x \in \ker e_\alpha$.

3.7. THEOREM. *Let $\lambda \in P_+$. Then we have:*

(i) *The crystal lattice $L(\lambda)$ is invariant under the operators* (3.8). *Correspondingly, these operators act in $K(\lambda)$.*

(ii) *The operators* (3.8) *are conjugated with respect to the form* $(\cdot, \cdot)_0$ *in the space $K(\lambda)$ (see 2.4).*

(iii) *The set $[B(\lambda)] \cup \{0\}$ is invariant under the operators* (3.8). *The equality $a = \tilde{e}_\alpha b$ for any elements a, $b \in [B(\lambda)]$ is equivalent to $b = \tilde{f}_\alpha a$.*

PROOF. It is sufficient to note that $e_\alpha = \hat{e}_i$, $f_\alpha = \hat{f}_i$ in the notation of 3.4 and to use Theorem 3.5. See also 2.4. $\square$

3.8. CONJECTURE. *For any $\lambda \in P_+$, $w \in W$ the equality $B'(\lambda) = B(\lambda)$ in the notation of* 3.5 *is valid. Correspondingly the set $[B(\lambda)]$ in assertion* (iii) *may be replaced by $B(\lambda)$.*

Using the methods of [Z2], it is easy to verify this conjecture for simple laced root systems, i.e., for the series A_n, D_n, E_n.

3.9. In the algebra $U_q(\mathfrak{g})$ the map $\Pi \mapsto w\Pi$ induces a new comultiplication structure, i.e., a new tensor product structure in the category of $U_q(\mathfrak{g})$-modules. We denote this new tensor product by the previous symbol $\otimes$.

We note (see 3.5) that the canonical bilinear form in $D(\lambda) \otimes D(\mu)$ does not depend on $w \in W$, i.e., this form coincides with the form $(\cdot, \cdot)$ defined in 2.7 for $w = 1$. Correspondingly, $L(\lambda) \otimes L(\mu)$ is a crystal lattice of $D(\lambda) \otimes D(\mu)$.

3.10. LEMMA. *Let $D(\lambda) \otimes D(\mu)$ be the tensor product of modules defined by the system $w\Pi$. Then we have:*

(i) *For any pair of elements $x \in X \cap wX$ and $v \in D(\lambda)$ the following relation is valid*:

$$(3.9) \qquad (x1_{\lambda+\mu}, v \otimes 1_\mu) = (x1_\lambda, v).$$

(ii) *If the elements $a \in B(\lambda)$, $b \in B(\mu)$ satisfy the conditions $\tilde{f}_\alpha a \neq 0$, $\tilde{a}_\alpha b = 0$ for $\alpha \in w\Pi$, then*

$$\tilde{f}_\alpha(a \otimes b) = \tilde{f}_\alpha a \otimes b. \tag{3.10}$$

(iii) *The operators $\tilde{e}_\alpha$, $\tilde{f}_\alpha$ for $\alpha \in w\Pi$ are conjugated with respect to the form $(\,\cdot\,,\,\cdot\,)_0$ in the space $K(\lambda) \otimes K(\mu)$.*

PROOF. Using the comultiplication rule for the generators f_α ($\alpha \in \Delta_+ \setminus \Delta_w$) of the algebra $X \cap wX$, we find:

$$x1_{\lambda+\mu} = x1_\lambda \otimes 1_\mu + \sum_i y_i 1_\lambda \otimes z_i 1_\mu,$$

where $z_i \in X \cap wX$. If $x = 1$ then (3.9) is evident. If x is an element of nonzero weight, then we have $(z_i 1_\mu, 1_\mu) = 0$ for any i, and (3.9) also follows.

The assertions (ii), (iii) follow from the general rules of action of the operators $\tilde{e}_\alpha$, $\tilde{f}_\alpha$ ($\alpha \in \Pi$) given in [Ka, Lemma 4.4.3]. □

§4. Extremal systems

4.1. Let us fix $\tau \in NO(\Delta_+)$ with initial segment Δ_w ($w \in W$). In [Z4], the following three families of subspaces in the algebra X are considered:

$$X_w = X \cap wX^+H, \qquad Y_w = X \cap wX, \tag{4.1}$$

$$Z_w = \{x \in X \mid a_\alpha x = 0 \text{ for any } \alpha \in \Delta_w\}, \tag{4.2}$$

where $a_\alpha \in \operatorname{End} X$ is the operator defined in [Z4] and coinciding with $wa_i w^{-1}$ for $\alpha = w\alpha_i$ on the subspace Y_w.

Let $F_{\tau,w}$ (respectively, $G_{\tau,w}$) be the part of Lusztig basis F_τ (see 3.3) generated by the elements f_α for $\alpha \in \Delta_w$ (respectively, for $\alpha \in \Delta_+ \setminus \Delta_w$). It is easy to see that [Z4] that $F_{\tau,w}$ (respectively, $G_{\tau,w}$) is a basis of the vector space X_w (respectively, Y_w). From this is clear that the subspaces (4.1) satisfy the following relations:

(α) X_w increases, Y_w decreases with the growth of w;
(β) $X = X_w Y_w \approx X_w \otimes Y_w$;
(γ) $Y_w = wX_{w'}$ for $w_{\max} = ww'$.

It is clear also that all the spaces (4.1), (4.2) are H-modules with respect to the adjoint action $x \mapsto q^h x q^{-h}$ ($h \in \mathfrak{h}$). In the terminology of [Z4], the family X_w (respectively, Y_w) is called an *increasing* (respectively, *decreasing*) *Schubert filtration* of the algebra X.

The family Z_w is called an *extremal filtration* of the algebra X. One of the principal results of the work [Z4] is the equality $Y_w = Z_w$ for any $w \in W$, which allows one to describe exactly all the solutions of the system of extremal equations (4.2).

In this section we shall give an analogous result (Theorem 4.5) under the corresponding filtrations of the module $D(\lambda)$.

REMARK. Definition (4.1) does not depend on the choice of $\tau \in NO(\Delta_+)$. Hence the equality $Y_w = Z_w$ yields a similar assertion for the definition (4.2).

4.2. We shall consider the following three filtrations of the module $D(\lambda)$:

(4.3) $$X_w(\lambda) = X_w 1_\lambda, \qquad Y_w(\lambda) = Y_w 1_\lambda.$$
(4.4) $$Z_w(\lambda) = \{x \in D(\lambda) \mid e_\alpha x = 0 \text{ for any } \alpha \in D_w\}.$$

The family $X_w(\lambda)$ (respectively, $Y_w(\lambda)$) is called an *increasing* (respectively, *decreasing*) *Schubert filtration* of the module $D(\lambda)$. The family $Z_w(\lambda)$ is called *extremal filtration* of the module $D(\lambda)$.

REMARK. The filtration $X_w(\lambda)$ in the classical case $q = 1$ is considered usually in connection with the theory of Demazure characters. In view of the relation (4.7) and following [Z4], we shall mainly consider the filtration $Y_w(\lambda)$.

4.3. PROPOSITION. *The canonical bilinear form in the space $D(\lambda)$ defines a duality of the following pair of vector spaces*:

(4.5) $$Y_w(\lambda)/I_w(\lambda), \qquad Z_w(\lambda),$$

where $I_w(\lambda)$ is the intersection of $Y_w(\lambda)$ with the sum of images of the operators f_α for $\alpha \in D_w$.

PROOF. Actually, $I_w(\lambda)$ is nothing but the orthogonal complement of $Z_w(\lambda)$ in the space $Y_w(\lambda)$. On the other hand, using the basis F_τ, we find $D(\lambda) = Y_w(\lambda) + J_w(\lambda)$, where $J_w(\lambda)$ is a sum of images of the operators f_α for $\alpha \in D_w$. If a vector $x \in Z_w(\lambda)$ is orthogonal to $Y_w(\lambda)$, then it is orthogonal to $D(\lambda)$, i.e., $x = 0$. □

In 4.5, a crystal version of duality for the spaces (4.5) will be given. Namely, we shall prove that the crystal images of these spaces are coinciding for all $\lambda \in P_+$, $w \in W$.

4.4. DEFINITION. Let $\lambda \in P_+$. Similarly to 4.2, we shall consider the following three filtrations of the lattice $L(\lambda)$:

(4.6) $$L_w(\lambda) = L_w 1_\lambda, \qquad M_w(\lambda) = M_w 1_\lambda,$$
(4.7) $$N_w(\lambda) = \{x \in L(\lambda) \mid \tilde{e}_\alpha x = 0 \text{ for any } \alpha \in D_w\},$$

where $L_w = L \cap X_w$, $M_w = L \cap Y_w$.

It is clear that L_w (respectively, M_w) is a free A-module with basis $F_{\tau, w}$ (respectively, $G_{\tau, w}$), for any $\tau \in NO(\Delta_+)$ with initial segment Δ_w. On the other hand, the equality $\ker e_\alpha = \ker \tilde{e}_\alpha$ (see 3.6) implies $N_w(\lambda) = L(\lambda) \cap Z_w(\lambda)$.

Here we may conclude that $L_w(\lambda)$ (respectively, $M_w(\lambda)$) is an increasing (respectively, decreasing) filtration of the lattice $L(\lambda)$. The family $N_w(\lambda)$ is the extremal filtration of the lattice $L(\lambda)$.

4.5. THEOREM. *Let $\lambda \in P_+$, $w \in W$. Then we have*:
(i) *The crystal images of $M_w(\lambda)$, $N_w(\lambda)$ coincide, i.e., $M_w(\lambda)^0 = N_w(\lambda)^0$.*
(ii) *Let $\omega_\lambda(x) = (x1_\lambda, x1_\lambda)_0$, where $x \in L$. The elements $x \in G_{\tau, w}$ (see 4.4) satisfying the relation $w_\lambda(x) \neq 0$ generate an orthonormal basis of the space $N_w(\lambda)^0$.*

(iii) *Let $m_w(\lambda, \mu)$ be the multiplicity of all solutions of weight μ of the system of equations (4.7) in the space $K(\lambda)$. Then we have*:

(4.8) $$m_w(\lambda, \mu) = \text{card}\{x \in G_{\tau, w, \mu-\lambda} \mid w_\lambda(x) \neq 0\},$$

where $G_{\tau, w, \nu}$ is the weight component of the weight μ for the subset $G_{\tau, w}$.

The proof is given in 4.7–4.10.

REMARK. The definition (4.7) depends on the choice of roof vectors e_α, i.e., on the choice of $\tau \in NO(\Delta_+)$. At the same time, according to (i), the space $N_w(\lambda)^0$ does not depend on the choice of $\tau \in NO(\Delta_+)$.

4.6. THEOREM. *For any set of elements $\lambda \in P_+$, $n \in \mathbb{Z}_+$, $\alpha \in \Delta_w$, $z \in M_w(\lambda)$, we have*

$$\tilde{f}_\alpha^n z \equiv f_\alpha^{(n)} z \mod qL(\lambda).$$

Correspondingly, for any $\tau \in NO(\Delta_+)$ elements of the basis $F_\tau(\lambda)$ may be written on the following form:

$$f(k) 1_\lambda \equiv \tilde{f}_{\alpha(1)}^{k_1} \cdots \tilde{f}_{\alpha(m)}^{k_m} 1_\lambda.$$

The proof is given in 4.9 below.

4.7. DEFINITION. Changing the operators $e_\alpha, f_\alpha \in \text{End}\, D(\lambda)$ in 3.6 to the operators $a_\alpha, f_\alpha \in \text{End}\, X$ (see 4.1), we define the corresponding operators $\tilde{e}_\alpha, \tilde{f}_\alpha \in \text{End}\, X$ (see [Ka, Z4] for the case $\alpha \in \Pi$).

Let $h_\alpha = wh_i$ for $\alpha = w\alpha_i$, (in the notation of 3.2), and let $\lambda_\alpha = \langle h_\alpha, \lambda \rangle$ for any pair $\lambda \in \mathfrak{h}^*, \alpha \in \Delta$.

The following assertion allows one to consider the operators $\tilde{e}_\alpha, \tilde{f}_\alpha \in \text{End}\, X$ as an asymptotic analog for the corresponding operators $\tilde{e}_\alpha, \tilde{f}_\alpha \in \text{End}\, D(\lambda)$.

4.8. LEMMA. *Let $\alpha = w\alpha_i$ and $x \in M_w$, where wr_i is a reduced decomposition. Then we have for $\lambda_\alpha \gg 0$*:

(4.9) $$(\tilde{e}_\alpha x) 1_\lambda \equiv \tilde{e}_\alpha(x 1_\lambda) \mod qL(\lambda).$$

PROOF. It is sufficient to consider the basis elements $x = f_\alpha^{(n)} y$, where $y \in M_{wr_i} = M_w \cap \ker a_\alpha$. Following (4.7), we also set $y = wz$, where $z \in M_{r_i} \cap \ker a_i$. Using the rule of action of the operators $\text{ad}\, e_i$ and the rule of permutation for the operators $a_i, \bar{a}_i$ (see 2.5), we find for $n \in \mathbb{Z}_+$:

$$e_i^n z \underset{\alpha_i}{\equiv} \bar{q}_i^{n(n-1)} \theta_i^{-n} t_i^n (\bar{a}_i^n z),$$

where the symbol $\underset{\alpha}{\equiv}$ means equivalence mod $U_q(\mathfrak{g}) e_\alpha$. Using the automorphism w, we get a similar relation for e_α, with factors $q_\alpha = q_i, \theta_\alpha = \theta_i, t_\alpha = wt_i$:

$$e_\alpha^n y \underset{\alpha}{\equiv} \bar{q}_\alpha^{n(n-1)} \theta_\alpha^{-n} t_\alpha^n (\bar{a}_\alpha^n y).$$

Applying both sides of this relation to the vector 1_λ (for which $e_\alpha 1_\lambda = 0$), we find an analog of Corollary 3.4.6 from [Ka] for the roof $\alpha = w\alpha_i$. The next proof is quite similar to that for the particular case $w = 1$ considered in [Ka] (Lemma 4.4.1). □

COROLLARY. *The elements $z \in M_w(\lambda)$ satisfy the relations $\tilde{e}_\alpha z \equiv 0 \mod qL(\lambda)$ for $\alpha \in \Delta_w$, $\lambda \gg 0$, i.e., $M_w(\lambda)^0 \subset N_w(\lambda)^0$ for $\lambda \gg 0$.*

PROOF. We set $\alpha = u\alpha_i$, where ur_i is a reduced decomposition and $ur_i \leqslant w$. We note that $M_w \subset M_{ur_i}$. Applying (4.9) and changing w to u, we find $\tilde{e}_\alpha M_{ur_i}(\lambda) \equiv 0$. In particular, we have $\tilde{e}_\alpha M_w(\lambda) \equiv 0$. □

REMARK. According to [Ka, Proposition 4.4.2] the analog of (4.9) with $\tilde{e}_\alpha$ replaced by $\tilde{f}_\alpha$ is valid for any $\lambda \in P_+$.

4.9. LEMMA. *$M_w(\lambda)^0 \subset N_w(\lambda)^0$ for any $\lambda \in P_+$.*

PROOF. The idea of the proof reduces to using Corollary 4.8 with λ replaced by $\lambda + \mu$ for $\mu \gg 0$. We set $x \in G_{\tau,w}$, and let $\tilde{e}_\alpha(x1_\lambda)^0 \neq 0$, where $\alpha = u\alpha_i$ in the notation of 4.8. Then we have $(x1_\lambda)^0 = \tilde{f}_\alpha z$ for some $z \in B(\lambda)$. Using Lemma 3.10 with w replaced by u, we find

$$w_\lambda(x) = (x1_\lambda, \tilde{f}_\alpha z)_0 = (x1_\lambda \otimes 1_\mu, \tilde{f}_\alpha z \otimes 1_\mu)_0 = (x1_{\lambda+\mu}, \tilde{f}_\alpha z \otimes 1_\mu)_0$$
$$= (x1_{\lambda+\mu}, \tilde{f}_\alpha(z \otimes 1_\mu))_0 = (\tilde{e}_\alpha(x1_{\lambda+\mu}), z \otimes 1_\mu)_0 = 0,$$

according to Corollary 4.8 for $\mu \gg 0$. Hence we have $(x1_\lambda)^0$. This is a contradiction. Hence we conclude that $\tilde{e}_\alpha M_w(\lambda)^0 = 0$ for any $\alpha \in \Delta_w$. □

COROLLARY. *Theorem 4.6 is valid for any $\lambda \in P_+$.*

PROOF. If $x \in M_w(\lambda)$, then $\tilde{e}_\alpha x \equiv 0 \mod qL(\lambda)$ for $\alpha \in \Delta_w$, hence $x \equiv y \mod qL(\lambda)$ for $y \in L(\lambda) \cap \ker \tilde{e}_\alpha$ (see [Ka, Lemma 2.6.2]). From this the assertion of Theorem 4.6 is evident. □

REMARK. The next assertion follows also from Remark 4.8.

4.10. PROOF OF THEOREM 4.5. According to Lemma 4.9, it is sufficient to verify that all the elements of the form $z = x1_\lambda$, where $x \in F_\tau$, satisfy the relations

(4.10) $$\tilde{e}_\alpha z \equiv 0 \mod qL(\lambda) \quad \text{for any } \alpha \in \Delta_w$$

only for $x \in G_{\tau,w}$.

PROOF. We set $x = f_\alpha^{(n)} y$, where $\alpha = u\alpha_i$ in the notation of 4.8, and $y \in M_{wr_i}$ such that $\tilde{e}_\alpha(y1_\lambda) \equiv 0$ (Lemma 4.9). Using Theorem 4.6, we obtain:

$$z = f_\alpha^{(n)} y1_\lambda \equiv \tilde{f}_\alpha^n(y1_\lambda), \qquad \tilde{e}_\alpha z \equiv \tilde{f}_\alpha^{n-1}(y1_\lambda) \equiv f_\alpha^{(n-1)} y1_\lambda,$$

where $\equiv$ means equivalence mod $qL(\lambda)$. If $n \neq 0$, this implies $z \equiv \tilde{f}_\alpha \tilde{e}_\alpha z$. Hence the relation (4.10) is possible only for $n = 0$. □

REMARKS. 1. If $\omega_\lambda(x) \neq 0$ in the conditions of Theorem 4.5, then $w_\lambda(x) = 1$.

2. The function $\omega_\lambda(x)$ coincides for $q = 0$ with the expression $\pi_\lambda(x^*x)$, where π_λ is the Harish–Chandra homomorphism specialized at the point $\lambda \in \mathfrak{h}^*$.

Hence for the computation of $w_\lambda(x)$ it is sufficient to use the contravariance relations given in 2.3.

§5. Reduction problem

5.1. A well-known reduction problem in the theory of finite-dimensional $\mathfrak{g}$-modules concerns the description of the reduction $\mathfrak{g} \downarrow \mathfrak{g}_0$, where $\mathfrak{g}_0$ is a reductively embedded subalgebra of the algebra $\mathfrak{g}$ (i.e., the adjoint action of $\mathfrak{g}_0$ in $\mathfrak{g}$ is reductive).

According to the general reductivity criterion (the Weyl theorem), one may reduce the reduction problem to the study of spectral decompositions of the following form:

$$(5.1) \qquad D(\lambda) = \bigoplus_{\mu} m(\lambda, \mu) D_0(\mu),$$

where $D_0(\mu)$ is a simple $\mathfrak{g}_0$-module with highest weight μ. The problem consists in the description of isotypical components $D(\lambda, \mu) = m(\lambda, \mu) D_0(\mu)$, in particular, in the computation of the multiplicities $m(\lambda, \mu)$.

The complete solution of the reduction problem is known only in particular cases (see, for example, [Z1]).

5.2. It should be recalled that the definition of highest weights in the decomposition (5.1) is connected with the choice of triangular decompositions of the algebra $\mathfrak{g}$ and the subalgebra $\mathfrak{g}_0$, i.e., with the choice of systems of positive roots $\Delta_+ \subset \mathfrak{h}^*$, $\Delta_{0+} \subset \mathfrak{h}_0^*$, where $\mathfrak{h}, \mathfrak{h}_0$ are some fixed Cartan subalgebras in $\mathfrak{g}, \mathfrak{g}_0$.

As is known (see, for example, [Di]), every Cartan subalgebra $\mathfrak{h}_0 \subset \mathfrak{g}_0$ is contained in some Cartan subalgebra $\mathfrak{h} \subset \mathfrak{g}$. The corresponding triangular decompositions in $\mathfrak{g}$ and $\mathfrak{g}_0$ may be chosen accordingly, i.e.,

$$(5.2) \qquad \mathfrak{g} = \mathfrak{n}' \oplus \mathfrak{h} \oplus \mathfrak{n}, \qquad \mathfrak{g}_0 = \mathfrak{n}_0' \oplus \mathfrak{h}_0 \oplus \mathfrak{n}_0,$$

where $\mathfrak{n}$ (respectively, $\mathfrak{n}'$) is the span of root vectors e_α (respectively, f_α) for $\alpha \in \Delta_+$, and $\mathfrak{n}_0 = \mathfrak{n} \cap \mathfrak{g}_0$, $\mathfrak{n}_0' = \mathfrak{n}' \cap \mathfrak{g}_0$ are similar subalgebras of $\mathfrak{g}_0$ defined by the system Δ_{0+}.

Hence we may assume that $\lambda \in P_+$, $\mu \in P_{0+}$, where P_+, P_{0+} are the corresponding weight lattices in $\mathfrak{h}^*, \mathfrak{h}_0^*$ (see 1.6).

5.3. The reduction problem (5.1) may be transformed in the standard way to the study of extremal subspaces with respect to $\mathfrak{g}_0$. The extremal subspace V^{ext} of a $\mathfrak{g}_0$-module V is defined as $\ker \mathfrak{n}_0$. The decomposition (5.1) is equivalent to the following:

$$(5.3) \qquad D^{\text{ext}}(\lambda) = \bigoplus_{\mu} m(\lambda, \mu) D_0^{\text{ext}}(\mu),$$

where $D^{\text{ext}}(\lambda) = D(\lambda)^{\text{ext}}$ (similarly for $D_0(\mu)$). Moreover, $\dim D_0^{\text{ext}}(\mu) = 1$ for any μ. Hence the spectral decomposition (5.1) is reduced to a weight decomposition (5.3). In other words, let $\operatorname{ch} V$ be the character of a weight $\mathfrak{h}_0$-module V. Then we have:

$$(5.4) \qquad \operatorname{ch} D^{\text{ext}}(\lambda) = \sum_{\mu} m(\lambda, \mu) e^{\mu},$$

where e^μ is the formal exponential (an element of the group ring $\mathbb{Z}[P_0]$).

We recall also that any reductive $\mathfrak{g}_0$-module V is of the form $\ker \mathfrak{n}_0 \oplus \operatorname{im} \mathfrak{n}'_0$. The projection operator on $V^{\text{ext}} = \ker \mathfrak{n}_0$ with respect to this decomposition is denoted by $p = p^{\text{ext}}$ and is called the *extremal projector a module V*. In particular, we have:

$$(5.5) \qquad D^{\text{ext}}(\lambda) = pD(\lambda).$$

The extremal projector p has a universal description as an element of some canonical extension $F(\mathfrak{g})$ of the algebra $U(\mathfrak{g})$ generated by a certain formal series over $U(\mathfrak{g})$ [Z2].

5.4. The embedding $\mathfrak{g}_0 \subset \mathfrak{g}$ is called *simple* if the system Π_0 of simple roots of the subalgebra $\mathfrak{g}_0$ with respect to $\mathfrak{h}_0$ is contained in the system Π of simple roots of the algebra $\mathfrak{g}$ (w.r.t. $\mathfrak{h}$), for its restriction to $\mathfrak{h}_0$. Hence we have:

$$(5.6) \qquad \Pi = \{\alpha_i\}_{i \in I}, \qquad \Pi_0 = \{\alpha_i\}_{i \in I_0},$$

where $I_0 \subset I$. Correspondingly, we have $\Delta_{0+} \subset \Delta_+$, and the triangular decompositions (5.2) are in agreement. Replacing the subalgebra $\mathfrak{g}_0$ by its central extension, we shall assume also, without loss of generality, that $\mathfrak{h}_0 = \mathfrak{h}$.

If is clear that any simple embedding $\mathfrak{g}_0 \subset \mathfrak{g}$ induces the corresponding embedding of the quantum algebras $U_q(\mathfrak{g}_0) \subset U_q(\mathfrak{g})$. The reduction problem (5.1) is completed by the quantum analog for the reduction $U_q(\mathfrak{g}) \downarrow U_q(\mathfrak{g}_0)$.

Moreover, the reduction problem (5.1) may be expressed in crystal terms, for an extremal subspaces $L^{\text{ext}}(\lambda)$, $K^{\text{ext}}(\lambda)$ defined as the intersections of kernels of the operators $\widetilde{e}_\alpha$ ($\alpha \in \Pi_0$) in $L(\lambda)$ and $K(\lambda)$ respectively. Finally, the relation (5.4) possesses the following crystal analog:

$$(5.7) \qquad \operatorname{ch} K^{\text{ext}}(\lambda) = \sum_\mu m(\lambda, \mu) e^\mu.$$

We note also that the relation (5.5) preserves for a quantum modules. The extremal projection p possesses a universal description in terms of the canonical extension $F_q(\mathfrak{g})$ of the algebra $U_q(\mathfrak{g})$ [Z2, Z5, Z7].

5.5. The relation (5.5) possesses the following detalization using the Poincaré–Birkhoff–Witt theorem for the algebra $U(\mathfrak{g})$.

We set $\mathfrak{n}' = \mathfrak{n}'_0 \oplus \mathfrak{m}$, where $\mathfrak{m}$ is the complementary $\mathfrak{h}_0$-submodule. Correspondingly, $X = U(\mathfrak{n}') = X_0 Y$, where $X_0 = U(\mathfrak{n}'_0)$ and Y is the span of the monomials generated by an ordered weight basis of the module $\mathfrak{m}$. We note that the operator p annulates im $\mathfrak{n}'_0$. Then we have

$$(5.8) \qquad D^{\text{ext}}(\lambda) = pX_0 Y 1_\lambda = pY 1_\lambda.$$

Correspondingly, the character $\operatorname{ch} D^{\text{ext}}(\lambda)$ is majorized by the character $\operatorname{ch} Y 1_\lambda$. In other words, $m(\lambda, \mu) \leqslant n(\lambda, \mu)$, where $n(\lambda, \mu)$ is the multiplicity of the weight μ in the $\mathfrak{h}_0$-module $Y 1_\lambda$.

In particular, let $\mathfrak{g}_0 \subset \mathfrak{g}$ be a simple embedding. The preceding argument works in this case with $U(\mathfrak{g})$ replaced by $U_q(\mathfrak{g})$. Moreover, we have

$$(5.9) \qquad X_0 = X_{w_0}, \qquad Y = Y_{w_0},$$

where $w_0 \in W$ is the greatest element generated by the elements r_i for $i \in I_0$. We note that $w_0 = w_{\max}$ for the subgroup $W_0 \subset W$ isomorphic to the Weyl group of the subalgebra $\mathfrak{g}_0$.

Consequently, the subspace $Y1_\lambda$ in (5.8) coincides with the corresponding component $Y_{w_0}(\lambda)$ of the Schubert filtration in $D(\lambda)$.

5.6. Theorem. *The spectral multiplicities $m(\lambda, \mu)$ for the reduction $U_q(\mathfrak{g}) \downarrow U_q(\mathfrak{g}_0)$ may be computed by formula (5.8) for $w = w_0$, i.e., we have*

(5.10) $$m(\lambda, \mu) = \operatorname{card}\{x \in G_{\tau, w_0, \mu-\lambda} \mid w_\lambda(x) \neq 0\}.$$

PROOF. It is sufficient to note that $K^{\mathrm{ext}}(\lambda) = N_{w_0}(\lambda)^0$, in the notation of 4.4, and to use Theorem 4.5. □

5.7. Theorem. *Let $\mathfrak{g}_0 \subset \mathfrak{g}$ be a simple embedding. Then the spectral multiplicities $m(\lambda, \mu)$ for the reduction $\mathfrak{g} \downarrow \mathfrak{g}_0$ coincide with the corresponding quantum multiplicities, i.e., may be computed by formula (5.10).*

PROOF. We recall [Z2] that the action of $p \in F(\mathfrak{g})$ is reduced to a finite subseries on any vector from $Y1_\lambda$. According to [Z6] (see also [Z5]), this action coincides as $q \to 1$ with the corresponding action of $p \in F_q(\mathfrak{g})$. From this we get $m(\lambda, \mu) \leqslant m_q(\lambda, \mu)$, where $m_q(\lambda, \mu)$ is the corresponding quantum multiplicity. Multiplying both sides of this inequality by $\dim \mathbb{D}(\lambda)_\mu$ and summing over μ, we get an identical expression for $\dim D(\lambda)$. Hence we have $m(\lambda, \mu) = m_q(\lambda, \mu)$ for any λ, μ. □

REMARK. The formula (5.10) gives a solution of the classical reduction problem in crystal (quantum) terms.

§6. Concluding remarks

6.1. In the classical case $q = 1$, the description of the characters $\operatorname{ch} X_w(\lambda)$ for the Schubert filtration $X_w(\lambda)$, in term of Demazure operators (Z2) acting in the group ring $\mathbb{Z}[P]$ is well known. The Demazure operators ∂_i ($i \in I$) are defined in the following way:

(6.1) $$\partial_i = (1 - e^{-\alpha_i})^{-1}(1 - e^{-\alpha_i} r_i),$$

where $r_i \in \operatorname{End} \mathbb{Z}[P]$ acts on the generators e^λ of the ring $\mathbb{Z}[P]$ by the rule $r_i e^\lambda = e^{r_i \lambda}$ ($\lambda \in P$). The expression $\operatorname{ch} X_w(\lambda)$ arises from the exponential e^λ by successively applying the chain $\partial_{i_1} \ldots \partial_{i_n}$, where $w = r_{i_1} \ldots r_{i_n}$ is an arbitrary reduced decomposition of the element $w \in W$.

On the other hand, P. Littelmann [Li] proposes a combinatorial construction of the crystal basis $B(\lambda)$ in terms of the following analogs of the Demazure operators acting in $B(\lambda)$:

(6.2) $$\partial_i = (1 - \tilde{f}_i)^{-1}(1 - \tilde{f}_i r_i),$$

where r_i maps the elements $b \in B(\lambda)_\mu$ to the element $\tilde{f}_i^n b$ for $n = \mu_i \geqslant 0$ or to the element $\tilde{e}_i^n b$ for $n = -\mu_i \geqslant 0$. Here $\mu_i = \mu_\alpha$ for $\alpha = \alpha_i$.

The next theorem is essentially a positive answer to the Littelmann conjecture [Li] proved by him for all root systems excluding F_4, E_7, E_8.

6.2. Theorem. *Let $\lambda \in P_+$, and let $K_w(\lambda) = L_w(\lambda)/qL_w(\lambda)$ for $w \in W$. Then we have*:
  (i) *The set $B_w(\lambda) = B(\lambda) \cap K_w(\lambda)$ is an orthonormal basis of the space $K_w(\lambda)$.*
  (ii) *For any reduced decomposition $w = r_{i_1} \ldots r_{i_n}$ we have*

$$(6.3) \qquad B_w(\lambda) = \partial_{i_1} \ldots \partial_{i_n} 1_\lambda,$$

where the operators ∂_i are defined by the formula (6.2).
  (iii) *Under the conditions of* (ii), *we also have*

$$(6.4) \qquad \operatorname{ch} K_w(\lambda) = \partial_{i_1} \ldots \partial_{i_n} e^\lambda,$$

where the operators ∂_i are defined by the formula (6.1).

A proof may be given similarly to the proof of Theorem 2.5.8 in [Z2], using a crystal version of the theory of induced modules. The details presented elsewhere.

Remark. A slight modification of the formulas (3), (4) also allows one to change the family $L_w(\lambda)$ to $M_w(\lambda)$.

6.3. It is interesting to compare the general scheme of reduction given in 5.5 with its crystal version (see 5.6). Setting $M = M_{w_0}$ in the notation of 4.4, we see that the equality $K^{\text{ext}}\lambda = N_{w_0}(\lambda)^0$ (see 5.6) may be rewritten as follows:

$$(6.5) \qquad K^{\text{ext}}(\lambda) = (M1_\lambda)^0.$$

Comparing this relation with formula (5.8), we find that the extremal projector $p = p(q)$ disappears in the relation (6.5). In other words, the relation (6.5) may be written symbolically in the form

$$(6.6) \qquad \lim_{q \to 0} p(q) = 1 \quad \text{on } (M1_\lambda)^0.$$

6.4. The use of quantum methods for general reductive embeddings $\mathfrak{g}_0 \subset \mathfrak{g}$ leads to obstacles in the general problems of describing the corresponding quantum embeddings $U_q(\mathfrak{g}) \subset U_q(\mathfrak{g})$.

See, for example [Z8]. The last problem is also connected with the general question of defining and studying "quantum symmetric spaces" associated with the algebra $U_q(\mathfrak{g})$.

References

[CK] C. De Concini and V. G. Kac, *Representations of quantum groups at roots of* 1, Progr. Math. **92** (1980), 471–506.
[Di] J. Dixmier, *Enveloping algebras*, North Holland, Amsterdam, 1977.
[D] V. G. Drinfeld, *Hopf algebras and the Yang–Baxter equation*, Dokl. Akad. Nauk SSSR; English transl., Soviet Math. Dokl. **32** (1985), 254–258.
[J] M. Jimbo, *A q-difference analog of $U(\mathfrak{g})$ and the Yang–Baxter equation*, Lett. Math. Phys. **10** (1985), 63–69.
[H] G. Heckmann, *Projections of orbits and asymptotic behavior of multiplicities for compact connected Lie groups*, Invent. Math. **67** (1982), 333–356.
[K] V. G. Kac, *Infinite-dimensional Lie algebras*, Cambridge Univ. Press, Cambridge, 1985.

[Ka] M. Kashiwara, *Crystalizing the q-analog of universal enveloping algebra*, Duke Math. J. **63** (1991), no. 2, 465–516.
[Li] P. Littelmann, *Crystal graphs and Young tableaux*, Preprint (1991), Basel Univ.
[Lu] G. Lusztig, *Canonical bases arising from quantized enveloping algebras*, J. Amer. Math. Soc. **3** (1990), 447–498.
[Z1] D. P. Zhelobenko, *Compact Lie groups and their representations*, "Nauka", Moscow, 1970; English transl., Amer. Math. Soc., Providence, RI, 1973.
[Z2] _____, *Representations of reductive Lie algebras*, "Nauka", Moscow, 1994; English transl., Oxford Univ. Press, Oxford (to appear).
[Z3] _____, *On quantum methods in the representation theory for reductive Lie algebras*, Funktsional. Anal. i Prilozhen. **28** (1994), no. 1, 5–14; English transl. in Functional Anal. Appl. **28** (1994), no. 1.
[Z4] _____ The algebra of quantum bosons, Schubert filtration and Lusztig bases, Izv. Ross. Akad. Nauk Ser. Mat. **57** (1993), no. 6, 3–32; English transl., Math. USSR-Izv. **43** (1994).
[Z5] _____ Extremal projectors and the problem of reduction in classical and quantum modules, Vestnik Ross. Univ. Druzhby Narodov **1** (1993), no. 1, 22–33. (Russian)
[Z6] _____, *Extremal projectors over the Drinfeld–Jimbo algebras*, Vestnik Ross. Univ. Druzhby Narodov **2** (1994) (to appear). (Russian)
[Z7] _____, *Constructive modules and reductivity problem in the category* $\mathcal{O}$, Transl. Series 2, Amer. Math. Soc., Providence, RI (to appear).
[Z8] _____, *Contragredient structures*, Sovremennye Problemy Matematiki. Fundamental'nye Mapreavleniya, VINITI, Moscow, 1994.

Other Titles in This Series

(*Continued from the front of this publication*)

130 **M. M. Lavrent'ev, K. G. Reznitskaya, and V. G. Yakhno,** One-dimensional Inverse Problems of Mathematical Physics

129 **S. Ya. Khavinson,** Two Papers on Extremal Problems in Complex Analysis

128 **I. K. Zhuk et al.,** Thirteen Papers in Algebra and Number Theory

127 **P. L. Shabalin et al.,** Eleven Papers in Analysis

126 **S. A. Akhmedov et al.,** Eleven Papers on Differential Equations

125 **D. V. Anosov et al.,** Seven Papers in Applied Mathematics

124 **B. P. Allakhverdiev et al.,** Fifteen Papers on Functional Analysis

123 **V. G. Maz'ya et al.,** Elliptic Boundary Value Problems

122 **N. U. Arakelyan et al.,** Ten Papers on Complex Analysis

121 **V. D. Mazurov, Yu. I. Merzlyakov, and V. A. Churkin, Editors,** The Kourovka Notebook: Unsolved Problems in Group Theory

120 **M. G. Kreĭn and V. A. Jakubovič,** Four Papers on Ordinary Differential Equations

119 **V. A. Dem'janenko et al.,** Twelve Papers in Algebra

118 **Ju. V. Egorov et al.,** Sixteen Papers on Differential Equations

117 **S. V. Bočkarev et al.,** Eight Lectures Delivered at the International Congress of Mathematicians in Helsinki, 1978

116 **A. G. Kušnirenko, A. B. Katok, and V. M. Alekseev,** Three Papers on Dynamical Systems

115 **I. S. Belov et al.,** Twelve Papers in Analysis

114 **M. Š. Birman and M. Z. Solomjak,** Quantitative Analysis in Sobolev Imbedding Theorems and Applications to Spectral Theory

113 **A. F. Lavrik et al.,** Twelve Papers in Logic and Algebra

112 **D. A. Gudkov and G. A. Utkin,** Nine Papers on Hilbert's 16th Problem

111 **V. M. Adamjan et al.,** Nine Papers on Analysis

110 **M. S. Budjanu et al.,** Nine Papers on Analysis

109 **D. V. Anosov et al.,** Twenty Lectures Delivered at the International Congress of Mathematicians in Vancouver, 1974

108 **Ja. L. Geronimus and Gábor Szegő,** Two Papers on Special Functions

107 **A. P. Mišina and L. A. Skornjakov,** Abelian Groups and Modules

106 **M. Ja. Antonovskiĭ, V. G. Boltjanskiĭ, and T. A. Sarymsakov,** Topological Semifields and Their Applications to General Topology

105 **R. A. Aleksandrjan et al.,** Partial Differential Equations, Proceedings of a Symposium Dedicated to Academician S. L. Sobolev

104 **L. V. Ahlfors et al.,** Some Problems on Mathematics and Mechanics, On the Occasion of the Seventieth Birthday of Academician M. A. Lavrent'ev

103 **M. S. Brodskiĭ et al.,** Nine Papers in Analysis

102 **M. S. Budjanu et al.,** Ten Papers in Analysis

101 **B. M. Levitan, V. A. Marčenko, and B. L. Roždestvenskiĭ,** Six Papers in Analysis

100 **G. S. Ceĭtin et al.,** Fourteen Papers on Logic, Geometry, Topology and Algebra

99 **G. S. Ceĭtin et al.,** Five Papers on Logic and Foundations

98 **G. S. Ceĭtin et al.,** Five Papers on Logic and Foundations

97 **B. M. Budak et al.,** Eleven Papers on Logic, Algebra, Analysis and Topology

96 **N. D. Filippov et al.,** Ten Papers on Algebra and Functional Analysis

95 **V. M. Adamjan et al.,** Eleven Papers in Analysis

94 **V. A. Baranskiĭ et al.,** Sixteen Papers on Logic and Algebra

93 **Ju. M. Berezanskiĭ et al.,** Nine Papers on Functional Analysis

92 **A. M. Ančikov et al.,** Seventeen Papers on Topology and Differential Geometry

91 **L. I. Barklon et al.,** Eighteen Papers on Analysis and Quantum Mechanics

(See the AMS catalog for earlier titles)